Building Governance and Climate Change

The contribution of buildings to climate change is widely acknowledged. This book investigates how building regulatory systems are addressing the current and future effects of climate change, and how these systems can be improved. After presenting a comprehensive overview of how the current building regulatory system developed, some of the inadequacies are identified. The largest part of the book examines the potential for innovative policy solutions to address the real-world problem of mitigating and adapting buildings to climate change. This publication contributes significantly to our understanding of the complexities of long-term energy efficiency in buildings.

The chapters in this book were originally published as a special issue of the *Building Research & Information* journal.

Richard Lorch is the Editor-in-Chief of *Building Research & Information* journal and a book series editor at Routledge/ Taylor and Francis. He is also a researcher, writer, and policy advisor on energy, buildings, and climate.

Jacques Laubscher is a practicing architect and academic. He is currently head of the technology research group in the Department of Architecture at the Tshwane University of Technology, South Africa.

Edwin H.W. Chan is a Chartered Architect (Authorized Person), Chartered Surveyor, and also a Barrister-at-Law called to the UK and Hong Kong Bars. He obtained his Ph.D. from King's College, University of London, UK.

Henk Visscher is Full Professor in Housing Quality and Process Innovation at Delft University of Technology, the Netherlands, where he is director of the Graduate School in the Faculty of Architecture and the Built Environment.

Building Governance and Climate Change

Regulation and Related Policies

Edited by
Richard Lorch, Jacques Laubscher, Edwin H.W. Chan, and Henk Visscher

Routledge
Taylor & Francis Group

LONDON AND NEW YORK

First published 2018
by Routledge

2 Park Square, Milton Park, Abingdon, Oxfordshire OX14 4RN
52 Vanderbilt Avenue, New York, NY 10017

Routledge is an imprint of the Taylor & Francis Group, an informa business

First issued in paperback 2020

British Library Cataloguing in Publication Data
A catalogue record for this book is available from the British Library

ISBN13: 978-0-8153-9520-1 (hbk)
ISBN13: 978-0-367-51939-1 (pbk)

Typeset in MinionPro
by diacriTech, Chennai

Publisher's Note
The publisher accepts responsibility for any inconsistencies that may have arisen during the conversion of this
book from journal articles to book chapters, namely the possible inclusion of journal terminology.

Disclaimer
Every effort has been made to contact copyright holders for their permission to reprint material in this book. The
publishers would be grateful to hear from any copyright holder who is not here acknowledged and will undertake
to rectify any errors or omissions in future editions of this book.

Contents

CONTENTS

Citation Information

The chapters in this book were originally published in *Building Research & Information*, volume 44, issues 5–6 (July–August 2016). When citing this material, please use the original page numbering for each article, as follows:

Introduction
Visscher, H., Laubscher, J. and Chan, E. (2016) Building governance and climate change: roles for regulation and related polices, *Building Research & Information*, 44(5–6), 461–466, DOI: 10.1080/09613218.2016.1182786

Chapter 1
Meacham, B.J., Sustainability and resiliency objectives in performance building regulations, *Building Research & Information*, 44(5–6), 474–489, DOI: 10.1080/09613218.2016.1142330

Chapter 2
Shapiro, S., The realpolitik of building codes: overcoming practical limitations to climate resilience, *Building Research & Information*, 44(5–6), 490–506, DOI: 10.1080/09613218.2016.1156957

Chapter 3
Eisenberg, D.A., Transforming building regulatory systems to address climate change, *Building Research & Information*, 44(5–6), 468–473, DOI: 10.1080/09613218.2016.1126943

Chapter 4
Mulville, M. and Stravoravdis, S., The impact of regulations on overheating risk in dwellings, *Building Research & Information*, 44(5–6), 520–534, DOI: 10.1080/09613218.2016.1126943

Chapter 5
Karatas, A., Stoiko, A. and Menassa, C.C., Framework for selecting occupancy-focused energy interventions in buildings, *Building Research & Information*, 44(5–6), 535–551, DOI: 10.1080/09613218.2016.1182330

Chapter 6
Visscher, H., Meijer, F., Majcen, D. and Itard, L., Improved governance for energy efficiency in housing, *Building Research & Information*, 44(5–6), 552–561, DOI: 10.1080/09613218.2016.1180808

Chapter 7
Rosenow, J., Fawcett, T., Eyre, N. and Oikonomou, V., Energy efficiency and the policy mix, *Building Research & Information*, 44(5–6), 562–574, DOI: 10.1080/09613218.2016.1138803

Chapter 8
Nishida, Y., Hua, Y. and Okamoto, N., Alternative building emission-reduction measure: outcomes from the Tokyo Cap-and-Trade Program, *Building Research & Information*, 44(5–6), 644–659, DOI: 10.1080/09613218.2016.1169475

Chapter 9
Zhang, J., Zhou, N., Hinge, A., Feng, W. and Zhang, S., Governance strategies to achieve zero-energy buildings in China, *Building Research & Information*, 44(5–6), 604–618, DOI: 10.1080/09613218.2016.1157345

Chapter 10

Sha, K. and Wu, S., Multilevel governance for building energy conservation in rural China, *Building Research & Information*, 44(5–6), 619–629, DOI: 10.1080/09613218.2016.1152787

Chapter 11

Janda, K.B., Bright, S., Patrick, J., Wilkinson, S. *and Dixon, T.J.,* The evolution of green leases: towards inter-organizational environmental governance, *Building Research & Information*, 44(5–6), 660–674, DOI: 10.1080/09613218.2016.1142811

Chapter 12

Conejos, S., Langston, C., Chan, E.H.W. and Chew, M.Y.L., Governance of heritage buildings: Australian regulatory barriers to adaptive reuse, *Building Research & Information*, 44(5–6), 507–519, DOI: 10.1080/09613218.2016.1142811

Chapter 13

Qian, Q.K., Fan, K. and Chan, E.H.W., Regulatory incentives for green buildings: gross floor area concessions, *Building Research & Information*, 44(5-6), 675–693, DOI: 10.1080/09613218.2016.1181874

Chapter 14

Van der Heijden, J., The new governance for low-carbon buildings: mapping, exploring, interrogating, *Building Research & Information*, 44(5–6), 575–584, DOI: 10.1080/09613218.2016.1159394

Chapter 15

Drummond, P. and Ekins, P., Reducing CO_2 emissions from residential energy use, *Building Research & Information*, 44(5–6), 585–603, DOI: 10.1080/09613218.2016.1180075

Chapter 16

Lord, S.-F., Noye, S., Ure, J., *Tennant, M.G. and Fisk, D.J.,* Comparative review of building commissioning regulation: a quality perspective, *Building Research & Information*, 44(5–6), 630–643, DOI: 10.1080/09613218.2016.1181955

For any permission-related enquiries please visit:
http://www.tandfonline.com/page/help/permissions

Notes on Contributors

Susan Bright is Professor of Land Law, based at New College, University of Oxford, UK.

Edwin H.W. Chan is a Chartered Architect (Authorized Person), Chartered Surveyor, and also a Barrister-at-Law called to the UK and Hong Kong Bars. He obtained his Ph.D. from King's College, University of London, UK.

Michael Y.L. Chew is Professor in the Department of Building, School of Design and Environment at the National University of Singapore, Singapore.

Sheila Conejos is a Research Fellow in the Department of Building, School of Design and Environment at the National University of Singapore, Singapore.

Timothy J. Dixon is Professor of Sustainable Futures in the Built Environment at the University of Reading, UK.

Paul Drummond is a Senior Research Associate in Energy and Climate Policy at the Institute for Sustainable Resources, University College London, UK.

David A. Eisenberg is the co-founder and Executive Director of the Development Center for Appropriate Technology in Tucson, USA.

Paul Ekins is Professor of Resources and Environmental Policy, and Director of the Institute for Sustainable Resources, at University College London, UK.

Nick Eyre is Senior Research Fellow and Professor of Energy and Climate Policy at the Environmental Change Institute's Energy Programme, and Oriel College, University of Oxford, UK.

Ke Fan is a doctoral researcher in the Department of Building and Real Estate, Hong Kong Polytechnic University, Hong Kong.

Tina Fawcett is a Senior Researcher in the Environmental Change Institute's Energy Programme at the University of Oxford, UK.

Wei Feng is Principal Scientific Engineering Associate, and a member of the China Energy Group, the International Energy Analysis Department, and the Energy Analysis and Environmental Impacts Division at the Lawrence Berkeley National Laboratory, USA.

David J. Fisk is Emeritus Professor in the Department of Civil and Environmental Engineering at Imperial College London, UK.

Adam Hinge manages Sustainable Energy Partnerships, a small consultancy specializing in energy efficiency program and policy issues. He is also an Adjunct Associate Professor at Columbia University, USA.

Ying Hua is Associate Professor and Director of Undergraduate Studies in the College of Human Ecology at Cornell University, USA.

Laure Itard is Associate Professor in the OTB – Research for the Built Environment at TU Delft, the Netherlands.

Kathryn B. Janda is a sociotechnical scholar focused on people, energy, and buildings. She is a Principal Research Fellow in Organizations and Non-Domestic Buildings at the UCL Energy Institute at University College London, UK.

Aslihan Karatas is Assistant Professor of Civil Engineering at Lawrence Technological University, Southfield, USA.

Craig Langston is Professor of Construction and Facilities Management at Bond University, Gold Coast, Australia.

NOTES ON CONTRIBUTORS

Jacques Laubscher is a practicing architect and academic. He is currently head of the technology research group in the Department of Architecture at the Tshwane University of Technology, South Africa.

Richard Lorch is the Editor-in-Chief of *Building Research & Information* journal and a book series editor at Routledge/ Taylor and Francis. He is also a researcher, writer, and policy advisor on energy, buildings, and climate.

Sue-Fay Lord is based at the Centre for Environmental Policy at Imperial College London, UK.

Daša Majcen is a Post-Doctoral Researcher at the Institute for Environmental Sciences at the University of Geneva, Switzerland.

Brian J. Meacham, formerly associate professor at Worcester Polytechnic Institute, USA, is Managing Principal at Meacham Associates. He conducts research and consults in the areas of building regulatory system structures and components, fire engineering, and risk-informed performance-based solutions to complex building and infrastructure challenges.

Carol C. Menassa is Associate Professor and John L. Tishman Faculty Scholar in the Department of Civil and Environmental Engineering at the University of Michigan, USA.

Mark Mulville is Senior Lecturer in Sustainable Built Environments and MSc Portfolio Manager (Construction) in the Faculty of Architecture, Computing and Humanities at the University of Greenwich, UK.

Yuko Nishida is the Manager of the Climate Change Group at the Renewable Energy Institute, Japan. Prior to that, she worked for the Tokyo Metropolitan Government on climate policies and international environmental cooperation projects.

Sarah Noye is a Researcher at TECNALIA Research and Innovation, and holds a Ph.D. from the Centre for System Engineering and Innovation, Imperial College London, UK.

Vlasis Oikonomou is based at the Joint Implementation Network, Groningen, the Netherlands. He is coordinating projects on the assessment of policy interrelationships and impacts on sustainability in Europe, mobilizing and transferring knowledge on post 2012 climate policy implications, and energy-saving policies and energy efficiency obligations schemes.

Naomi Okamoto is based in the Bureau of Environment for the Tokyo Metropolitan Government, Japan.

Julia Patrick is an Honorary Research Associate in the Environmental Change Institute at the University of Oxford, UK.

Queena K. Qian is an Assistant Professor in the OTB - Research for the Built Environment Department at TU Delft, the Netherlands.

Jan Rosenow is Director of European Programmes at the Regulatory Assistance Project, a global team of highly skilled energy experts. He is also a Senior Research Fellow at the Centre on Innovation and Energy Demand at the University of Sussex, UK, and runs his own company, EnPolicy Consulting Ltd.

Kaixun Sha is based in the School of Management Engineering at Shandong Jianzhu University, Shandong, China.

Shari Shapiro is a Research Affiliate at the University of Pennsylvania Law School Program on Regulation, and Head of Public Affairs in Pennsylvania and Delaware for Uber.

Allisandra Stoiko holds a B.S. in Chemical Engineering at the University of Michigan, USA.

Spyridon Stravoravdis is Senior Lecturer in the Department of Built Environment at the University of Greenwich, UK.

Mike G. Tennant is a Lecturer in Business and the Environment at Imperial College London, UK.

Jim Ure is Director of Allied Project Consultants, based in Hampshire, UK. He has spent his career developing and promoting the through-life approach to facilities operation and design, focused on the priority of occupant well-being.

Jeroen Van der Heijden is an Associate Professor at the Australian National University, Canberra, and at Wageningen University, the Netherlands. He works at the intersections of regulation and governance, policy change, and urban development and transformation.

Henk Visscher is Full Professor in Housing Quality and Process Innovation at Delft University of Technology, the Netherlands, where he is director of the Graduate School in the Faculty of Architecture and the Built Environment.

Sara Wilkinson is Associate Professor in the School of the Built Environment at the University of Technology, Sydney, Australia. She is also a Chartered Building Surveyor, a Fellow of the Royal Institution of Chartered Surveyors, and a member of the Australian Property Institute.

Shaoyan Wu is Associate Professor in the Management School at Tianjin University of Technology, China.

Jingjing Zhang is a Postdoctoral Researcher in the China Energy Group, the International Energy Analysis Department, and the Energy Analysis and Environmental Impacts Division at the Lawrence Berkeley National Laboratory, USA.

Shicong Zhang is Deputy Director of the Research Center for Development and Strategy, Institute for Building Environment and Energy, China Academy of Building Research, Beijing, China.

Nan Zhou is Leader of the China Energy Group and Head of the International Energy Analysis Department at the Lawrence Berkeley National Laboratory, USA.

Building governance and climate change: roles for regulation and related polices

Henk Visscher, Jacques Laubscher and Edwin Chan

The contribution of buildings to climate change has become widely acknowledged. On 3 December 2015, the United Nations Environment Programme (UNEP) held the first 'buildings day' at COP 21 (the UN Climate Change Conference) devoted to the decarbonization of the building stock. There are several forms of negative contributions that buildings make to climate change, but high on the list are embodied and operational energy demands, which largely depend on fossil fuels and result in greenhouse gas emissions.

Given the urgency of the risks associated with climate change, the mandate to contain global warming to 1.5–2.0°C and the urgency to reduce energy demand and decarbonize radically, a key challenge is what actions can be taken across a whole suite of areas relating to the building stock. Over the past 18 years, there have been several *Building Research & Information* special issues exploring this theme in terms of technical, social, environmental and economic aspects as well as numerous papers in regular issues.

This special issue explores the governance options and regimes for addressing climate change in the building stock. Specifically, it investigates how building regulatory systems and related polices are addressing the current and future effects of climate change. It considers alternative governance approaches. Key questions are whether the focus and scope of building regulations are adequate, what can be done to improve them, how to accelerate change, what other regimes and policies are needed, whether more flexibility in regulation is needed and whether today's regulations sufficiently anticipate tomorrow's problems and needs. In posing these questions, one must ask what is a clear, efficient, transparent policy (and process) for ensuring radical reductions in greenhouse gas emissions from both individual buildings and the building stock as a whole and when/how the regulations are applied over the various stages of a building's life.

Although notable exceptions exist, the current approach to regulatory codes in the built environment largely originates from the past. It continues to be influenced by past events where adverse conditions often resulted in loss of human life and much financial damage. The result of this approach is not only a long time lag with the current requirements of the built environment but also a failure to anticipate and plan for the current challenges of mitigation and adaptation to climate change. It is evident that building regulations have been slow to engage with mitigating climate change and their design may not be suited to the task.

Many traditional regulatory systems tend to be mandatory and prescriptive. One way to address these challenges is for the governance of buildings to embrace more integrative solutions that may include voluntary alternative approaches. These voluntary approaches are initially established to provide some incentives or recognition for buildings that meet higher requirements. Some of these approaches are market based, others are led by altruistic concerns.

The safety and health of building users are still the main influence on the present regulatory system for the built environment, although regulation has considerably tightened over the past 30–40 years on some issues involving energy efficiency. When considered against the background of the Brundtland Report: *Our Common Future* (World Commission on Environment and Development, 1987), the lack of progress in the governance of the building stock is notable. For example, current building regulations take little account of dwindling resource availability and the impact of future climate change on buildings. These challenges necessitate a different regulatory approach in terms of the scope of concerns, and the points of intervention over the life cycle of a building.

However, some Western governments want to deregulate the regulation of buildings by reducing the role of the state and assigning more responsibility to market forces. A key challenge for building regulations is how to expand the focus from issues of inhabitant health and safety to include a wider remit of environmental health and safety. This central question challenges most traditional regulatory roles and responsibilities. It also necessitates alternatives for the governance of these issues.

It is worth remembering that governance consists of a broad landscape entailing many actors. It is not limited to only governmental regulation (and the mix of many different policies that comprise regulation) and is also composed of many other influences: standards, professional obligations, private contracts, insurance and finance requirements, judicial decisions, market conditions, etc. These (and the engagement of the inhabitants) all have a role to play in determining the actual outcomes for buildings.

Historical development of building regulations

The early origins of building regulations can be traced back to the first signs of urbanization in most civilizations (Visscher, Meijer, & Costa Branco De Oliveira Pedro, 2012). The oldest actual building regulations can be found in Babylon's Codex Hammurabi from 1700 BCE. This contains rules that set out the responsibilities of architects and builders with respect to structural safety. Fire safety was a basic issue in European towns during the Middle Ages. Buildings should not collapse and a fire in a single building should not easily spread to other buildings. Therefore, some towns at this time developed rules (*e.g.* for the spacing between individual buildings, a minimum width for streets, the use of fire-resistant materials). Aside from these local building regulations, owners, designers and builders discharge of their responsibilities was also a strong influence on the quality of buildings. In the Middle Ages the guilds developed rules for good craftsmanship that were of great importance in creating structural soundness and durability.

In the 19th century, industrialization prompted fast development of very small dwellings, built at high densities to accommodate labourers migrating from rural locations into towns. Concerns were raised about the unhealthy conditions in these dwellings. Most were built back to back, so very little daylight could enter and the internal air quality was poor. Some progressive factory owners and philanthropists created houses that offered better living conditions for labourers. Doctors realized that the living conditions of labourers in the towns had an increasingly important influence on their health. In the Netherlands, the government responded by developing laws and regulations to provide better dwellings. This led to the introduction of the Housing Law in 1901, providing a legal basis for municipalities to introduce minimum standard regulations and the right to enforce them through building permit procedures and powers to enforce improvements to dwellings that are responsible for dangerous conditions. Other countries experienced similar developments to protect the health of the individual inhabitants but also as a wider responsibility to protect 'public health' and reduce the potential spread of disease. A similar response to the poor standards of health and hygiene in Hong

Kong led to the enactment of the Buildings and Nuisances Ordinance 1856, which was one of the few earliest ordinances in the then British colony. Until the beginning of 20th century, the public regulation of building standards was mainly in the hands of municipalities. Throughout the 20th century there emerged an increasingly uniform and centralized regulation of building standards that are often independent of local circumstances. From the early 20th century, the influence of national and municipal governments on the quality of housing production grew steadily, until the early 1990s. By that time concerns about overregulation stifling development encouraged governments to reduce the size of public administration, and this has now became a dominant factor in some countries.

While structural failure and fire safety motivated the early development of building regulations, growing concern about poor living conditions and health (both for inhabitants and the wider spread of disease) prompted new regulations. Concerns for safety and health were also reasons for more detailed regulations concerning the dimensions, building materials and amenities of dwellings. Also 'usability' became a starting point to require minimum spatial dimensions of living rooms, bedrooms, kitchens, toilets and bathrooms. During the 20th century there has been a nearly continuous development in which the regulations were refined and extended to assure good basic minimum standard living conditions. Building physics in terms of damp, light, air, noise etc. were better controlled by requirements with respect to the quality of the building envelope and foundations. The provision of water, sewage, rainwater drains, and the supply of electricity and gas for heating and cooking were all subject to detailed regulation. However, heat insulation requirements with respect to the exterior of dwellings were only introduced in many Western countries in the 1970s as a response to the oil crises. In the Netherlands, energy performance regulations were introduced only in 1995, which moved away from a prescriptive basis and allowed a variety of options for saving energy in space and water heating. A broader anxiety about environmental sustainability began to emerge during the 1980s in some European countries. Until now it has been perceived as too complicated to define public regulations to limit the overall environmental burden of buildings (*e.g.* through the use of building materials).

Impact of building regulations on the built environment

Meacham's paper 'Sustainability and resiliency objectives in performance building regulations' starts with a comprehensive overview of how the current building regulatory system developed and evolved. This identifies the mounting pressure on the building regulatory system to respond to specific climate-related issues.

The precarious position of existing buildings is highlighted for their limited regulatory oversight. Based on the regulatory policies of the 12 countries participating in the Inter-jurisdictional Regulatory Collaboration Committee (IRCC), Meacham concludes that the issues of sustainability and resiliency are not receiving the same level of attention as health and safety. A better understanding of the holistic building performance is needed if regulatory regimes are to achieve safe, healthy, sustainable and resilient buildings.

The governance of buildings and the building stock embraces the notion of time. In most developed countries, the building stock changes slowly over time. With buildings lasting between 50 and 300 years, they have to respond to current and future needs. Effective governance of the building stock needs to account for change over time and incorporate future scenarios into the requirements for today's buildings.

In 'The realpolitik of building codes: overcoming practical limitations to climate resilience', Shapiro highlights the degree of hostility towards the adoption of codes that moderate the impacts of climate change in the US. The political resistance (often from lobbyists on behalf of particular vested interests, *e.g.* the house-building industry) is based on a claim that the associated actions would raise the cost of individual entrance into the property market, specifically the housing market. These claims are debunked as the actual increase is a small amount which also has a reasonable payback period. Nevertheless, these are political influences that need to be acknowledged and countered. Following the disasters of precipitated by Hurricane Katrina in 2005 and Hurricane Sandy in 2012, various reports called for a more resilient built environment that protects communities from all types of hazards and disasters. Shapiro puts matters in perspective by stating that 'the real cost impacts of climate change are borne by society as a whole, not the individual building owner'. Shapiro argues that this opportunity will remain unrealized until 'participation, pricing, and policing' are addressed. The governance of buildings (*e.g.* building regulations) depends on those participating in its framing – which needs to include an explicit acknowledgement of the professional and ethical obligations to protect the public good by taking a more holistic assessment of the larger impacts (*cf.* Bordass & Leaman 2013; Hill, Lorenz, Dent & Lützkendorf, 2013). At the other end of the spectrum, the enforcement of regulations is equally important – both having the necessary capabilities and the will to enforce.

Inadequacy of current building regulatory systems

The regulation of energy in buildings is still undertaken on a gateway basis – usually through simulations or 'deemed to satisfy' solutions when the building is at the design stage. This fails to recognize two significant factors that affect energy demand: namely the performance gaps between anticipated and actual consumption, as well as the intensity of use.

Buildings will increasingly encounter changing environmental conditions in the future. In order to address this, Eisenberg suggests a process that identifies, evaluates and considers hazards across different categories, timeframes, locations of impact and impacted populations. In 'Transforming building regulatory systems to address climate change', he compellingly argues for an expansion of the regulatory role to enable regenerative outcomes. This is required because the inherent characteristics of built environment regulatory systems ignore the significant large-scale hazards associated with climate change and environmental degradation. A call is made for new regulatory practices in the US (and by extension, elsewhere) that are based on systems thinking. These practices can be built on current knowledge while integrating it with the larger framework of risk and responsibility.

Climate change will impact extensively on the regulation of both new and existing buildings. As noted by Meacham, the emergence of new building regulations and standards focusing on climate change mitigation, can result in unintended consequences. Mulville and Stravoravdis' paper, 'The impact of regulations on overheating risk in dwellings', explores how regulations focus mainly on the 'point of handover' or the pre-occupancy phase. They successfully show that at this moment the building may not be fit for its intended purpose. The creation of a highly insulated, energy-efficient building can result in a building that overheats and becomes uninhabitable from its first day of habitation. This suggests that regulations may also be failing in one of their traditional roles: to protect the health and safety of the inhabitants. It points to a larger, looming public health crisis of a failure to anticipate future conditions (*e.g.* increased summer temperatures and occupant heat stress).

Addressing the performance gap

The actual results of energy policy tools are often less than anticipated. According to Karatas, Stoiko and Menassa, one reason for this is caused by policies focusing on the cost and ease of installation while neglecting the role of occupant behaviour. Their paper, 'Framework for selecting occupancy-focused energy interventions in buildings', emphasizes factors contributing to occupants' sustainable behavioural patterns. To include these concerns, the authors propose a multilevel intervention strategy tailored to specific occupant characteristics. With this approach it is expected that large-scale energy savings for buildings (maintained over time) will be achieved. The significance for governance and policy-making is clear: control of the traditional gateway (building design and build quality) is insufficient. Effective governance will also need to consider the behaviours, demands and management of the building during occupancy.

The aim of a carbon-neutral building stock by 2050 as set by the European Union is highly unlikely without strict and supportive governmental policies on energy saving. In accordance with the Energy Performance of Buildings Directive (EPBD) and the Energy Efficiency Directive (EED), various member states have developed and strengthened their respective energy performance regulations. In 'Improved governance for energy efficiency in housing', Visscher, Meijer, Majcen and Itard analyse the effectiveness of current European energy governance tools as applied in the Netherlands to determine its actual impact on CO_2 reduction. Currently, the actual energy consumption in the residential sector in the Netherlands differs significantly from the required or modelled energy use. Understanding and influencing occupant behaviour is found to be an essential link for achieving the predicted energy performance of buildings. Other contributing factors include improving the capabilities for prediction by using actual performance measurements; increasing the knowledge base and skills of building professionals as well as rethinking the current building control and enforcement processes. Given these observations, schemes offering an energy performance guarantee are presented as an alternative approach to realize the energy saving potential of the existing building stock.

Innovative policy solutions

The governance of buildings is not limited to building regulations. Many different policies, instruments, barriers and incentives influence the creation and operation of buildings and the wider built environment. In this special issue, various alternatives approaches that have contextual relevance are presented. Learning from these is essential in achieving a sustainable and resilient built environment that is appropriate in responding to the challenges of climate change.

In 'Energy efficiency and the policy mix', Rosenow, Fawcett, Eyre and Oikonomou present an empirical study of the current policy mixes that focus on building efficiency in 14 European Union countries. In contrast to single-policy evaluation, the effectiveness concept is employed to observe the interaction of 55 policy pairs. The study identifies energy savings policy mixes 'likely to deliver more'. Its contribution is to show how the different mixes of building policies could be made more effective as well as identifying current knowledge gaps and highlighting key research needs. The authors also emphasize the need for policy and instruments to use actual data from buildings' operational phase.

This will ensure that potential energy savings from interventions are actually realized and maintained.

In 'Alternative building emission-reduction measure: outcomes from the Tokyo Cap-and-Trade Program', Nishida, Hua and Okamoto reflect on five years of implementing the Tokyo Cap-and-Trade Program (TCTP). The results indicate that this alternative policy instrument has a unique capacity to reduce energy consumption and greenhouse gas emissions. Unlike other instruments, this focuses on measured performance in relation to actual reduction targets and therefore is able to bridge the issues of design, build quality, commissioning, operation and energy source. It also allows the owner to ascertain how to best meet the emissions target, which explicitly acknowledges the diversity of circumstances affecting buildings and their uses. By doing so, it provides for options for different solution paths. The cap-and-trade approach allows incremental, tighter targets to be imposed over time. It has successfully created significant reductions and offers a new approach to governance based on actual performance.

In response to climate change, some governmental policies propose 'zero-energy' buildings (ZEBs) as the ideal. The absence of a universally accepted definition of a ZEB creates difficulties. Zhang, Zhou, Hinge, Feng and Zhang highlight the fact that traditional building regulations cannot address all key ZEBs issues. At the same time, different policy measures, regulatory codes and governance structures for ZEBs exist. Using Denmark and the US state of California as examples, their paper, 'Governance strategies to achieve zero-energy buildings in China', reviews current ZEBs policies, programmes and governance approaches with the prospect of implementation in China. For possible implementation in China, Zhang et al. suggest a participatory governance process in which all relevant societal actors actively participate. An argument is made for the inclusion of hierarchical, market and network elements as part of the policy development and implementation phases.

Another paper about China, 'Multilevel governance for building energy conservation in rural China', highlights the need for contextual consideration when developing regulatory strategies. As mentioned above, in many countries there has been a tendency to develop a national framework for building regulations. According to the authors, Sha and Wu, the current Chinese building governance system distinguishes between urban and rural areas, resulting in an almost 'dual' building economy. Using a perspective of economic governance, this paper examines different mechanisms and strategies that would be appropriate for promoting building energy conservation in the rural areas of Shandong province, China. The authors develop a three-level analytic framework in response where markets, governments and the third parties (professionals and others) work together regulating stakeholder behaviour. A case is made for building governance to respond to local conditions by understanding what particular drivers would (in this case) promote energy efficiency.

Several aspects of governance are determined by the private sector, whether by intent or happenstance. In 'The evolution of green leases: towards inter-organisational environmental governance', Janda, Bright, Patrick, Wilkinson and Dixon examine the interplay between different property interests, obligations and outcomes. The rental agreement for a building or piece of land may be contributing directly to a split incentive problem between the landlord and tenants. A private instrument called a 'green lease' is shown to be a constructive solution enabling both landlords and tenants to achieve specific environmental targets. Green leases represent a new form of private environmental governance that could promote polycentric governance, ultimately resulting in a more resource efficient built environment. Is there potential for the public sector to encourage or incorporate this into part of the policy mix?

Historic buildings

The adaptive reuse of existing buildings is a practical substitute to demolition and landfill. Additionally, the need for sustainable conservation of heritage buildings is increasingly acknowledged and accepted. In 'Governance of heritage buildings: Australian regulatory barriers to adaptive reuse', Conejos, Langston, Chan and Chew argue that historic buildings have continually existed in a changing climate. A central question is whether and how building regulations (as applied to heritage buildings) have the flexibility to accommodate green building technologies and heritage conservation principles. As a viable alternative, the authors propose a protocol that could guide the conservation of the built heritage according to sustainability and adaptability aspects. This can provide insights into reducing the unintended consequences created by policies and regulations that seek to upgrade the building stock, but result in increased demolition and increased use of resources.

Financial incentives

The costs of voluntarily implementing green building strategies are often perceived to be prohibitive and therefore present a barrier to some developers and investors. Shapiro noted that the real costs associated with climate change adaptation should be borne by society which, in turn, raises the question of what constitutes fair allocation of costs, risks and responsibilities. Simply stated: who should pay for the additional costs and who benefits from the reduced risks? Whether this is resolved through a regulatory system that demands higher standards for all stakeholders or a voluntary system that provides incentives is a significant question for civil society to consider. In addition to the question of fairness and social justice, one might ask which system actually implements the broadest change in a short period of time.

One interesting financial instrument is a gross floor area (GFA) concession scheme, currently implemented in Hong Kong and Singapore. This voluntary scheme enacted by government provides a financial incentive to developers and has contributed significantly to the market uptake of green building practices. These concession schemes link government land sale conditions with developmental control measures. In 'Regulatory incentives for green building: gross floor area concessions', Qian, Fan and Chan find that developers are willing to pay a premium for a concession that allows them to develop the property to a higher GFA. Although the apparent costs of introducing green building systems and technologies are evident, there are other hidden costs and benefits, and their fair allocation to different stakeholders are often overlooked. The use of any incentive scheme relies on maintaining a balance between the public good and the private benefits – which are likely to shift over time. By examining the GFA concession schemes in Hong Kong and Singapore, this paper identifies other costs and benefits to both the industry and society that could inform their future evolution and the creation of other similar schemes elsewhere.

Future governance

A more sophisticated system of governance is needed which, at the very least, sets clear limits on building energy use in practice. A strong body of evidence exists for what can be reasonable limits in different contexts. The governance must also shift from being based only on hypothesized consumption, to actual evidence of energy use (not a simulation).

In 'The new governance for low-carbon buildings: mapping, exploring, interrogating', Van der Heijden emphasizes the limitations of traditional, mandatory governance instruments to achieve low-carbon buildings. The paper evaluates over 50 'new governance' instruments for low-carbon buildings. These 'new governance' approaches developed in reaction to the confines of traditional governance instruments that were struggling to actually achieve low-carbon buildings. The central question addressed in this paper is how different forms of governance can accelerate a large scale transition to low-carbon buildings. A change in the application of new governance instruments is recommended by Van der Heijden. Firstly, the current leadership and showcasing purposes have to be questioned during the development of the second generation instruments. Secondly, these instruments have to be positioned carefully within the local, national and international regulatory frameworks. Ideally, it should more mandatory and/or or less voluntary. Thirdly, the contribution and interaction between new governance instruments and existing policy mixes need careful consideration. An experiential model is proposed where the mandatory bottom line of the future is informed by experience gained from today's leading practices.

With an ambitious 2050 greenhouse gas emissions reduction target of 80–95% below 1990 levels, the European Union (EU) developed a range of policy instruments. Drummond and Ekins argue that EU member states need to target the residential sector as a significant contributor to CO_2 emissions. Their paper, 'Reducing CO_2 emissions from residential energy use', examines current EU and UK policies by grouping them into three 'pillars of policy'. The UK serves as an illustrative member state. Using instrument coverage and effectiveness as metrics, the strengths and weaknesses of each pillar are evaluated. The authors recommend a range of priority reforms. These include improved enforcement of regulatory requirements, the expansion of carbon pricing to include heating fuels (coupled with the removal of residential subsidies where applicable), requiring a net-zero energy performance of new buildings and a post-2020 commitment to address the energy consumption of existing buildings.

Enforcement

Regulations without enforcement are useless. Eisenberg states that effective enforcement is essential for regulations to provide their intended outcomes and benefits. In the US the lax enforcement of energy codes have been an ongoing issue partly because they were perceived to be outside the traditional remit of building codes. Shapiro also stresses the significance that enforcement has for the energy codes. Visscher et al. note inadequate enforcement as one of the reasons for the energy performance gap of new buildings. Within a broader landscape where responsibilities in some countries are shifting from the public sector (national and local governments) to the private sector (building owners and private organizations), the accurate enforcement of regulations will become increasingly important.

In 'Comparative review of building regulation: a quality perspective' Lord, Noye, Ure, Tennant and Fisk focus on the commissioning of energy related installations (e.g. building services). They successfully argue that a thorough commissioning is needed to ensure the energy efficiency of the installations. A convincing argument is made that regulations (for non-domestic buildings) need to extend their remit to include this aspect which means that enforcement would cover the commissioning of equipment and systems. This would result in the optimization of building performance and a reduction in energy demand and CO_2 emissions.

In the traditional systems of building regulations the link between regulations, control and enforcement actions was evident: full responsibility resided with municipal legal authorities. In the past decades more and more of the actual control actions have been exported to private parties. The process of enforcement under these changed (i.e. private) legal conditions creates new problems and dangers: the new approach might not work satisfactorily, especially when compared with the traditional situation. In the privatized organization models, the legal authorities may still need to be empowered in order to enforce the regulations particularly if there is reluctance by private organizations to do so. The system should ensure that the authorities remain sufficiently empowered and are also provided with enough information to enforce.

Conclusions

Although automobiles are a mass market commodity (unlike most buildings which are bespoke, unique objects), autos are tested for their performance in two ways: (1) before they are marketed (for actual fuel consumption, emissions, crash tests, etc.) for their hypothetical performance (although this is based on standardized test measurements); and (2) in many jurisdictions vehicles must undergo annual, independent checks for their emissions and other safety features to ensure their actual performance is within a prescribed limit.

In comparison, there is little or no governance to ascertain a building's actual annual performance and therefore little opportunity to ascertain whether a building's performance is optimized (i.e. commissioned correctly, working correctly, understood and operated correctly). The exceptions to this are Tokyo's exemplary TCTP and, to some extent, the European EPBD. The success of Tokyo's scheme depends on having a rich dataset for the buildings, which includes historical and current performance.

This sets a number of challenges for regulators and, in turn, wider civil society and the construction industry:

- Setting norms for buildings (which may need to vary according to morphology, location, and type) in terms of kWh/m^2 and CO_2e/m^2 or a similar metric for (1) space heating/cooling, (2) lighting and (3) hot water. Many of these already exist for low carbon buildings, but there is not yet a system for regulating and validating this in use.

- Consideration about the intensity of building use and whether additional metrics (*e.g.* kWh/person or CO_2e/person) are required.

- Providing clear, understandable feedback to building occupants (and others) on building performance and its operation (which need to include factors such as indoor environmental quality, outside weather conditions, etc.). This would identify incorrect or profligate use. (Smart meters alone will not do this.)

- Creating a database for the energy consumption of buildings – correlating energy bills to the geographical information system (GIS) coordinates and actual premises.

- Mindset change for building regulatory systems to embrace innovative approaches that integrate alternative solutions and voluntary measures.

- Scale: developing new forms of governance using both incentives and deterrents to accelerate energy reductions in the building stock. This may be on a building or neighbourhood basis (as not all old buildings can be expected to meet new criteria so synergies between buildings can be encouraged). It may also involve aspects of planning and place new demands on buildings when they are exchanged.

- Enforcement: any form of regulation stands or falls with the enforcement of the regulations. The low level of enforcement strength in privatized building control systems is a serious concern.

The challenges are not only for governments. The construction industry (supported by its research base) has a significant role to not only create the technical solutions but also to rethink how this can be effectively implemented, monitored and governed. The research community has a wider role beyond supporting the construction industry. Given that civil society and policy-makers (and not the construction industry) will determine the policies, strategies and instruments for the governance of buildings, the research community should broaden its engagement to actively involve these key stakeholders. It is important for the research community to be seen as an independent agent able to assess policy formulation and outcomes for all the stakeholders involved.

The baselines for many low energy buildings in different parts of the world now exist. The challenge is moving from these prototypes to a system of governance that ensures the built environment plays a significant role in climate mitigation and adaptation.

Although much research already exists on occupant behaviours, social practices, energy demand and thermal comfort, an urgency now exists to incorporate this knowledge in the formulation of smarter policies and instruments.

Henk Visscher
Jacques Laubscher
Edwin Chan

Acknowledgements

This special issue originated from the efforts of the CIB Task Group 79: Building Regulations in the Face of Climate Change. This task group focuses on the themes of governance and policies, regulatory practices, existing buildings, life cycle approach and resulting effects. The special issue is the product of a two-year collaboration between Henk Visscher, Edwin Chan and Jacques Laubscher with the journal's Editor-in-Chief Richard Lorch. It is based on an international call for papers and the journal's independent, double-blind peer-review process.

References

Bordass, B., & Leaman, A. (2013). New professionalism: Remedy or fantasy? *Building Research & Information*, *41*(1), 1–7.

Hill, S., Lorenz, D., Dent, P., & Lützkendorf, T. (2013). Professionalism and ethics in a changing economy. *Building Research & Information*, *41*(1), 8–27.

Visscher, H. J., Meijer, F. M., & Costa Branco De Oliveira Pedro, J. A. (2012). Housing standards: Regulation. In S. J. Smith, M. Elsinga, L. Fox O'Mahony, O. Seow Eng, S. Wachter, & G. Wood (Eds.), *International encyclopedia of housing and home* (pp. 613–619). Oxford: Elsevier Science.

World Commission on Environment and Development. (1987). *Our common future* (The Brundtland Report). Oxford: Oxford University Press for the United Nations World Commission on Environment and Development.

Sustainability and resiliency objectives in performance building regulations

Brian J. Meacham

Building regulatory agencies worldwide are grappling with how to define and implement appropriate mandatory and voluntary measures for new and existing buildings that address societal and political demands for increased environmental and resource sustainability and resiliency to the effects of climate change without lessening the historical building regulatory focus on health, safety and welfare of building occupants. It can be argued that a transition from prescriptive to performance-based building regulatory regimes, coupled with the introduction of new policy objectives for sustainability and resiliency, in a rather short period of time, without full assessment of how they interact with existing building regulatory objectives, and without broadly agreed holistic solutions, has led to the introduction of new objectives that have the potential to result in increased hazards and risks to occupants. To explore the current situation and future needs associated with performance building regulatory regimes and the inclusion of sustainability and resiliency objectives for new and existing buildings, the literature was reviewed and a survey of building regulatory bodies and institutions in 12 countries was conducted to obtain perspectives on whether and how sustainability and resiliency objectives are being incorporated into their building regulations and if any challenges have been identified.

Introduction and context

Historically, building regulations have focused on the health, safety and welfare of building occupants. They emerged in response to widespread illness, death and destruction that occurred in urban centres as a result of unsanitary conditions and significant hazard events, and the social and political mandate to mitigate these hazards as part of urban redevelopment. A social compact (contract) developed between government and the citizenry in which it became acceptable for the government to collect taxes in return for the regulation and enforcement of minimum levels of safety and sanitary conditions. As such, early building regulations addressed such issues as minimum requirements for fire separation and resistance of materials, structural resiliency to natural hazards, and safe heating and sanitation systems for occupants. Over time, needs such as standardized testing and listing to assure minimum performance, industry standards for compatibility of systems and components, minimum competency of practitioners, and mechanisms to assure compliance of constructed buildings with stated designs gave rise to the other components within the building regulatory system, all based around a culture of facilitating a built environment that delivered minimum levels of safety and health to building occupants (Cheit, 1990; Field & Rivkin, 1975; Geschwind, 2001; Hemenway, 1975; ICC, 2007; Imrie & Street, 2011; Meacham, 2014; Wermeil, 2000; Yatt, 1998).

Generally speaking, changes to the regulations and supporting regulatory system occurred slowly as new materials, products, systems and methods of analysis, design and construction were developed, tested, introduced and adopted as acceptable into the regulatory system. The process can take many years, as with the development and implementation of the single burning item (SBI) test (van Mierlo, 2005) or even decades, such as with the development and implementation of the Eurocodes for structural design (Eurocode, 2015). In other cases, change occurred rapidly, but nearly always in response to some significant hazard event which resulted in unacceptable damage, injury or death. In these situations change was often made to the building regulations first, with changes to standards and other instruments following, with the intent of avoiding similar losses in the future. When this type of change occurred, however, it was primarily focused on new construction, and even when upgrade of some of the existing building stock was required, it was often not to the same level of new construction: a trend that continues today (*e.g.*, IREM, 2013; Stannard, 2014). There are obvious reasons for this, primarily the cost of retrofitting all existing building stock in comparison with the overall risk and benefits. In order to provide options to legislated provisions, voluntary standards or guidelines (*e.g.*, NFPA 101A, 2013; FEMA 356, FEMA 114) and/or market instruments (*e.g.*, insurance industry requirements, such as FM Global Data Sheets (*e.g.*, FM Global 1–28) are used to encourage building owners to increase the safety performance of existing buildings.

While there have always been some challenges in addressing issues within the historical building regulatory system as outlined above, the focus on health and safety, and the development of regulatory infrastructure and generally common understanding of those issues amongst stakeholders around these topics, had created an environment where many of the issues were able to be addressed proactively and cooperatively. In recent years, however, three concurrent activities have emerged within building regulatory systems that have placed pressure on the ability of the systems to adapt and respond in as successful a manner as in the past: a relatively rapid transition to performance-based regulations, new policy objectives in the area of sustainability, some of which appear to be difficult to harmonize with traditional health and safety objectives, and a desire to address, but lack of clarity as to how, climate change impacts on buildings.

When the transition to performance-based (function-based) building regulatory systems and regimes began, health and safety was still the focus. The motivation for change was related to reducing regulatory burden, reducing costs to the industry and the public, increasing innovation and flexibility in design, and being better positioned to address emerging issues (BRRTF, 1991; May, 2003; Meacham, 2009; Meacham, Moore, Bowen, & Traw, 2005; Meijer & Visscher, 1998; 2006). All this was to be achieved while maintaining tolerable levels of safety and performance.

In some cases the transition has worked reasonably well: in other cases there have been issues, including some significant system failures (Lundin, 2005; May, 2003; Meacham, 2010; Mumford, 2010). With respect to failures, contributing factors include lack of agreed performance measures (criteria) and means to predict performance in use, lack of test methods that yield data that can be used in engineering analysis, limited availability and quality of data, inadequate

competency and accountability in the market and of those in oversight (compliance checking) roles, insufficient product certification/means to assure the performance of products, lack of transferring knowledge between regulatory regimes, lack of having persons with the right expertise and experience involved, and challenges with insurance, liability assignment and limitation, and consumer protection mechanisms (Lundin, 2005; May, 2003; Meacham, 2010; Mumford, 2010). These shortcomings become particularly important when one considers that buildings have been constructed under these systems for some 25 years, and now might reflect a significant contribution to the existing building stock in many countries. As such, if performance issues exist, the impact could be significant. The 'leaky building' scenario that unfolded in New Zealand is probably the highest profile example of what could happen (May, 2003), but there may be other latent problems as well.

Part of the challenge is that performance-based building regulatory systems, while similar in structure, can vary significantly in implementation. While the structure of the building code (regulation) itself follows the general format outlined by the Nordic Building Code Committee (NKB) in the 1970s (Meacham 2004; Meacham et al., 2005), there can be significant variation in the means for demonstrating compliance with the mandatory provisions (Meacham, 2009) (Figure 1).

In some countries there are no mandatory means of demonstrating compliance, and it is the designer's choice to follow 'deemed to comply' documents, which may reference national or international standards, or to take a 'first principles' (performance-based) approach. England and New Zealand are examples. In other countries there are mandatory compliance documents, which in turn reference national or international standards, and variance from the requirements is by exception. There are also situations in which performance criteria and means of verification are built into the code (regulation). A good example here is Japan.

In addition to the transition to performance-based systems, building regulations and regulatory systems have increasingly become complicated by the establishment of policy mandates and the introduction of voluntary assessment and rating instruments originating from environmental sustainability and climate change resiliency concerns, which have historically been outside the realm of building regulation. While a social compact between government and the populace has developed around the understanding that we need to do more to protect the environment, resulting in the proliferation of environmental regulation and enforcement legislation which nominally began in the 1970s, it is only recently that the connection of the

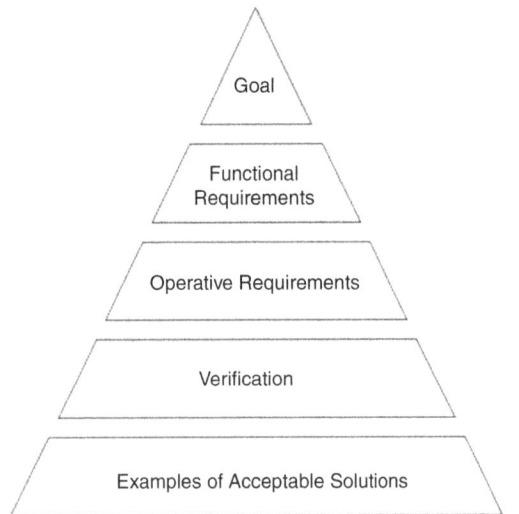

Figure 1 NKB hierarchy
Source: Meacham et al. (2005)

built environment to negative impacts on the environment, and the need to make buildings more resilient to climate change, have come onto the political agenda. A result is new pressures on building regulations from new actors, inside and outside of government.

These new pressures pose a significant challenge, not just because the traditional building regulatory environment is itself undergoing change and has structural challenges to overcome, but because the success of recent governmental policies and market approaches aimed at increasing the sustainability of the built environment has arguably been limited (Simmons, 2015; Van Bueren & de Jong, 2007). Whereas a robust approach to engaging stakeholders in issues of health and safety developed over decades in the historical building regulatory system, new stakeholders have emerged around sustainability objectives, and the different groups are fragmented and not working effectively together (Cole, 2011; du Plessis & Cole, 2011; Foxell & Cooper, 2015; IPCC, 2007; Simmons, 2015; Van Bueren & de Jong, 2007). In addition, the introduction of voluntary measures have resulted in inconsistent levels of performance being realized (Newsham, Mancini, & Birt, 2009; Scofield, 2009), in part because it rests outside the realm of regulatory oversight. The situation is further complicated because there are incomplete building performance measures, monitoring and enforcement mechanisms for sustainability (Van Bueren & de Jong, 2007) as well as increasing liability concerns (Brinson & Dolan, 2008). As noted by Foxell and Cooper (2015, p. 403),

> The gap between policy intent and effective solutions remains difficult to close. This serves to increase the number of challenges that the built environment sector needs to address successfully.

This fragmented regulatory approach, not necessarily having the right persons involved, and the introduction of new social objectives has led to unintended consequences being introduced, some of which could present considerable risk to building occupants. This includes structural hazards due to moisture-related failures of enclosed structural systems (May, 2003; Mumford, 2010), health hazards related to mould and indoor air quality due to weather-tight buildings (Mudarri, 2010), fire and health hazards due to the flammability of thermal insulating materials (Babrauskas et al., 2012; Simonson McNamee, Blomqvist, & Andersson, 2011), fire and smoke spread potential through the use of double-skinned facades (Chow, Hung, Gao, Zou, & Dong, 2007), and fire hazards and impediments to emergency responders associated with interior and exterior use of vegetation, photovoltaic panels, and other 'green' features and elements (Meacham, Poole, Echeverria, & Cheng, 2012).

Finding a suitable balance between sustainability and fire safety objectives can be particularly complex due to the multidimensional aspects of each. Timber is 'sustainable' but also is combustible, so if not addressed appropriately can present a significant fire safety hazard (Meacham et al., 2012; UL, 2008). High-strength concrete requires less material and is more sustainable than regular-strength concrete, but can be highly susceptible to spalling during a fire (Kodur & Phan, 2007). Insulation and alternative energy sources are good for sustainability, but photovoltaic panels that can cause an ignition, and flammable insulation material, can be a catastrophic combination (Meacham et al., 2012).

Tightly coupled with goals to increase environmental sustainability through better energy efficiency, reduction in/reuse of construction materials, incorporation of alternative energy sources and related measures in the built environment are goals to increase the resiliency of the built environment to climate change effects. In recent years we have seen devastating hurricane, cyclone, flood, snowfall, drought and wildland fire events and seasons. Each has resulted in significant impacts on the built environment, including widespread building damage associated with moisture, wind, snow load and fire. This is not unexpected, as it has been predicted as a real possibility for some time (*e.g.*, IPCC, 2007; Steenbergen, Geurts, & van Bentum, 2009; Wilby, 2007).

However, there are no easy answers in terms of developing comprehensive resiliency strategies, since while the problems are easy to recognize, the solutions are difficult to agree and implement. In many cases there is not a single policy area that has responsibility for avoiding or mitigating the impacts. Planning, zoning, environmental and resource legislation have a significant effect on the susceptibility of buildings to flooding and wildland fire (as is said in real estate (property) valuation: location, location, location). In some cases, policy-makers wish to avoid moving people or restricting expansion into hazard-prone areas if that has an impact on economic development. That places a burden on building regulation. Some of this can be addressed in regulations for new construction; however, affordability then becomes a concern. The challenges become even more amplified when addressing existing buildings, as there is less regulatory oversight and often less economic capacity to manage from the ownership side (*i.e.*, older buildings, particularly residential, house a higher percentage of lower-income families).

In summary, the literature suggests that advancing sustainability and resiliency objectives in the built environment is necessary, in addition to traditional occupant health and safety objectives. However, challenges exist in that sustainability and resiliency are not yet being viewed as having the same level of importance, or equivalent level of social compact between government and the public, as providing for minimum levels of health and safety in buildings. As such, the rather sudden entry of these new policy objectives has created a wide range of challenges, from regulatory development to enforcement, with potentially the most significant issues around existing building stock and trying to assure regulatory and market instruments adequately address the spectrum of policy objectives without increasing hazards, risks or costs, or decreasing building performance. As a means to begin assessing just how significant the issues are in practice, a survey was conducted of representatives of lead building regulatory agencies and research institutions in 12 countries to benchmark the situation in a number of areas. Select outcomes from the survey and preliminary analysis, along with select outcomes from a 2009 survey of the same group, are presented below. It should be noted that the information provided represents the author's interpretation of the survey responses and materials provided in the countries cited, and do not reflect official government or organizational policies or positions.

Performance, sustainability and resiliency in IRCC member countries

In 1997, six countries that were developing or had implemented performance-based building regulations created a forum where they, and countries with similar interests, could discuss experiences, issues and opportunities associated with performance-based building regulatory systems. This forum, the Inter-jurisdictional Regulatory Collaboration Committee (IRCC), has grown considerably since then and now has members from 12 countries – Australia, Austria, Canada, China, Japan, New Zealand, Norway,

Scotland, Singapore, Spain, Sweden and the US – with affiliated organizations participating from England and the Netherlands (see www.IRCC.info). Over the years, the IRCC has hosted several workshops and published a number of documents addressing a wide range of topics related to performance building regulatory systems (*e.g.*, Meacham, 2010; Meacham et al., 2005). In 2009, a report was prepared that reinforced the principles underlying the need for building regulation, the benefits that IRCC members see with performance-based approaches, and the building regulatory situation in the IRCC member countries at that time (Meacham, 2009). A general conclusion is that all IRCC member countries support performance-based building regulatory regimes, acknowledging that there are challenges with current systems and that more needs to be done to facilitate the full potential of such systems, including the following:

- Mechanisms are needed to define and quantify better levels of tolerable building performance, be they in terms of health, safety, welfare, risk, sustainability or other measures.

- Quantified performance metrics must be developed and incorporated into regulations. Recognizing that some metrics may be best addressed prescriptively (*e.g.*, rise and run of a stair), there remains significant scope for performance measures, for which associated verification methods are needed.

- Tools and methods for helping with the enforcement of performance-based building regulations are still lacking. In part related to the lack of quantified performance measures, those responsible for approval of designs and enforcement of regulations are faced with the challenge of making decisions in the face of significant uncertainty.

In 2014, several IRCC members presented papers at the World Sustainable Building Conference SB14 in Barcelona, Spain, on topics related to sustainability and resiliency objectives and issues for building regulations (*e.g.*, Dodds, 2014; McDonald, 2014; Neng Kwei Sung, 2014; Serra, Tenorio, Echeverria, & Sanchez-Ostiz, 2014). These were complemented by contributions of members of CIB Task Group 79, Building Regulations and Control in the face of Climate Change (*e.g.*, Chan, Chan, & Quin, 2014; Laubscher, 2014; Visscher & Meijer, 2014). Collectively, these papers, as well as other literature which was reviewed, identified several issues worthy of additional investigation around this topic.

To begin exploring some of the issues in more detail, from a building regulatory perspective, a survey was conducted of the IRCC members to obtain additional information about current policies, activities and concerns around sustainability, resiliency and building regulation. The survey consisted of 30 questions within three primary topic areas:

- The relative importance/priority of sustainability and resiliency principles and policies to the government – in general and with respect to the built environment.

- How concepts of sustainability and resiliency are addressed within building regulations, and within the built environment, for new and existing buildings.

- How well the performance- (objective-, function-) based building regulatory system has been working, with a particular emphasis on aspects that have been identified as somehow lacking (or failed, or could be significantly improved), why the challenges exist, whether actions are underway to change the system going forward, and what 'the next generation' of the regulatory system might include.

An open-ended question to provide for additional feedback was also provided.

Surveys were sent by e-mail to each IRCC member and affiliated organization and responses were consolidated. The survey questions are given in Appendix A in the supplemental data online. Due to the small sample size (12 reporting entities), no statistical analysis of the responses is provided. Rather, summary data in terms of whether or not a country addresses the issue at question are presented.

Due to space limitations, the focus of the discussion that follows is limited to feedback received as related to top-level government policy/perspective on sustainability and resiliency and how these topics are addressed within the building regulatory system. Furthermore, not all responses from each country are included, and only examples from some countries are provided. It is anticipated to present more outcomes in future.

It should be noted that sustainability and resiliency were not defined in the survey, and countries were free to interpret the terms. This was felt to be important to gauge commonality and differences. In many countries, sustainability encompasses the management of resources (water, energy, materials, etc.), and resiliency reflects the ability to withstand and recover from impacts – from earthquake to climate change impacts, such as drought, increased storm frequency and severity, changing snow locations and amounts, and similar. While the entire paper could be focused

on the definitions and broad government policy responses, it was decided to keep the focus limited to which countries have policies to address sustainability and resiliency concerns, and how it manifests within the building regulatory system.

From a top-level (overarching) national government policy perspective, is sustainability a guiding principle for legislative/regulatory development?

Yes	No	Variable
Australia, China, England, Japan, Netherlands, Norway, Scotland, Singapore, Spain, Sweden	New Zealand	US

The clear message here is that sustainability has become embodied as a guiding principle in most countries at the national government level. While sustainability is embodied in many legislated areas in New Zealand, there is no 'top of government' sustainability guidelines or policy. The reason that the US is listed as 'variable' is that the priorities can change with administrations, and while the Barack Obama administration has a strong climate change and sustainability focus, subsequent administrations may not. This is not to say that other countries are not also susceptible to shifts in policy and politics, but in many countries policy guidance has been implemented, which reflects a firmer position, at least for the moment.

In Australia, it was recognized that significant benefits could be achieved through improvements in energy efficiency, so in 2004 the Australian Ministerial Council on Energy (which comprises representatives from all levels of Australian government) endorsed the National Framework for Energy Efficiency (NFEE) and agreed to the implementation of a number of energy efficiency packages. The Ministerial Council on Energy anticipated that the NFEE would comprise more than one stage, and activities under the NFEE continued for the next several years. In July 2009, the Council of Australian Governments (COAG) agreed to the comprehensive, 10-year National Strategy on Energy Efficiency (NSEE), to accelerate energy efficiency improvements and deliver cost-effective energy efficiency gains across all sectors of the Australian economy. The overarching National Partnership Agreement on Energy Efficiency (NPAEE) is the Intergovernmental Agreement that gives effect to the NSEE and sets out specific action to be taken by the commonwealth, state and territory governments to maximize cost-effective energy efficiency gains across the economy.

The NSEE aims to streamline roles and responsibilities across government by providing a nationally consistent and coordinated approach to energy efficiency. The NSEE is framed around the following four key themes: assisting households and businesses to transition to a low-carbon future, reducing impediments to the uptake of energy efficiency, making buildings more energy efficient, and government working in partnership and leading the way, which was published in 2009 (NSEE, 2009).

In Japan, the Basic Environment Law sets out the overall mandate for environmental conservation and sustainability, the Purpose of which is stated in Article 1:

> The purpose of this law is to comprehensively and systematically promote policies for environmental conservation to ensure healthy and cultured living for both the present and future generations of the nation as well as to contribute to the welfare of mankind, through articulating the basic principles, clarifying the responsibilities of the State, local governments, corporations and citizens, and prescribing the basic policy considerations for environmental conservation.
>
> (BEL, 1993)

The stated purpose of the Scottish government is to focus government and public services on creating a more successful country, with opportunities for all Scotland to flourish, through increasing sustainable economic growth (Scotland, 2015). The Scottish parliament passed the world-leading Climate Change (Scotland) Act 2009 (CCSA, 2009), committing to reduce Scotland's greenhouse gas emissions by at least 42% by 2020 and by 80% by 2050 compared with 1990 levels. In addition, the Scottish government has committed to the equivalent of 100% of Scotland's electricity demand to be generated from renewable energy by 2020. Sections 63 and 64 specifically relate to making regulations for improving the energy efficiency of existing non-domestic and domestic buildings respectively (CCSA, 2009).

Sweden was the first country to adopt the goals of a global development policy in 2003 (Sweden, 2003). In brief, global development policy encompasses all political decisions that directly or indirectly affect poor people in developing countries or countries undergoing transition to a market economy, one of which is sustainable development, which encompasses sustainable use of natural resources and environmental concern, economic growth, and social development and security. Beginning in spring 2016, all Swedish ministries will develop action plans around the global development policy.

From a top-level (overarching) national government policy perspective, is resiliency a guiding principle for legislative/regulatory development?

Yes	No
Australia, China, England, Japan, Netherlands, New Zealand, Norway, Scotland, Singapore, Sweden, US	Spain

While the majority response to this question is also yes, the definition and interpretation of 'resiliency' varies much more than with sustainability, as well as resiliency against what hazards or events. In some cases there is a focus on specific natural hazards (*e.g.*, earthquake), in others all natural hazards, and in some cases all disasters, including deliberate events. Only in some countries, such as Norway, is climate change resiliency a clear focus. Even in Spain, it can be argued sustainability policy is a climate change resiliency policy, even if not stated precisely as such.

In Australia, the Council of Australian Governments (COAG) agreed in 2007 to the National Climate Change Adaptation Framework (NCCAF). The NCCAF covered a range of cooperative actions between all Australian governments to begin to address key demands from business and the community for targeted information on climate change impacts and adaptation options (NCCAF, 2007). In 2009, the COAG agreed to adopt a whole-of-nation resilience-based approach to disaster management, which recognizes that a national, coordinated and cooperative effort is needed to enhance Australia's capacity to prepare for, withstand and recover from disasters. The National Emergency Management Committee subsequently developed the National Strategy for Disaster Resilience (NSDR, 2009), which was adopted by the COAG in 2011. The purpose of the NSDR is to provide high-level guidance on disaster management, with a focus on priority areas necessary for building disaster resilient communities across Australia.

In China there are several policies and laws related to resiliency against natural hazards, including protection and mitigation of earthquake disasters and meteorological disasters, and on disaster prevention and planning. While earthquake and meteorological disasters are important in Japan as well, the Fundamental Plan for National Resilience speaks to being committed to promoting initiatives for building national resilience with the aim of creating safe and secure national lands, regions and economic society that have strength and flexibility in the event of any disaster (NRPO, 2015). The guiding principles include prevent human loss by any means, avoid fatal damage to important functions for maintaining administration as well as social and economic systems, mitigate damage to property of the citizenry and public facilities, and achieve swift recovery and reconstruction.

New Zealand has a long standing focus on resiliency to earthquakes, with renewed thinking and discussion following the Canterbury earthquakes, and is also a signatory to the international Sendai framework (formerly Hyogo), monitoring and reporting on the disaster risk reduction objectives of the framework. In the Netherlands, climate change resiliency has been a focus for a long time, given the potential susceptibility to sea level rise. There is also a focus on infrastructure and security resiliency. This is also the case in the US, which under the Obama administration there are Executive Branch initiatives on climate change resiliency and on infrastructure resiliency, with several government departments and agencies having legislation and guidance on these topics.

A national assessment of the impacts of climate change on Norway has been produced and published as an Official Norwegian Report, Adapting to a changing climate (NOU, 2010). As noted in the report, the Norwegian government appointed a committee to study society's vulnerability and the need to adapt to the effects of climate change. Various climate change effects were identified, including higher temperatures, increased precipitation and potential sea level rise and effects on the coasts and fjords. Impacts to infrastructure and buildings were identified as needed to be addressed, and the committee recommended that adaptation considerations be incorporated into new regulations associated with the Planning and Building Act.

From a building regulatory policy perspective, is sustainability addressed within the building regulations?

All countries report some aspects of sustainability being embodied in building regulations. In Australia, sustainability was included as a specific objective of the Building Code of Australia (BCA) in 2007. In the European Union (EU), the Energy Performance of Buildings Directive (EPBD, 2010) and the Energy Efficiency Directive (EED, 2012) provide common direction and expectations, which in general are adopted in member states' legislation, although other legislation may be promulgated as well.

In New Zealand, even though sustainability is not an overarching objective of the current government, it is one of four purposes of the Building Act, so that 'buildings are designed, constructed, and able to be used in

ways that promote sustainable development'. The act also requires certain principles to be taken into account, including: the need to facilitate the efficient use of energy and energy conservation and the use of renewable sources of energy in buildings, the need to facilitate the efficient and sustainable use in buildings of material [...] and material conservation, the need to facilitate the efficient use of water and water conservation, and the need to facilitate the reduction in the generation of waste during the construction process. The Building Code has a specific code clause related to energy efficiency, H1 (DBH, 2011).

In Scotland, sustainability was first explicitly addressed in the Building (Scotland) Act 2003, which speaks to 'furthering the achievement of sustainable development'. However, sustainability has been a focus of the national building regulations since the Building (Scotland) Act 1959, predating the term 'sustainable development'. The current approach to development of standards and guidance within the Scottish building regulations framework looks to embed sustainability into the minimum measures that apply to new building work. However, this is always tempered by cost, as it could be considered unsustainable to have new buildings that are very environmentally friendly and accessible but the capital construction cost is elevated to such a level that means they become unaffordable. As such, within the Scottish system, a sustainability labelling system (with clearly defined measures) exists that enables those who wish to go further and demonstrate their 'green' credentials to do so (Dodds, 2014).

In 2005, Singapore's Building and Construction Authority (BCA), a statutory board under the Ministry of National Development, kick-started its drive to green Singapore's physical landscape by launching the BCA Green Mark: a rating system to evaluate a building's environmental impact and recognize its sustainability performance (Neng Kwei Sung, 2014). The BCA Green Mark scheme forms the backbone of the first Green Building Masterplan (GBMP) which was introduced in 2006 to encourage, enable and engage industry stakeholders to step up efforts in environmental sustainability, with a first focus on greening new buildings. This included legislation, financial incentives, industry training programmes and a public outreach campaign to place green buildings in the forefront of industry and consumer awareness. Under the legislation, Singapore' Building Control Act was amended in April 2008 to impose minimum standards on environmental sustainability for buildings (BCESR, 2008). In January 2008, an Inter-Ministerial Committee on Sustainable Development (IMCSD) was also set up to formulate a national strategy for Singapore's sustainable development. A Sustainable Development Blueprint was developed by the IMCSD and released in 2009 to outline key targets and initiatives to improve resource efficiency and enhance Singapore's urban environment for the next 10–20 years. One of the key targets set by the IMCSD for Singapore's built environment, is to green at least 80% of the buildings in Singapore by 2030. This would reduce the city-state's carbon footprint in the long term, whether in terms of energy and water efficiency, waste reduction or the use of sustainable materials.

In the US, 13 states and the District of Columbia have adopted the International Green Construction Code (IgCC) (2015). The IgCC is the first model code to include sustainability measures for the entire construction project and its site – from design through construction, certificate of occupancy and beyond. This code is expected to make buildings more efficient, reduce waste, and have a positive impact on health, safety and community welfare. It is in addition to other codes, such as the International Energy Conservation Code (IECC) (2015), which focuses of energy-efficiency measures in new buildings.

From a building regulatory policy perspective, is resiliency, as reflected in general government policy, addressed within the building regulations?

Yes

Australia, China, England, Japan, Netherlands, New Zealand, Norway, Scotland, Singapore, Spain, Sweden, US

While it is noted that 'resiliency' is addressed in building regulations in all countries, the situation here is somewhat complex, as with the case of national resiliency policy. In all countries, resiliency against 'normal' loads such as wind, snow, earthquake and the like, as appropriate to the country, is included. However, fewer countries have explicit infrastructure and security resiliency clauses beyond these, and countries addressing resiliency to climate change impacts through building regulations are even fewer. In some cases, this is not because of lack of interest, but more related to lack of knowledge and data from the experts.

In Australia, for example, the Australian Building Codes Board (ABCB) recognizes that they are not the experts on climate change, but need input from experts and other stakeholders, and released a Discussion Paper in 2014 to inform and seek feedback from stakeholders on the resilience of new buildings to extreme weather events (ABCB, 2014). Some of the issues are identified by McDonald (2014) as including the fact that the National Construction Code of

Australia (NCC) currently does not cover hail, storm tide or have specific requirements relating to heat stress, but that some of the largest insurance property losses result from hail damage, and that storm tide is potentially a very high risk in low-lying coastal communities, especially those subject to the risk of cyclones. However, it is acknowledged that any proposed changes would need to pass regulation impact analysis, and it is unlikely that requiring all external building materials to resist hail impact would be cost effective, and it would be very costly and restrictive to design and construct buildings to resist storm surge because of the significant water forces involved. Ultimately, the feedback to be obtained from stakeholders is expected to inform the ABCB on strategic advice it provides to governments in identifying future areas to focus its activities.

In the Netherlands, which has a significant coastal flooding concern, flooding is not regulated within the Building Regulations, but is part of the Act 'Deltawet waterveiligheid en zoetwatervoorziening,' which addresses flood protection and care of fresh water in relation to the expected climate change, under the responsibility of the Ministry of Infrastructure and the Environment. However, potentially surprising to some, the Building Decree will be addressing earthquake in the future, to address man-induced earthquakes which occurs frequently in the Northern part of the Netherlands due to gas production for many decades.

The examples from Australia and the Netherlands are reflective of challenges that other countries face as well – with numerous stakeholders providing input and various regulatory bodies having jurisdiction over resiliency, it can be difficult to craft holistic policies and legislation for the built environment. Climate change science is still developing. People live in areas of changing hazards. As will be highlighted later, it is challenging to upgrade existing buildings. These are but a few areas which need to be addressed.

Are building regulations in your country applicable to existing buildings?

Building regulations have traditionally been developed to apply to new construction. As a growing percentage of the construction activity is with the renovation, transformation, extension and upgrading of existing buildings more countries are engaged in developing regulatory tools for existing buildings (Meacham, 2009). In 2007, the IRCC conducted a survey of member countries with respect to building regulation and existing buildings. A summary of the 2007 survey results was published by Bergeron (2008).

Building regulation not applicable to existing buildings	Building regulation has provisions for new and existing buildings	Application to existing buildings determined by the regional government	Application to existing buildings for major alteration or renovation	Separate regulation for existing buildings
Austria	China, England, the Netherlands, Singapore, Sweden[1]	Australia	Canada, Japan, New Zealand, Scotland, Spain,[2] Sweden[1], US	US

Some of the key outcomes of the survey include the following:

- An important aspect of regulation for existing buildings is the determination of the performance level required from building upgrades. In many countries this is achieved by comparing existing building upgrades to the performance levels required for new construction.

- A fundamental difficulty encountered is the unavailability of sufficient knowledge to express the performance target of building regulations in quantified measurable terms that can be verified. Developing tools and methods to help develop these performance parameters is essential to the success of this approach.

- In some countries the regulations make allowance for risk-based approaches for determining what constitutes an acceptable level of performance. With the rapid expansion of the scope of regulations from the traditional fire and safety issues, to emerging social objectives such as accessibility and resource conservation, new decision-making tools need to be developed to support this approach.

From the 2015 survey, the situation appears to be generally unchanged. It was reiterated by several countries that regulations for new buildings are applicable to existing buildings, but whenever any intervention takes place in an existing building (expansion, renovation or change of use) due consideration should be given to proportionality, flexibility and no worsening of the conditions. How this is done, however, varies, and tools are sought to help make reasonable risk–cost–performance decisions.

The challenges have become highlighted when considering building retrofit for energy performance, which can have an impact on other building performance objectives (*e.g.*, Meacham et al., 2012; Sanchez-Ostiz, Meacham, Echeverria, & Pietroforte, 2014). This is especially true given that most countries have voluntary (market-based) mechanisms aimed at helping to increase sustainability in buildings, for which regulatory oversight may not occur.

Challenges are also highlighted for existing buildings within areas impacted by climate-related events, such as drought and subsequent wildland (bush) fire (*e.g.*, Australia and parts of the US), coastal storm surge and flooding (*e.g.*, Netherlands et al., Australia, US), increased snow load (*e.g.*, New Zealand) and more. In concept the challenge is not new – we have to deal with earthquake and other hazards, and adjust the building regulations as needed. However, in this case, numbers of buildings/people at risk may be higher, while ability to predict impact is not as reliable. Data, tools and methods for predicting a broad range of impacts, at a regional level to a building level, remain needed.

In your country, have any sustainability measures been mandated for existing buildings through regulation?

Yes	No	Variable
Netherlands, Scotland, Singapore, Spain, Sweden	China, England, New Zealand, Norway	Australia, Japan, US

While the summary indicates significant movement in addressing sustainability measures in existing buildings, many of the legislated requirements are associated with energy performance certification, or are only triggered if significant renovation is being undertaken (see the above section on existing buildings).

In Australia the Department of Industry, Innovation and Science delivers the Commercial Building Disclosure (CBD) Programme which mandates the disclosure of energy efficiency information for commercial office spaces of 2000 m^2 or more (CBD, 2015). Disclosure of this information before sale or lease assists prospective buyers and tenants to make informed decisions. The CBD Programme is an initiative of the Council of Australian Governments (COAG). The Australian Capital Territory (ACT) introduced under the ACT's House Energy Rating Scheme (ACTHERS) in 1999 a requirement for the disclosure of an existing dwelling's energy rating in all advertisements for sale, and that the contract of sale includes information about the buildings Energy Efficiency Rating. Minimum Energy Performance Standards (MEPS) Regulations apply in Australia for individual domestic appliances. MEPS is regulated through Commonwealth legislation, *i.e.*, the Greenhouse and Energy Minimum Standards (GEMS) Act 2012. It allows consumers to make informed decisions regarding the performance of individual domestic appliances.

As noted above, for EU member countries, the 2010 Energy Performance of Buildings Directive and the 2012 Energy Efficiency Directive are principal pieces of legislation. Under the Energy Performance of Buildings Directive:

- Energy performance certificates are to be included in all advertisements for the sale or rental of buildings.

- EU countries must establish inspection schemes for heating and air conditioning systems or put in place measures with equivalent effect.

- All new buildings must be nearly zero energy buildings by 31 December 2020 (public buildings by 31 December 2018).

- EU countries must set minimum energy performance requirements for new buildings, for the major renovation of buildings and for the replacement or retrofit of building elements (heating and cooling systems, roofs, walls, *etc.*).

- EU countries have to draw up lists of national financial measures to improve the energy efficiency of buildings.

Under the Energy Efficiency Directive:

- EU countries make energy efficient renovations to at least 3% per year of buildings owned and occupied by central government

- EU governments should only purchase buildings which are highly energy efficient

- EU countries must draw-up long-term national building renovation strategies which can be included in their National Energy Efficiency Action Plans

In Japan, Article 11 of the Act for the Improvement of the Energy Saving Performance of Buildings states that the owner of a building undergoing major renovations exceeding a certain size is obliged to make them comply with the energy consumption performance standard. In the US, existing buildings undergoing major renovations are required to meet current building codes, including for energy performance.

Are there market-based approaches to enhance sustainability in existing buildings?

Yes

Australia, China, England, Japan, Netherlands, New Zealand, Norway, Scotland, Singapore, Spain, Sweden, US

As one might expect, all countries have voluntary mechanisms to help enhance sustainability in existing buildings. A representative sampling includes:

- Australia: *Green Star* is a national, voluntary, industry-developed sustainability rating system for buildings and communities. The National Australian Built Environment Rating System (NABERS) is a government-developed national rating system that measures the environmental performance of Australian buildings, tenancies and homes. NABERS is mandatory where required to comply with the Commercial Building Disclosure (CBD) Programme.

- England: *BREEAM* is voluntary, sustainability rating system for buildings, and is used mainly in commercial buildings. The Code for Sustainable Homes ceases to have effect in October, to be replaced by the Home Quality Mark system.

- Japan: Energy saving performance can be indicated in *BELS: Building Energy-efficiency Labelling System*, overall environmental performance can be indicated in *CASBEE: Comprehensive Assessment System for Built Environment Efficiency*, and energy-saving performance of housing can be indicated in the housing performance indicator system.

- New Zealand: The New Zealand Green Building Council (NZGBC) promotes several green star rating systems, including *greenstar, homestar* and *NABERS/NZ*.

- Singapore: The BCA *Green Mark* is a rating system to assess a building's environmental

impact and recognize its sustainability performance, designed specifically for buildings in the
tropics.

- Sweden: *Miljöbyggnad* is a voluntary Swedish
  environmental assessment system developed by
  Boverket a few years ago and now managed by
  the Sweden Green Building Council.

- US: There are several systems, including *LEED*, a
  green building certification programme that recognizes best-in-class building strategies and practices
  and the *Energy Star* system for homes and
  businesses.

Have any resiliency measures been mandated for existing buildings through regulation?

Yes	No	Variable
Australia, China, Japan, Netherlands, New Zealand, Scotland,	England, Norway, Singapore, Spain, Sweden	US

The answer to this question closely tracks the
responses with national resiliency policy and existing
building policy outlined above. In most cases, mandated resiliency upgrades are for specific hazards,
such as earthquake, with upgrades associated with
resiliency requirements associated with climate
change effects less prevalent. This seems likely due
to the complexity of the problem. As noted by
Thompson, Cooper, and Gething (2015; as cited in
Foxell & Cooper, 2015, p. 403), 'climate change is
the prime example of a well-recognized but still
intractable policy issue for almost all sectors and
certainly for the built environment'.

As an example of earthquake upgrade requirements
for existing buildings, New Zealand focuses on
'earthquake-prone buildings', which are defined as
those having one-third the capacity of a new building
at that location. While some changes are underway
to the legislation, most seismic assessment of buildings is conducted by following New Zealand
Society for Earthquake Engineering (NZSEE) guidelines (NZSEE, 2014; 2015). These too are currently
undergoing a major revision based on the new proposed legislation and the lessons from Canterbury.
In addition, New Zealand is also reviewing requirements for snow loading and flooding, based on
recent events.

Are there voluntary regulatory measures, or voluntary market mechanisms, to enhance resiliency in existing buildings?

Yes	No
New Zealand, Norway, Scotland, Sweden	Australia, China, England, Japan, Netherlands, Singapore, Spain

Here again, the mechanisms are largely related to long-
known natural hazards, such as earthquake. However,
there are some mechanisms for flooding, which is a
potential climate change effect.

For example, in Japan, voluntary seismic retrofitting
under the Law for Promotion of Seismic Retrofitting
of Buildings exempts the owners from application of
other requirements. In New Zealand, NZSEE has a
rating scheme to encourage building owners to
upgrade to a higher level, which has seen an uptake
in use since the Canterbury earthquakes. This is
having a significant impact on the market, particularly
in Wellington where the seismic hazard is greater.

In Scotland there are a number of guidance documents
available for flood protection, as well as industry guidance and advice. The same is the case in the US,
where guidelines on seismic rehabilitation of buildings,
flood protection and protection against other hazards
is available from the Federal Emergency Management
Agency (FEMA), the American Society of Civil Engineers (ASCE) and others (e.g., FEMA 114, 1986; FEMA
356, 2000; and FM Global 1–28, 2012).

Have there been any failures of/gaps in/challenges to your system, particularly as related to sustainability or resiliency policies or issues?

Yes	No
Australia, England, Netherlands, New Zealand, Scotland, Spain, US	China, Japan, Norway, Singapore, Sweden

This question is actually quite complex, and a yes or no
answer is not that simple. Many countries have identified existing buildings as presenting building regulatory policy challenges for sustainability and resiliency.
Existing buildings are not specifically a new challenge,
but the new objectives, and assessment of actual versus
intended performance, is providing new focus.

In Australia, the Commonwealth Scientific and Industrial Research Organization (CSIRO) was contracted to evaluate the effectiveness of the 5-star energy efficiency standard for houses introduced in 2006, compared with the previous 3.5–4-star standard. The resulting report, *The Evaluation of the 5-Star Energy Efficiency Standard for Residential Buildings*, is the first evaluation of energy efficiency standards for houses in Australia based on comparison of actual energy use (Ambrose, James, Law, Osman, & White, 2013). The study monitored the energy use of more than 400 houses (with around half the houses undergoing more detailed monitoring of heating and cooling) in Brisbane, Melbourne and Adelaide, from June 2012 to February 2013. One of the findings was that the expected energy ratings of the houses in the sample have not increased in line with the changes in building regulation. This tracks with published research about the how well voluntary rating schemes do at actually delivering cost savings and expected performance (*e.g.*, Newsham et al., 2009; Oates & Sullivan, 2012; Scofield, 2009), the point being that both mandatory and voluntary approaches appear to have difficulty in delivering on projected outcomes.

Comments from England and Scotland suggest growing concern that increasing the energy efficiency of buildings will lead to problems of poor air quality in homes, moisture damage to fabric, and overheating in summer. The moisture and indoor air quality issues were seen with New Zealand's 'leaky building' problem (May, 2003; Meacham, 2010), and similar issues have been observed in Sweden and the US as well (Hagentoft, 2011; Odom et al., undated). In Scotland, survey feedback indicates that there is evidence that either poorly designed or extremely well built air tight buildings are causing poor indoor air quality and leading to the increase in respiratory diseases. With the ever increasing drive to improve energy efficiency this is likely to acerbate the IAQ issue. There is also evidence that poorly executed energy efficiency retrofitting projects have led to fire and moisture issues with existing buildings. There are numerous reports of fires related to energy retrofits and alternative energy installations on buildings (Meacham et al., 2012), which supports these concerns.

For several countries, issues associated with how to address sustainability and resiliency in existing buildings – to what level should they perform, what is the basis of the performance analysis, and what is an appropriate regulatory instrument – was identified as a significant concern. Tied closely to this is the need for quantitative performance metrics in the building regulations, which several countries identified as well. This is not only for sustainability and resiliency, but across regulation. It becomes focused when new objectives, such as energy performance of buildings, compete with safety objectives, such as fire performance of buildings.

Conclusions

A review of the literature and responses to survey of IRCC members (Table 1) reflect that challenges exist with performance-based building regulatory systems due to lack of clear performance criteria, verification methods and related issues. Outcomes also indicate that challenges exist in incorporating sustainability and resiliency into building regulation, for both new and existing buildings, because these social objectives are not yet being viewed as having the same level of importance, or equivalent level of social compact between government and the public, as providing for minimum levels of health and safety in buildings.

While a review of the literature and responses from the survey reflect sustainability objectives are being included in building regulations, it has also been found that their inclusion has been triggered from outside of the 'traditional' building regulatory system actors, and it appears that holistic or integrated performance has not been fully assessed (*i.e.*, making sure that adding a new objective does not result in an unanticipated impact somewhere else). This failure fully to assess potential impacts has resulted in observable, if not fully measured, reduction in performance for some areas, such as fire safety, structural safety and occupant health.

While the literature suggests that resiliency to climate change effects, narrowly viewed as resulting in hazards such as drought, increased rain, wind and snow loads, sea level rise and storm surge, should be addressed at all levels of government, survey results indicate that such objectives are generally not explicitly addressed in the building regulations of the responding countries.

Overall, the literature review and survey responses suggest that addressing sustainability and resiliency objectives for existing buildings is the greatest challenge, since regulatory oversight is low, building modifications associated with increasing sustainability (*e.g.*, more insulation, alternative energy sources) is becoming a concern, and it is extremely costly to increase the overall performance of existing buildings to that expected for new construction.

To address these issues, research suggests that a better understanding of holistic building performance is needed, along with the data and tools to assess performance, and more integrated regulatory and market measures are needed to achieve societal expectations for safe, healthy, sustainable and resilient buildings.

Table 1 Summary responses to selected survey questions

	Australia	Canada	England	Japan	Netherlands	New Zealand	Norway	Scotland	Singapore	Spain	Sweden	US
1	From a top-level (overarching) national government policy perspective, is sustainability a guiding principle for legislative/regulatory development?	✓	✓	✓	✓	✓	X	✓	✓	✓	✓	✓	V
2	From a top-level (overarching) national government policy perspective, is resiliency a guiding principle for legislative/regulatory development?	✓	✓	✓	✓	✓	✓	✓	✓	✓	✓	X	✓
3	From a building regulatory policy perspective, is sustainability, as reflected in general government policy, addressed within the building regulations?	✓	✓	✓	✓	✓	✓	✓	✓	✓	✓	✓	✓
4	From a building regulatory policy perspective, is resiliency, as reflected in general government policy, addressed within the building regulations?	✓	✓	✓	✓	✓	✓	✓	✓	✓	✓	✓	✓
5	Are building regulations in your country applicable to existing buildings?	✓	✓	✓	✓	✓	✓	✓	✓	✓	✓	✓	✓
6	In your country, have any sustainability measures been mandated for existing buildings through regulation?	V	X	X	V	✓	X	X	✓	✓	✓	✓	V
7	Are there market based approaches to enhance sustainability in existing buildings?	✓	✓	✓	✓	✓	✓	✓	✓	✓	✓	✓	✓
8	In your country, have any resiliency measures been mandated for existing buildings through regulation?	✓	✓	X	✓	✓	✓	X	✓	X	X	X	V
9	Are there voluntary regulatory measures, or voluntary market mechanisms, to enhance resiliency in existing buildings?	X	X	X	X	X	✓	✓	✓	X	X	✓	✓
10	Have there been any 'significant' failures of/gaps in challenges to your system? If so what are/were they, how did they come about?	✓	X	✓	X	✓	✓	X	✓	X	✓	X	✓

Notes: ✓ = yes; X = no; V = varies.

Acknowledgements

The author sincerely thanks the members and affiliates of the Inter-jurisdictional Regulatory Collaboration Committee (IRCC) for their assistance in responding to the survey referenced in this paper, and for their continued collaboration in sharing experiences, challenges and future directions related to the building regulatory systems in their countries. The author also thanks the referees for their insightful comments, which helped to improve the quality of this paper.

Disclosure statement

No potential conflict of interest was reported by the author.

Supplementary data

Supplemental data for this article can be accessed at: 10.1080/09613218.2016.1142330.

References

ABCB. (2014). *Resilience of buildings to extreme weather events – ABCB discussion paper*. Canberra, Australia: Australian Building Codes Board. Retrieved November 15, 2015 http://www.abcb.gov.au/work-program/Natural%20Disaster%20Mitigation.aspx

Ambrose, M. D., James, M., Law, A. Osman, P., & White, S. (2013). *The evaluation of the 5-star energy efficiency standard for RESIDENTIAL buildings – final report*. Canberra, Australia: Department of Industry. Retrieved from http://www.industry.gov.au/Energy/Documents/Evaluation5StarEnergyEfficiencyStandardResidential Buildings.pdf

Babrauskas, V., Lucas, D., Eisenberg, D., Singla, V., Dedeo, M. and Blum, A. (2012). Flame retardants in building insulation: A case for re-evaluating building codes. *Building Research & Information*, 40(6), 738–755. doi:10.1080/09613218.2012.744533

BCESR. (2008). *Building Control (Environmental Sustainability) Regulations 2008*. Building Control Act, Chapter 29, Minister for National Development, Singapore.

BEL. (1993). *The basic environment law*. Tokyo: Ministry of the Environment, Government of Japan. Retrieved November 15, 2015, from http://www.env.go.jp/en/laws/policy/basic/index.html

Bergeron, D. (2008). *Codes for existing buildings: Different approaches for different countries*. Proceedings. 7th International Conference on Performance-Based Codes and Fire Safety Design Methods. Society of Fire Protection Engineers, Bethesda, MD, pp. 15–23.

Boverket. (2011). *Boverket's mandatory provisions on the amendment to the Board's building regulations (2011:6) –mandatory provisions and general recommendations*. Sweden: Author. Retrieved November 15, 2015, from http://www.boverket.se/globalassets/publikationer/dokument/2012/bbr-engelsk/bfs-2011-26_introduktion.pdf

Brinson, R. A. and Dolan, J. B. (2008). Emerging risks of green construction. *Structural Engineering Magazine*, NCSEA/SEI, June 2008.

BRRTF. (1991). *Microeconomic reform building regulation* (Report of the Building Regulation Review Taskforce). Dr. John Nutt, Chair, Canberra, ACT, Australia, 1991.

CBD. (2015). *Commercial Building disclosure – A national energy efficiency program*. Canberra, ACT, Australia: Department of Industry and Science, Australian Government. Retrieved November 15, 2015, from http://www.cbd.gov.au/

CCSA. (2009). *Climate change (Scotland) Act 2009*. Edinburgh, Scotland: Scottish Government.

Chan, C.-K., Chan, E. H. W., & Quin, Q. K. (2014). *Building regulatory control in facing the challenge of climate change: A case of Hong Kong*. Proceedings, WSB14. Green Building Council Espana, Madrid, Spain.

Cheit, R.E. (1990). *Setting safety standards: Regulation in the public and private sectors*. Berkeley, CA: University of California Press.

Chow, W. K., Hung, W. Y., Gao, Y., Zou, G., & Dong, H. (2007). Experimental study on smoke movement leading to glass damages in double-skinned façade. *Construction and Building Materials*, 21, 556–566. doi:10.1016/j.conbuildmat.2005.09.005

Cole, R. J. (2011). Motivating stakeholders to deliver environmental change. *Building Research & Information*, 39(5), 431–435. doi:10.1080/09613218.2011.599057

DBH. (2011). *Compliance document for New Zealand building code, clause H1, energy efficiency* (3rd ed.). Wellington, New Zealand: Department of Building and Housing (now incorporated into the Ministry of Business, Innovation and Employment).

Dodds, B. (2014). *Sustainability labelling for building standards*. Proceedings, WSB14, Green Building Council Espana, Madrid, Spain.

du Plessis, C., & Cole, R. J. (2011). Motivating change: Shifting the paradigm. *Building Research & Information*, 39(5), 436–449. doi:10.1080/09613218.2011.582697

Energy Efficiency Directive (EED). (2012). *Directive 2012/27/EU of the European Parliament and of the Council of 25 October 2012 on energy efficiency*. Brussels.

Energy Performance of Buildings Directive (EPBD). (2010). *Directive 2010/31/EU of the European Parliament and of the Council of 19 May 2010 on the energy performance of buildings*. Brussels.

Eurocode. (2015). Retrieved November 15, 2015, from http://www.eurocode-online.eu/en/eurocode-information/history-of-eurocodes

FEMA 114. (1986). *Design manual for retrofitting floodprone residential structures*. Washington, DC, US: Federal Emergency Management Agency, Department of Homeland Security.

FEMA 356. (2000). *Prestandard and commentary for the seismic rehabilitation of buildings*. Washington, DC, US: Federal Emergency Management Agency, Department of Homeland Security.

Field, C. G., & Rivkin, S. R. (1975). *The building code Burden*. Lexington, MA: Lexington Books, D.C. Heath and Company.

FM Global 1–28. (2012). *Wind design*. Johnston, RI, US: FM Global Property Loss Prevention Data Sheet 1–28, FM Global.

Foxell, S., & Cooper, I. (2015). Closing the policy gaps. *Building Research & Information*, 43(4), 399–406. Retrieved from http://dx.doi.org/10.1080/09613218.2015.1041298 doi:10.1080/09613218.2015.1041298

Geschwind, C. H. (2001). *California earthquakes: Science, risk and the politics of hazard mitigation*. Baltimore, MD, US: The Johns Hopkins University Press.

Hagentoft, C.-E (2011). *Probabilistic analysis of hygrothermal conditions and mould growth potential in cold attics: Impacts of weather, building systems and constructions design characteristics*. International Conference on Durability of Building Materials and Components, Porto, Portugal, 12–15 April. Retrieved November 15, 2015, from http://www.irbnet.de/daten/iconda/CIB22349.pdf

Hemenway, D. (1975). *Industrywide voluntary product standards*. Cambridge, MA: Ballinger Publishing Company.

ICC. (2007). The purpose of controls. Chapter 1. *Building department administration* (3rd ed.). Washington, DC: International Code Council.

IECC. (2015). *International energy conservation code* (2015 ed.). Washington, DC: Author.

Imrie, R., & Street, E. (2011). *Architectural design and regulation*. Chichester, UK: Wiley-Blackwell.

IPCC. (2007). Levine, M., Ürge-Vorsatz, D., Blok, K., Geng, L., Harvey, D., Lang, S., Levermore, G., Mongameli Mehlwana, A., Mirasgedis, S., Novikova, A., Rilling, J., & Yoshino, H. Residential and commercial buildings. In B. Metz, O. R. Davidson, P. R. Bosch, R. Dave, & L. A. Meyer (Eds.), *Climate change 2007: Mitigation*. Contribution of Working Group III to the Fourth Assessment Report of the Intergovernmental Panel on Climate Change. Cambridge, United Kingdom and New York, NY, US: Cambridge University Press.

IgCC. (2015). *International green construction code* (2015 ed.). Washington, DC: Author.

IREM. (2013). *Life safety laws: Sprinklers in high-rise buildings*. Chicago, IL: Institute of Real Estate Management (IREM) Legislative Report. Retrieved November 15, 2015, form (https://www.irem.org/File%20Library/Public%20Policy/ LifeSafety.pdf

Kodur. V., & Phan, L. (2007). Critical factors governing the fire performance of high strength concrete systems. *Fire Safety Journal*, 42(6–7), 482–488. doi:10.1016/j.firesaf.2006.10. 006

Laubscher, J. (2014). *Reviewing challenges between the need for government-subsidised housing in South Africa and the sustainability requirements of the National Building Regulations*. Proceedings, WSB14, Green Building Council Espana, Madrid, Spain.

Lundin, J. (2005). *Safety in case of fire: The effect of changing regulations* (PhD Dissertation). Department of Fire Safety Engineering, Lund University, Lund, Sweden.

May, P. J. (2003). Performance-based regulation and regulatory regimes: The saga of leaky buildings. *Law and Policy*, 25(4), 381–401.

McDonald, M. (2014). *Resilience of Australian buildings to extreme weather events*. Proceedings, WSB14. Green Building Council Espana, Madrid, Spain.

Meacham, B.J. (2004). Performance-based building regulatory systems: Structure, hierarchy and linkages, *Journal of the Structural Engineering Society of New Zealand*, 17(1), 37–51.

Meacham, B. J. (2009). Editor, performance-based building regulatory systems: principles and experiences. IRCC. Retrieved from www.ircc.info

Meacham, B. J. (2010). Accommodating innovation in building regulation: Lessons and challenges. *Building Research & Information*, 38(6), 686–698. doi:10.1080/09613218. 2010.505380

Meacham, B. J. (2014). A brief overview of the building regulatory system in the united states. Chapter 9. In P. Stollard (Ed.), *Fire from first principles* (4th ed.) (pp. 138–158). London: Routledge.

Meacham, B. J., Moore, A., Bowen, R., & Traw, J. (2005). Performance-based building regulation: Current situation and future needs. *Building Research & Information*, 33(1), 91–106. doi:10.1080/0961321042000322780

Meacham, B., Poole, B., Echeverria, J., & Cheng, R. (2012). *Fire safety challenges of green buildings*. Springer Briefs in Fire. doi:10.1007/978-1-4614-8142-3, J. Milke, Series Editor, Springer.

Meijer, F.M. & Visscher, H.J. (1998). The deregulation of building controls: a comparison of Dutch and other European systems. *Environment and Planning B: Planning and Design*, 35, 617–629.

Meijer, F. M. and Visscher, H. J. (2006). Deregulation and privatisation of European building-control systems?. *Environment and Planning B: Planning and Design*, 33, 491–501. doi:10.1068/b3109

van Mierlo, R. (2005). The single burning item (SBI) test method – A decade of development and plans for the near future. *HERON*, 50(4). Retrieved November 15, 2015, from http://heronjournal.nl/50-4/1.pdf

Mudarri, D. H. (2010). *Building codes and indoor air quality, report prepared for the U.S.* Arlington, VA: Environmental Protection Agency, Cadmus Group. Retrieved November 30, 2015, from http://www2.epa.gov/sites/production/files/ 2014-08/documents/building_codes_and_iaq.pdf

Mumford, P. J. (2010). *Enhancing performance-based regulation: Lesson's from New Zealand's building control system* (PhD Thesis). Victoria University, Wellington, New Zealand.

NCCAF. (2007). *National climate change adaptation framework*. Canberra, Australia: Australian Government, Department of Climate Change and Energy Efficiency.

NDSR. (2009). *National strategy for disaster resilience*. Canberra, Australia: Council of Australian Governments. Retrieved November 15, 2015, from https://www.coag. gov.au/sites/default/files/national_strategy_disaster_ resilience.pdf

Neng Kwei Sung, J. (2014). *Lifetime environmental sustainability of buildings under the Singapore building control act*. Proceedings, WSB14, Green Building Council Espana, Madrid, Spain.

Newsham, G. R., Mancini, S., & Birt, B. (2009). Do LEED-certified buildings save energy? Yes, but . . . . *Energy and Buildings*, 41(8), 897–905. doi:10.1016/j.enbuild.2009.03.014

NFPA 101A. (2013). *Guide on alternative approaches to life safety*. Quincy, MA, US: National Fire Protection Association.

NOU. (2010). *Adapting to a changing climate: Norway's vulnerability and the need to adapt to the impacts of climate change*. Official Norwegian Reports NOU 2010:10, Recommendation by a committee appointed by Royal Decree of 5 December 2008, submitted to the Ministry of the Environment on 15 November 2010.

NRPO. (2015). National resilience promotion office, Cabinet Secretariat, 3-1-1 Kasumigaseki. Chiyoda-ku, Tokyo, Japan 100-8970.

NSEE, (2009, July). *National strategy on energy efficiency*. Canberra, Australia: Council of Australian Governments, First Printing.

NZSEE (2014). *Assessment and improvement of the structural performance of buildings in earthquake*. Wellington, New Zealand: New Zealand Society for Earthquake Engineering.

NZSEE (2015). *Assessment and improvement of the structural performance of buildings in earthquake, Corrigenda 4*. Wellington, New Zealand: New Zealand Society for Earthquake Engineering.

Oates, D., & Sullivan, K. (2012). Postoccupancy energy consumption survey of Arizona's LEED new construction population. *Journal of Construction Engineering and Management*, 138(6), 742–750. doi:10.1061/(ASCE)CO. 1943-7862.0000478

Odom, J. D., Scott, R., & DuBose, G. H. (undated). *The hidden risks of green buildings: Avoiding moisture & mold problems*. NCARB Mini-Monograph. Washington, DC: National Council of Architectural Registration Boards. Retrieved November 15, 2015, from http://www.ncarb.org/ Publications/Mini-Monographs/~/media/ 514202A644794ED3B734F04D78721705.ashx

Sanchez-Ostiz, A., Meacham, B. J., Echeverria, J. B., & Pietroforte, R. (2014). *Towards a risk-informed decision approach for optimizing retrofit of existing buildings for energy performance*. Paper 259, Proceedings of WSB14 – World Sustainable Building Conference, CIB, the Netherlands.

Scotland. (2015). Retrieved November 15, 2015, from http:// www.gov.scot/About/Performance/scotPerforms/purpose

Scofield, J. H. (2009). Do LEED-certified buildings save energy? Not really . . . . *Energy and Buildings*, *41*(12), 1386–1390. doi:10.1016/j.enbuild.2009.08.006

Serra, J., Tenorio, J. A., Echeverria, J. B., & Sanchez-Ostiz, A. (2014). *Perspective of the building sustainability regulatory evolution in Spain: From prescription to performance*. Proceedings, WSB14. Green Building Council Espana, Madrid, Spain.

Simmons, R. (2015). Constraints on evidence-based policy: Insights from government practices. *Building Research & Information*, *43*(4), 407–419. Retrieved from http://dx.doi.org/10.1080/09613218.2015.1002355   doi:10.1080/09613218.2015.1002355

Simonson McNamee, M., Blomqvist, P., & Andersson, P. (2011). *Evaluating the impact of fires on the environment*. Proceedings, 10th International Association of Fire Safety Science.

Stannard, M. (2014). *Sustainability, resilience and risk in earthquake-prone areas: Lessons for building regulators from the Canterbury earthquakes*. Proceedings, WSB14. Green Building Council Espana, Madrid, Spain.

Steenbergen, R. D. J. M., Geurts, C. P. W., & van Bentum, C. A. (2009). Climate change and its impact on structural safety. *HERON*, *54*(1), 3–36.

Sweden. (2003). Policy on global development, Swedish parliament (Riksdag). Retrieved November 15, 2015, from http://www.regeringen.se/contentassets/e30c371390234406acee1a396ba8aed9/ud-info—fact-sheet-swedens-policy-for-global-development

Thompson, M., Cooper, I., & Gething, B. (2015). The business case for adapting buildings to climate change: Niche or mainstream, Legacy Document, Design for Future Climate Programme, UK Innovate (formerly Technology Strategy Board) Swindon, UK (as cited in Foxell and Cooper, 2015, p. 403).

UL. (2008). Report on structural stability of engineered lumber in fire conditions. Project Number 07CA42520, Underwriters Laboratories, Northbrook, IL, US. Retrieved November 15, 2015 http://ul.com/global/documents/offerings/industries/buildingmaterials/fireservice/NC9140-20090512-Report-Independent.pdf

Van Bueren, E., & de Jong, J. (2007). Establishing sustainability: Policy successes and failures. *Building Research and Information*, *35*(5), 543–556. doi:10.1080/09613210701203874

Visscher, H. J., & Meijer, F. M. (2014). *Impact of energy efficiency goals on systems of building regulations and control*. Proceedings, WSB14. Green Building Council Espana, Madrid, Spain.

Wermeil, A. E. (2000), *The fireproof building: Technology and public safety in the nineteenth-century American city*. Baltimore, MD, US: The Johns Hopkins University Press.

Wilby, R. L. (2007). A review of climate change impacts on the built environment. *Built Environment*, *33*(1), Climate Change and Cities, 31–45. doi:10.2148/benv.33.1.31

Yatt, B. D. (1998). *Cracking the codes: An architect's guide to building regulations*. New York, NY: Wiley.

Endnotes

[1]In Sweden, essentially the same requirements are applied for both the construction of a new building as well as the alteration. For alterations, however, one should always take into account the scope of the alteration in accordance with Chapter 8, Article 7 of the Planning and Building Act and with Chapter 3, Article 23 of Planning and Building Ordinances as well as the building's conditions when the requirements apply. The requirements for new constructions are never directly applicable to alterations. However, one can often obtain some guidance from these for assessing the implications of the requirements for the alterations. For alterations, however, the requirements are often met through other solutions than for the construction of new buildings (Boverket, 2011).

[2]To comply with EU regulations on energy efficiency as part of the 20/20/20 plan, a system of energy certification of buildings for both new construction and for existing buildings that are sold or rented was implemented in 2013.

INFORMATION PAPER

The realpolitik of building codes: overcoming practical limitations to climate resilience

Shari Shapiro

Scientists predict a future with more natural disasters due to climate change. Up-to-date building codes can reduce carbon emissions and make the built environment more resilient. However, the hostile environment for code adoption and lax enforcement in many jurisdictions in the United States impede the full potential of building codes to moderate the impacts of climate change. There are three 'realpolitik' reasons that building codes are not as effective as they could be in moderating the impacts of climate change. First, most building code review and approval boards are comprised of construction industry professionals who rarely take climate change into account. Second, the homebuilding industry is waging an effective advocacy campaign against updating building codes in general, particularly objecting to improved energy efficiency through building energy codes. Finally, enforcement of building codes, especially energy codes, is uneven. Overcoming these political and practical challenges requires greater activity in the areas of participation, pricing and policing. Interest groups and policy-makers must become more engaged in code advocacy. Price signals must internalize the costs imposed on society by damage from climate change. Finally, improving compliance and enforcement through training and additional resources should be a priority.

Introduction

Many have imagined republics and principalities which have never been seen or known to exist in reality; for how we live is so far removed from how we ought to live, that he who abandons what is done for what ought to be done, will rather bring about his own ruin than his preservation.

(Machiavelli, *The Prince*)

Hurricane Katrina made landfall on 25 August 2005, and in the days that followed 80% of New Orleans was flooded, over 1000 people died and total economic losses exceeded US$125 billion (Federal Emergency Management Agency (FEMA), 2006). Many of the communities in areas that were heavily impacted by Katrina 'had either not adopted up-to-date model codes that incorporate flood and wind protection, or had no building codes at all' (FEMA, 2006, p. vi). The purposes of building codes are, of course, to protect individuals and society from hazards and to create equal conditions (rules) that all actors (developers, builders, designers, etc.) must adhere to. The principle is well established and is easily understood in terms of regulations for automobiles – to be fit for sale, each must meet a minimum of requirements: have headlamps and tail lights, seat belts, passive systems (air bags) for impacts, etc. and, of course, meet certain fuel and emissions standards. These (and other requirements) are accepted as necessary for the 'public good'.

Due to increased carbon emissions and the resulting climate changes, scientists predict a future with more Katrinas – hurricanes, flooding, storm surges, drought, fires and more. The experience of the past 10 years has already begun to show the accuracy of these predictions.

To address this outcome, society must take two paths simultaneously – reducing carbon emissions and making the built environment more resilient to the impacts of climate change. Up-to-date building codes can help to achieve both these goals, but the hostile environment for code adoption and lax enforcement in many states and local jurisdictions in the United States make these promises difficult to realize.

There are three 'realpolitik' reasons that building codes are not as effective as they could be. First, most state (or local) building code review and approval boards are comprised of construction industry professionals, who rarely take climate change into consideration in making their decisions. This raises two important issues: (1) what is the professional responsibility of the members of these boards, particularly the design professionals and government officials, in terms of taking larger societal considerations like climate changes and resilience into account?; and (2) if the existing makeup of these boards is not effective in taking these issues into consideration, what changes should be made to the range of interests represented on these boards? For example, should code adoption boards include insurance, emergency management, utilities, utility regulators and others more directly responsible for addressing climate change?

Second, the homebuilding industry is waging an effective advocacy campaign against updating building codes in general, particularly objecting to the enhancements in energy efficiency in the 2009 and 2012 versions of the International Energy Conservation Code (IECC). The homebuilding industry argues that enhanced code requirements make housing unaffordable, and that there is little demand from home buyers for more efficient and resilient dwellings. This paper will examine the evidence in the published literature regarding the relationship between enhanced code requirements and the affordability of housing, the potential change in the cost-benefit assessment when climate change is taken into consideration, and how incentives and other regulatory tools may change this calculus.

Finally, enforcement of building codes, especially energy codes, is uneven. Thus, even if a jurisdiction has enacted up-to-date codes, actual construction may not achieve the higher levels of energy efficiency and resiliency included in the codes. This paper provides an analysis of some of the contributing factors to the lack of enforcement, and efforts being made to provide tools and incentives to improve compliance and enforcement.

To overcome these political and practical challenges, activity in each of the 'Three Ps' – participation, pricing and policing – is needed. Interest groups and policy-makers involved in addressing climate change and creating a more resilient built environment must become more engaged in the code adoption and enforcement process. Price signals to building owners and the construction industry must internalize the costs imposed on society resulting from a changing climate. To do so, insurance companies and government actors can use a combination of incentives and risk-based pricing to shift the market towards more resilient codes and overcome political opposition. Finally, improving compliance and enforcement through training and additional resources should be a priority.

The first section of the paper provides an overview of the research regarding the opportunity to decrease the negative impacts from climate change and improve the resilience of the built environment through building codes, describes the most common regulatory mechanisms for building code adoption and enforcement in the US, and provides some specific examples of the jurisdictions where the political and practical challenges are hampering building code adoption and enforcement. The second section provides more detail on the 'Three P' model, including recommendations for realizing the benefits of up-to-date building codes in the face of a challenging political reality.

Towards a more resilient built environment

Society has two separate, yet interconnected, challenges in responding to the threat of climate change: reducing the extent of change (mitigation) and preparing for or preventing the negative impacts (adaptation) that climate change will inevitably bring. Improving the 'resilience' of the built environment is central to addressing both of these challenges.

Resilience is a term that can 'collapse into meaninglessness […] from having too many meanings' (Hassler & Kohler, 2014, p. 119). Here, resilience incorporates two concepts: built-in resilience and disaster risk reduction. Lee Bosher defines 'built-in resilience' as 'a quality of a built environment's capability (in physical, institutional, economic and social terms) to keep adapting to existing and emergent threats' (Hassler & Kohler, 2014, p. 123). By contrast, 'disaster risk reduction' is preventing threats from happening in the first place (Hassler & Kohler, 2014).

The built environment and climate change are intimately interconnected. In the US, buildings consume approximately 40% of US energy, more than either the transportation or industrial sectors, and are responsible for 40% of US carbon emissions (Vaughan & Turner, 2013). In addition, much of the threat to human life and property from climate change is building related: homes and businesses lost to flooding, power brown- and blackouts due to increased demand for heating and cooling, roofs blown off in hurricane-force winds.

There are steps that can be taken to reduce both the risk reduction and response sides of the resilience equation to improve the resilience of the built environment. Even a small reduction in the energy use and emissions from buildings would help with 'disaster risk reduction' by mitigating the extent of climate change. Prohibiting construction in flood-prone areas is also a risk-reduction strategy. On the other hand, tactics to improve the strength of roof fastenings would make buildings more resistant to damage from hurricane-force winds.

There are time-related aspects of resilience as well. As the most cost-effective time to implement resiliency features is usually when the building is first built, this underscores the importance of resilient building codes, which generally apply to new buildings and major renovations. However, resilience is not a one-time fix. New hazards will arise, as will new technologies and building practices. Strategies that were anticipated to work may be shown to be ineffective, and approaches designed to address one hazard may actually make buildings more vulnerable in other ways. As a result, older structures will need to be retrofitted, as the trend towards retrofitting older buildings with energy-efficiency features demonstrates. Even buildings built with resilience in mind may need to be tweaked or retrofitted over time.

Finally, the built environment is a living system. Human psychology is always a factor that affects the patterns of use and demands. Even if a building is designed and built to be more energy efficient, occupants increase their plug-load electricity or leave windows open, causing an overall *increase* in energy use and environmental damage – a rebound effect. Even the 'disaster' component of 'natural disaster' can have a human element. For example, a lack of political will to prohibit building in a flood-prone area can change a natural hazard into a community-devastating disaster through loss of life and property. Thus, changing building practices is by no means a panacea for eliminating or even reducing the climate impacts of the built environment – addressing the interrelationship between human behaviour and policy is also critical to success.

Ultimately, however, improving the 'resilience' of buildings provides an opportunity to reduce some of the damage from climate change measured both in dollars and in lives. As a result, scientists and policy-makers are exploring different ways of achieving this goal, and building codes are often cited as a component of a regulatory response to climate change.

US regulation of construction

The model code structure
In the US, most state and local building codes are based on model codes developed by non-profit, third-party code developers, like the International Code Council (ICC) and the National Fire Protection Association (NFPA). The model codes are updated every three years.

For the most part a state (or, in some cases, municipal) committee of construction industry experts review the code changes and make recommendations to the administrative agency that regulates construction or to the legislature. That regulatory agency or state legislature then enacts the new code through regulation or legislation.

In addition to codes addressing issues of health and safety, most jurisdictions' building codes also include a code, primarily the IECC, setting the minimum level of energy efficiency of buildings allowable by law.

Building codes and climate change
Building codes set the baseline for structures that are intended to last for decades, so they can be a major part of the solution to improve the resilience of the building stock in the face of an increasingly hostile weather environment (Insurance Institute for Business & Home Safety (IBHS), 2015b).

One of the Federal Emergency Management Agency's (FEMA) Mitigation Assessment Team's (MAT) primary recommendations to reduce damage similar to Katrina going forward was the 'adoption of modern building codes' (FEMA, 2006). Likewise, in the *Hurricane Sandy Rebuilding Strategy Report*, the Hurricane Sandy Rebuilding Task Force recommended to President Barack Obama that states use the most up-to-date building codes:

> to ensure that buildings and other structures incorporate the latest science, advances in technology and lessons learned [... and] that more resilient structures are built and that communities are better protected from all types of hazards and disasters [...].
> (Hurricane Sandy Rebuilding
> Task Force, 2013, p. 82)

In addition to making the building stock better able to withstand the effects of climate change, energy codes like the IECC can help to mitigate climate change by reducing building energy use and carbon emissions. Energy codes address building materials and practices, including lighting, insulation, fenestration, etc. Although potentially beneficial, renewable energy is not included as part of the energy codes.

From 1992 to 2006, the standard for building energy efficiency required by code largely remained static. At some point in the mid-2000s, however, energy efficiency advocates realized that building energy codes were an avenue for increasing building energy efficiency significantly without requiring legislative change. As a result of their efforts, the 2009 codes increased building energy efficiency by 15% over the 2006 codes, and the 2012 codes increased efficiency another 15% over the 2009 codes, for a 30% boost in energy efficiency in six years.

These increases in energy efficiency of building codes are estimated to deliver significant reductions in carbon emissions and energy savings over the next 30 years. If adopted *and enforced* nationwide, the Department of Energy (DOE) estimates that, from 2013 to 2040, building energy codes will cumulatively reduce full fuel cycle energy consumption by 41.6 quads (43.888 exajoules) (equal to 42% of total US energy use in a year), decrease electricity use by 3643 billion kWh (approximately equal to the total amount of electricity used annually in Illinois), and avoid 3178 million metric tonnes in CO_2 emissions (equal to taking 66 million cars off the road for 10 years) (US DOE, 2015).

A note of caution is needed on the estimates of future energy savings and reductions in carbon emissions. Most of these estimates rely on prospective estimates based on design modelling. Retrospective studies show lower levels of energy savings than the design estimates (Jacobsen & Kotchen, 2010; Kotchen, 2011), with some studies challenging whether energy codes result in overall energy savings at all (Levinson, 2015). Given the scale of the built environment's contribution to energy use and carbon emissions, the savings from energy codes will be worthwhile, even if they are not equivalent to the design-based estimates.

Thus, adoption and enforcement of up-to-date codes can serve two important purposes: mitigating climate change and ameliorating some of its economic and human effects.

Perhaps most significantly, unlike developing new programmes to address climate change, states and jurisdictions already have the necessary legal, educational and enforcement mechanisms in place to regulate buildings through building codes. In contrast to a carbon tax or other regulatory mechanism, no additional legal tools are required in the vast majority of jurisdictions to realize the resilience benefits of up-to-date energy codes.

However, as discussed in the remainder of the paper, political and practical barriers are currently preventing society from realizing the full potential of building codes to address climate change and its impacts. Greater action on the part of the private and public sector to overcome political opposition to the adoption of up-to-date codes, and additional investment in awareness, education, research and enforcement resources are necessary to obtain the greatest benefit from building codes.

Costs of up-to-date building codes

The residential homebuilding industry is the primary opponent of updated building codes. This industry is primarily represented by the National Association of Home Builders (NAHB), representing small to mid-size builders, and the Leading Builders of America (LBA), representing large-production homebuilders. The homebuilders' primary argument is that up-to-date codes, especially energy codes, cost more, and that potential homebuyers are not demanding more energy-efficiency homes and will not pay for the increased cost. Other organizations representing the commercial building sector, like the Building Owners and Managers Association (BOMA), have also expressed concern regarding the costs of increasing the energy efficiency of building codes (BOMA, 2015).

The homebuilders argue that increased costs 'price out' potential homebuyers from the market, thus negatively impacting low- and middle-income Americans (Siniavskaia, 2014), frequently citing the statistic that 'a $1000 increase in the home price leads to pricing out about 206 269 households from the home ownership market' (Siniavskaia, 2014).

To address the homebuilders' opposition to updated codes based on cost, it is useful to know what up-to-date codes are anticipated to cost. To provide some context for assessing these numbers, the NAHB estimates that the average sales price of a home is US$399 352 (Home Innovation Research Labs, 2015), and that Americans remain in their homes for an average of 13 years (Emrath, 2009).

In performing a high-level cost analysis, it appears that there *is* an increase in first costs for more resilient and efficient codes.

The increase in costs for the structural changes of up-to-date building codes (as opposed to energy efficiency provisions) is negligible. With respect to the updates to the International Residential Code (IRC), the code that governs the construction of one- and two-family dwellings, the Home Innovation Research Labs (the research arm of the NAHB itself) estimates that the costs of the 2015 IRC code changes range from US$203 to US$757 (Home Innovation Research Labs, 2015).

While a price increase of approximately US$500 seems like a minimal investment for a more resilient home, the NAHB estimates that a US$1000 increase in home prices will exclude more than 200,000 families from homeownership. Thus, even an increase this small could be politically problematic, if not cost prohibitive. But, as discussed further below, the 'priced out' analysis is fundamentally flawed. It is reasonable to assume that such a minimal increase, if it exists, would not be cost prohibitive. Indeed, there is evidence that consumers would pay for an increase in the price of a home if it resulted in more resilient structures (Dumm, Sirmans, & Smersh, 2009) as this increases the value of their property.

By comparison, the increase in first costs for the most recent editions of the IECC are more significant. According to the DOE, incremental costs for the 2009 IECC over the 2006 IECC range from US$525 to US$2014. Incremental costs for the 2012 IECC over the 2009 IECC range from US$1566 to US$2797 (Mendon, Lucas, & Goel, 2013). The NAHB has more costly estimates, particularly for the 2012 IECC. It estimates incremental construction costs for the 2012 IECC over the 2009 IECC ranging from US$3224 to US$7203 (NAHB Research Center, 2012a).

However, as indicated in Table 1, using the average of the cost estimates provided by different interest groups, the first costs are more than recouped over the life of the building, and pay for themselves within the 13-year average time that a person owns his or her home (Emrath, 2009). After the initial payback, any energy cost savings are profit to the homeowner.

Despite the apparent cost-effectiveness of the 2009 and 2012 IECC, caution must be taken in relying on either the NAHB or the DOE cost estimates. These estimates are design based and use an estimate of future energy savings. Retrospective studies have shown lower levels of energy savings, with some studies challenge whether energy codes result in overall energy savings at all (Jacobsen & Kotchen, 2010; Levinson, 2015).

Moreover, human factors must also be taken into consideration. 'Rebound effects' are well documented (Sorrell, 2007). People have a tendency to be more profligate with energy use (through increased plug load, for example) after energy-efficiency measures are implemented. Therefore, in spite of more energy-efficient buildings, energy use may stay the same or increase. As a result, estimates of energy savings based on design models alone tend to overestimate actual results.

With the dramatic costs of damage from climate change, as further explored below, even less impressive carbon and energy savings are likely to have societal benefits, and to be cost-effective when these societal benefits are internalized. Nonetheless, the uncertainty regarding energy savings and payback will play an important role in terms of consumer discounting of the value of energy-efficient codes.

The costs and benefits of improving building resilience

Thus far, the cost-effectiveness analyses have compared the increased first costs of building a house with benefits to the individual homeowner. But the real cost impacts of climate change are borne by society as a whole, not the individual building owner. These externalities, like increased taxes, healthcare costs, insurance rates, food prices, etc., are not currently internalized into property (real estate) prices to any significant extent.

Overall, the effects of climate change on the built environment are estimated to be staggering. Taking only loss of property and infrastructure into consideration,

> within the next fifteen years, higher sea levels combined with storm surge will likely increase the *average annual cost* along Eastern Seaboard and Gulf Coast from $2 to $3.5 billion. Adding in potential changes from hurricane activity, annual losses will grow to up to $7.3 billion, bringing the total annual price tag of storms to $35 billion.
> (Risky Business Project, 2014, p. 3, emphasis added)

In addition to the risks from increased storm activity, there is also increased risk of fire from hotter and drier conditions. Between 2000 and 2009, the property damage from wildfires averaged US$665 million per year, and wildfires cost states and the federal government billions in fire-suppression management (Union of Concerned Scientists, n.d.).

Increased resilience and energy savings from building code changes might be a loss or marginal benefit to an individual building owner at the time of purchase, but on a national (or even international) scale over many years, society would reap benefits many times the initial cost. Thus, these larger societal costs need to be acknowledged in public policy and regulation. As the role of public policy and regulation is to protect the individual and wider society, it is reasonable for public policy to require that the larger societal costs be shared and internalized by the builder and building owner. By shifting the price signals, the long-term and societal benefits will be weighed by the individual actors making decisions about up-to-date codes.

Method

Building codes are often cited by scholars as a strategy for addressing climate change impacts. However, as a practitioner and scholar, the author has witnessed a gap between the theoretical benefits of building codes and the practical and political challenges to realizing these benefits.

The goal of this paper is to provide a high-level overview of the relationship between building codes and

Table 1 Comparison of costs and energy savings between the 2009 and 2012 International Energy Conservation Codes (IECCs)

Climate zone	NAHB energy cost savings (US$)	NAHB incremental cost (US$)	NAHB simple payback	DOE energy cost savings (US$)	DOE incremental cost (US$)	DOE simple payback	Average of energy cost savings (US$)	Average of incremental cost (US$)	Simple payback
1	206	3224	15.7	344	1659	4.8	275.00	2441.50	8.9
2	294	3330	11.3	197	1995	10.1	245.50	2662.50	10.8
3	470	7203	15.3	290	2528	8.7	380.00	4865.50	12.8
4	410	7091	17.3	355	2035	5.7	382.50	4563.00	11.9
5	505	4653	9.2	410	1693.50	4.1	457.50	3173.25	6.9
6	397	6399	16.1	525	2797	5.3	461.00	4598.00	10.0
7	609	6465	10.6	592	2797	4.7	600.50	4631.00	7.7
8	725	6465	8.9	1360	2797	2.1	1042.50	4631.00	4.4

Sources: NAHB Research Center (2012b), Mendon et al. (2013).

politics, using theoretical literature, examples from the popular press and the author's personal experience.

The research was compiled from several sources, including:

- Westlaw

- Social Science Research Network

- HeinOnline

- Google Scholar

- Google Search

The information related to specific organizations, like the American Institute of Architects (AIA), the American Society of Heating, Refrigeration and Air-conditioning Engineers (ASHRAE), the NAHB, etc. was sourced from the organizations' websites or other official publications wherever possible.

The anecdotal information on the political environment for building codes in North Carolina and Pennsylvania was informed by the author's personal experiences and source material.

Challenges to adoption and enforcement of up-to-date building codes

Building resiliency advocates face four primary challenges to realizing the benefits of up-to-date codes: (1) failure to internalize societal costs and benefits; (2) political resistance and regulatory capture; (3) lack of participation by advocates of improved building resilience in decision-making about code adoption and enforcement; and (4) lack of funding and training for building code enforcement.

Failure to internalize societal costs and benefits
As discussed above, interest groups most frequently cite increased costs and lack of consumer demand as the primary arguments against more stringent building code requirements.

Examining first costs to builders and building owners only, the cost–benefit calculation is generated by comparing the amount that a potential building buyer is willing to pay versus the first cost of the building requirements. The building owner's willingness to pay is affected by both personal preferences *and* the price signals that the owner is receiving from outside sources. These price signals include assessment of the value of the building by mortgage companies and other institutions providing financing for the purchase or sale, property taxes, insurance costs and many other factors.

Currently, builder and buyer receive skewed price signals, particularly in the areas of insurance costs, property valuation, property taxes and energy costs. They do not effectively internalize the external costs to society of a less resilient building stock and increased carbon emissions.

For example, flood insurance is subsidized by the federal government through the National Flood Insurance Program (NFIP). Historically, the premiums that homeowners paid for flood insurance bore little relationship to the likelihood of flooding and the potential personal and property damage. As a result, homeowners have substituted insurance for investment in more resilient homes (Dumm et al., 2009).

Through the Biggert–Waters Flood Insurance Reform Act of 2012, however, NFIP premiums are coming more into line with actual risk. The results are striking – people are building more resilient homes, or choosing not to build in areas of excessive risk from natural disasters (Klibanoff, 2015). These results indicate that changes in insurance costs will, in fact, nudge buyers towards more resilient building practices, and presumably raise the value of structures built to more resilient building codes.

From a financing standpoint, financial institutions do not place much, if any, value on resilience and energy efficiency in determining the value of property or the amount that a potential buyer can borrow (Institute for Market Transformation, 2015). As a result, the resiliency and efficiency features are not factored directly into real estate prices. Sellers cannot recoup their investment at the time of sale, and buyers cannot borrow additional funds to pay for more resilient buildings. Therefore, buyers have little incentive beyond personal preference to invest in these features (Institute for Market Transformation, 2015).

In addition, financial institutions generally only require that a building is in compliance with the code in effect in the jurisdiction at the time of construction, not that the building is built to the most current code edition. If constituents were prevented from qualifying for mortgages unless buildings were built to the current code edition, the political attitudes towards updating building codes would change. Unfortunately, the potential for using mortgage price signalling has thus far gone largely unrealized.

Buyers have also devalued resilient structures because of the assumption that the government will provide funding to help rebuild structures if they are damaged (Dumm et al., 2009).

State and local property taxes do not generally reflect the increasing costs of recovery from natural disasters. In addition, much of the disaster recovery costs are paid for by the Federal government, and there is no federal mechanism for tying disaster recovery costs with property prices.

Finally, as has been discussed at length in scholarly and popular publications, energy costs in the US do not currently internalize climate impact (Union of Concerned Scientists, 2015). As a result, energy prices are lower than they would be if these impacts were taken into account. This skews the cost–benefit analysis of improved energy efficiency. If energy prices are made higher, the energy savings benefits would increase, decreasing the payback time for energy efficiency features and increasing the savings once the first costs have been recouped.

In summary, property-related price signals currently do not nudge consumers to value resilient and energy efficient building stock, and, in fact, devalue these features. However, these same mechanisms provide fertile opportunities to change the support for resilient and energy efficient codes.

Political resistance and regulatory capture

The Insurance Institute for Business & Home Safety (IBHS) publishes a periodic assessment of residential building code and enforcement systems for life safety and property protection in hurricane-prone regions. In its 2015 report, the IBHS notes, with diplomatic understatement, '[i]deally, states keep pace with new building science research and technology, but occasionally, they respond to political pressures that reduce or delay safety standards' (IBHS, 2015b, p. 3).

Lobbying and regulatory capture (domination of a regulatory body by the regulated community), are the primary methods for applying political pressure.

The NAHB and its state and local chapters, the primary opponents of updating building codes, are formidable interest groups (NAHB, 2014; Top Lobbyist of 2014, 2014). First, there is a homebuilder in every district, and many legislative representatives are or were involved in the construction industry. Second, the homebuilding industry is uniquely focused on building code policy. Many other groups have building codes as a part of their interests, like manufacturers of building products, environmental groups, first responders, trade unions, etc., but the homebuilders are the only group for which building codes are a primary policy focus. Third, homebuilders are well organized and their advocates are well funded. Finally, homebuilder advocates use all levers of political power, including lobbying, campaign contributions, grassroots organizing and litigation.

In addition, building code councils are subject to the type of regulatory capture that recent scholarship and two 2015 Supreme Court decisions demonstrate frequently plagues 'private' public boards: lack of government oversight and protecting private interests of their members over serving the public good (Volokh, 2014).

Building code review councils are staffed by construction industry professionals tasked with the responsibility of evaluating codes and advising administrative agencies and legislatures on updating the jurisdiction's building codes. The homebuilding industry is often disproportionately represented on these councils. Thus, although the enabling statutes require the members of the building code councils to protect the health, safety and welfare of the public, these boards are more likely to protect the business interests of their members.

Political challenges to updated building codes are being waged nationwide by homebuilder groups with increasing frequency. Two states – North Carolina and Pennsylvania – best illustrate the impact of intense lobbying and regulatory capture on adoption of up-to-date codes.

North Carolina

Historically, North Carolina has been up to date in its adoption of new codes. After the political transformation of the state's political environment from 'blue' to 'deep red' between 2008 and 2010 (Kromm & Sturgis, 2013, p. 11), the IBHS now warns that 'North Carolina's legislative and regulatory actions to weaken building code protection requirements during the past few years is a major concern' (IBHS, 2015b).

North Carolina's Building Code Council (BCC) adopted changes to the model codes that reduce the resilience of its building stock, including 'weakening wall bracing provisions [… and] eliminat[ing] permanent anchors around glazed openings' (IBHS, 2015b, p. 11).

North Carolina failed to adopt the latest codes on wind-borne debris, such as that which is blown around when there is a hurricane or major windstorm. '[T]he N.C. Building Code Council, a board dominated by development interests, said doing so could price many middle- and lower-income residents out of the housing market' (McGrath, 2007, p. 1).

In addition, legislation enacted in 2013 changed the adoption cycle of the model codes from every three to every six years. As a result, the BCC 'does not intend to update the statewide residential code until 2018, which would mean the code would be on a nine year cycle' (IBHS, 2015b, p. 11).

Finally, a combination of anti-regulatory sentiment and regulatory capture led to enactment of a law, H.B. 255, on 13 July 2015 dividing the BCC into two sub-committees – residential and commercial – which have veto power over whether new code provisions are even considered by the BCC.

The residential subcommittee is disproportionately weighted in favour of the residential homebuilding industry. The seven-member subcommittee is comprised of two residential general contractors, two subcontractors, an engineer, a fire service representative and a municipal or county building inspector. The statute mandates that the residential subcommittee must be chaired by a residential homebuilder. In the

original version of the law, there were no code officials represented on the subcommittee at all.

Representative Marc Brody, a homebuilder himself, introduced the bill. The editorial board of the Charlotte *News & Observer* observed that the 'implied intent of his legislation [was] to make things easier for one industry without enough input from others involved in that industry, and that the bill:

> meddles in the rights and powers of local government, fixes something that doesn't appear to be broken and helps one specific group to which he personally happens to be connected.
> (Code Breaker, 2015)

Specifically, the editorial highlighted:

> a couple of highly questionable items in the bill, including provisions that list violations that could put a building inspector's certification in jeopardy, things like 'habitual failure' to do timely inspections (in other words, when the builder wants them) or trying to enforce requirements 'more stringent' than the building code.
> (Code Breaker, 2015)

Indeed, these provisions serve to reduce the level of enforcement by making the building code officials personally liable for negligence if these ephemeral requirements are violated.

Although building codes are technical documents, as the North Carolina building code landscape demonstrates, the review, adoption and even enforcement of codes is subject to the larger political winds of the states, and even the country as a whole.

Pennsylvania

Like North Carolina, Pennsylvania's building code review process has been politicized and strongly influenced by the state's residential homebuilding lobby.

The Review and Advisory Council (RAC) was established in 2006 to review and adopt the provisions of the ICC model codes that were appropriate to protect the health safety and welfare of Pennsylvanians.

The RAC is comprised of 19 gubernatorial appointees representing various sectors of the construction industry, with seats for architects, engineers, code officials, municipal officials, etc. As in North Carolina, the residential homebuilding industry is disproportionately represented on the RAC. In addition to a seat for a residential general contractor, a manufactured housing representative, a modular housing representative, a residential remodelling contractor and a residential architect also have designated seats, totalling five of the 19 positions.

There is no government oversight of the RAC's adoption decision by any state authority. By statute, the RAC's decision becomes law without any opportunity for input or change.

In 2009, under a Democratic governor, the RAC voted to adopt all the 2009 codes, including the IECC and the requirement for fire sprinklers in one- and two-family dwellings. These changes raised the ire of the Pennsylvania Builders Association (PBA), which sued the commonwealth, i.e. the government of the state of Pennsylvania (Swift, 2010).

The PBA lost the suit, and proceeded to lobby the General Assembly to change the adoption process to make it much more difficult to adopt new codes (Swift, 2010). Prior to 'Act 1', as the legislation was ultimately coined, the codes were updated every three years except for provisions the RAC excluded by a majority vote. Act 1 switched the adoption process from 'opt-out' to 'opt-in' – the old code remained in place except for any provisions a two-thirds majority of the full 19 member RAC voted into the new code.

In addition to making the all-volunteer RAC's review process *much* more burdensome, the Act 1 structure runs a high risk of creating a code that would be impossible to comply with or enforce. Any new provisions added to the codes must be harmonized with the rest of the 'old' code provisions.

The RAC's job would be difficult, but not impossible, if Pennsylvania adopted the majority or all of the latest model code provisions on a regular basis. However, if the RAC skips a code cycle, the process becomes untenable, which is the situation currently in effect in Pennsylvania.

Pennsylvania adopted the 2009 codes, but skipped the 2012 codes. In 2015, the RAC was charged with reviewing the changes from the 2012 codes to the 2015 codes *only*. They were not permitted, by statute, to look at the changes from the 2009 codes to the 2012 codes. So, the RAC could not adopt any 2012/2015 code changes that were affected by the 2009/2012 code changes. Any changes that the RAC *did* decide to adopt had to be inserted and harmonized with the 2009 codes to ensure that the resulting code would be technically feasible.

As an all-volunteer body, the RAC does not have the technical expertise, time or resources to do this type of technical review. The statute does not provide any money for the RAC to contract for these services or require Pennsylvania's Department of Labor and Industry, the regulatory agency responsible for building codes, to do so.

In May, 2015, the RAC's technical subcommittees recommended adoption of more than 90% of the 2012/2015 changes. As a result of the flawed statutory structure, the RAC members' feared that major changes to the 2009 codes would be technically unworkable, and decided to reject all but 16 of the 1900 2012/2015 code changes (Fair, 2015).

The General Assembly was aware of the problems with Act 1 as soon as 2012. On 25 February 2015, the RAC itself wrote to the Pennsylvania legislature stating:

> in our professional opinion as construction industry experts, the RAC is unable to effectively fulfil its obligation to review and update the Uniform Construction Code without legislative change' (PA Review and Advisory Council (RAC), 2015). Two bills in two successive legislative cycles were introduced to reform the RAC process. The first bill failed because a rider proposed by the PBA in exchange for RAC reform was politically unacceptable to the municipalities and code officials. The second bill failed because the PBA insisted on extending the code cycle from three to six years.

In the wake of the General Assembly's inaction and the RAC's decision on the 2015 codes, the Clean Air Council, a Pennsylvania environmental group, sued the commonwealth, claiming that Act 1 is unconstitutional and that the RAC's decision to reject the vast majority of the 2015 code provisions is invalid.

Although Pennsylvania and North Carolina represent extreme examples, in many other jurisdictions the residential home building industry has championed delaying or weakening energy codes and extending the code cycle to six years from three years. The fraught political environment of building code adoption must be addressed or circumvented in order to achieve climate resilience through building codes.

Lax enforcement

Enforcement is the obvious and significant but unspoken problem when it comes to realizing the benefits of up to date codes. 'If builders do not build to the code and inspectors do not enforce the provisions, the resulting buildings will not actually be any more resistant [to natural disasters]' (Sutter, 2008, p. 2). The same is, of course, true regarding realizing the energy efficiency benefits.

Professor Daniel Sutter reviewed 'Building Code Effectiveness Grading Schedule' (BCGES) data produced by insurers through the Insurance Services Office (ISO) which evaluates the effectiveness of building code adoption in selected communities (Sutter, 2008). Communities agree to evaluation of their building code enforcement efforts in exchange for discounts on insurance for residents. The scores range from 1 (best) to 10 (worst).

Even among this self-selecting group of approximately 9000 communities, Sutter found that 'only five communities have the best rating of 1 and less than 7% have one of the top three ratings' (Sutter, 2008). He concludes 'overall, the provision of quality certification by the local public sector cannot be characterized as exemplary' (Sutter, 2008, p. 9).

If general code enforcement needs improvement, energy codes are particularly neglected in terms of enforcement. Many code officials have not been sufficiently trained to evaluate compliance with the more stringent energy efficiency requirements of the 2009 and 2012 codes. Building code officials also tend to believe that energy code enforcement is not important to protecting the health, safety and welfare of the public (Keystone Municipal Services Inc., 2010). Therefore, they spend the least amount of time it any on reviewing compliance with the energy codes.

Finally, a drastic reduction in code officials and financial resources contribute to the attitude and education challenges (Mallach, 2013). If there are an insufficient number of code officials, building inspections will be infrequent and cursory.

Thus, it is not merely the political environment surrounding adoption of building codes that is challenging, but also the practical application of the codes to the projects reviewed by code officials and design professionals every day. Both of these factors must be addressed for the full potential of building codes to improve the resilience of the built environment.

Improving code adoption and enforcement through the 'three P' model

The potential benefits for reducing loss of life and property from building codes are significant. However, the political and administrative realities described above make realizing this potential difficult. But, using tools from both the public and private sector, and at the state and federal levels, it is possible to overcome the political barriers to building code adoption and enforcement.

These tools can be summarized in the 'Three P' model: participation, pricing and policing. Below is a discussion of the specific actions in each of these areas that

will provide immediate benefits in terms of raising the profile of using building codes to improve resiliency, and longer-term benefits in terms of shifting the political climate to favour adoption and enforcement of resilient and efficient codes.

Participation

Building codes have changed over time. Initially, they were focused on the most basic safety and welfare considerations. Now, building codes incorporate provisions that effect resiliency and building energy efficiency.

Unfortunately, the perception of codes and, therefore, the makeup of building code council has not caught up with the present and future roles of codes. Most building code approval boards are comprised of construction industry professionals, who rarely take climate change into consideration when deciding whether to update building codes.

This unfortunate situation persists despite the obligation of licensed professionals, particularly design professionals, to take larger societal and environmental considerations into account.

Design professionals are subject to codes of ethics and professional conduct, which focus on the importance of protecting public safety and the impact of the design professionals' work on the environment. The code of ethics and professional conduct of the AIA specifies that architects must 'consider the social and environmental impact of their professional activities and exercise learned and uncompromised professional judgment' (AIA, 2012, p. 1). Likewise, the ASHRAE code of ethics requires that the 'efforts of the society, its members, and its bodies shall be directed at all times to enhancing the public health, safety and welfare' (ASHRAE, 2013, p. 1).

In addition to protecting public welfare generally, both the AIA and ASHRAE require their members to take the impact of their work on the environment into consideration:

- 'Members should promote sustainable design and development principles in their professional activities' (AIA, 2012, p. 4).

- 'Members and organized bodies of [ASHRAE] shall be good stewards of the world's resources including energy, natural, human and financial resources' (ASHRAE, 2013, p. 1).

ASHRAE also has policy statements on climate change specifically, committing to take a 'leadership role in responding to and reducing building climate change footprints' (ASHRAE, 2009, p. 1). It recommends that its members and staff 'need to become actively involved worldwide with policy-setting entities to encourage sound, balanced, and innovative actions to address long-range environmental problems and objectives' (ASHRAE, 2009, p. 7).

These standards of professional conduct seem to dictate that the architects and engineers on building code councils have a special responsibility to take the impacts of climate change on the built environment into consideration when making their decisions on adoption of or changes to building codes. In practice, however, these conversations rarely occur.

These ethical obligations also provide motivation for the national professional and advocacy organizations representing design professionals to play a direct role in advocating for climate resiliency through building codes, and to encourage their members to take similar action.

The NAHB has made building codes a high priority of its advocacy investment, and used strategies like state-level political action committees (PACs), campaign contributions and litigation strategies to advance their position on building codes.

National organizations invested in climate change, design, energy efficiency, next-generation manufacturing, insurance, first responders, housing advocates, disability rights representatives and other related areas should put similar emphasis and investment in building code adoption and enforcement efforts. In order to be successful, engagement with code issues cannot be a crisis-driven or reactive effort. Instead, it must be viewed as a long-term effort, requiring robust, long-lasting coalition building, sustained presence in the political arena and taking legal action where appropriate.

Over the past few years, there has been greater state-level advocacy by some of these groups. For example, most of the environmental advocacy groups in Pennsylvania have embraced code reform as one of their top advocacy priorities. The Delaware Valley Green Building Council included a question about RAC reform in its questionnaire of candidates during the 2014 gubernatorial campaign, and were successful in getting now-Governor Tom Wolf to commit to resolving the code adoption problems (Delaware Valley Green Building Council, 2014). Likewise, in the wake of the RAC's decision to reject all but 16 of the 2015 code changes, the Clean Air Council sued the Commonwealth of Pennsylvania challenging the validity of Act 1 and the RAC's decision (Clean Air Council v. Department of Labor and Industry of the Commonwealth of Pennsylvania and Uniform Construction Code Review and, No. 1041 C.D, 2015).

In addition, national organizations should engage their state chapters and members to become more politically active and provide advocacy training and resources to support their efforts. By helping to create advocacy capacity in their chapters and membership, the national organizations can create self-sustaining model that will achieve the long-term success.

The design professionals must uphold their ethical responsibilities regarding environmental impacts, but they should not bear the entire burden of knowledge of and advocacy for a more resilient built environment. Many different aspects of governmental policy should work in concert to support a more resilient built environment. These policies include utility policy, emergency preparedness, land development, fire protection, insurance and many others. Policy-makers and advocates in these sectors should be prominent participants in code adoption and enforcement proceedings as well.

Ideally, to ensure coordination of policy and proper consideration of climate resilience, building code councils should include representatives from these agencies and sectors. Most importantly, insurance, emergency management and utility regulators must be included in the building code decision-making.

In many cases, adding or substituting code council board members requires amendments to statues or regulations. In political environments where such legal changes are not possible, advocates and policy-makers from these sectors can have influence simply by participating in the building code review and adoption processes, contributing their perspectives and expertise to the conversation.

Increased engagement in building code politics by experts in climate change and resilience will shift the conversation from the cost and benefits to the construction industry alone towards a more holistic assessment of the larger impact of the built environment and the societal costs associated with climate change impacts.

Pricing

The residential building industry is the primary opponent of updating building codes. Largely, their opposition relates to perceived increases in the cost of building and increased complexity of codes as a whole. Many homebuilder representatives take the position that the costs outweigh the benefits (Dornfeld, 2012). They argue that the latest code editions fail to deliver real benefits in terms of safety and energy efficiency for the increased cost and compliance obligations required (Galbraith, 2009). Moreover, they argue, consumers are not demanding more efficient and resilient codes, and are unwilling to pay a premium for safer, more energy-efficient buildings. As a result, the homebuilders have invested heavily in influencing the political process, as they did in Pennsylvania and North Carolina.

There are at least two factors involved in responding to the cost argument: perception and reality. In politics, of course, there is no distinction.

From a perception point of view, while the official NAHB estimates of the IRC and IECC costs are within a reasonable distance of estimates from other sources, policy-makers are rarely exposed to these data. The leaders of the Builders Association of Minnesota claimed that 'possible changes in the state building energy codes could add US$8000–US$12 000 to the cost of a typical home' (Dornfeld, 2012). The St. Louis Homebuilders Association estimated that implementing the 2015 IECC energy efficiency standards would 'cost county homebuyers between *US$32,420 and US$42,345* more than what the energy efficiency measures required by the county existing code' (Lacapra, 2015, emphasis added). It is understandable that policy-makers would hesitate to undertake policy changes that would add US$35,000 to the price of a home!

Moreover, the NAHB and its chapters frequently cite the statistic that 'a $1000 increase in the home price leads to pricing out about 206,269 households from the home ownership market' (Siniavskaia, 2014). If true, this statistic would be very problematic for any policy-maker. However, the analysis that led to the 'priced out' finding is fundamentally flawed.

First, an increase in price because of building codes will not result in 'pricing out' homebuyers from the market. Rather, if true, these 'priced out' buyers will substitute purchases of existing homes built to older code standards which would presumably be less expensive. So, the homebuilders' alleged concern for consumers is actually self-interest. An increase in the cost of a *new* home may make it less competitive with existing homes, not that consumers will be eliminated from the housing market entirely.

Second, the 'priced out model' uses 'ability to qualify for a mortgage as the affordability standard' (Siniavskaia, 2014). Based on the assumption that mortgages are not provided to those households where housing costs exceed 28% of the buyer's gross monthly income, the 'priced out' analysis assumes that any cost increase will disqualify all homebuyers where the increase results in the housing costs exceeding this threshold.

However, the 'priced out' analysis includes people who would *never* qualify for mortgages, before or after the increase, in its calculation. The NAHB paper includes

'over 9 million households with annual income of less than $10,219. [...] A household of less than $10,219 would never qualify for a home loan!' (Dewar, 2015).

The NAHB should have used the marginal cost, which would determine which homebuyers would be forced into unaffordability with the addition of a US$1000 increase in costs. Since the US$1000 increase translates into an increase of '$4.17 a week over the course of mortgage' or a '$217 increase in the required annual income' the actual number of households allegedly priced out would be negligible (Dewar, 2015).

In addition, the price out analysis assumes a 20% down payment. At the low end of housing purchase price, the down payments are generally 10% or less. So, the actual outlay for a purchase price increase of US$1000 is extremely marginal (Dewar, 2015).

Finally, any increase in first costs assumes that the improvements provide no value to the homeowner during the life of the home. For example, a more energy-efficient home reduces a homeowners' utility costs. Thus, the buyers' 'housing costs', which typically include utility costs, would be reduced as well. Reduced losses from natural disaster-related damage also provides economic benefits to the homeowner.

Given the flaws in the NAHB's analysis, it would be valuable for dispelling the persuasive force of the 'priced out model' to perform more economic research to determine accurately the impact of increases in housing construction costs due to updating building codes on the behaviour of homebuyers.

With respect to the argument that consumers neither want nor will pay for updated codes, there is some empirical evidence indicating that this is not necessarily the case. The NAHB's publishing arm, BuilderBooks, performed a consumer research study entitled 'What Homebuyers Really Want'. The study found 'first and foremost, energy efficiency. [...] Nine out of 10 buyers would rather buy a home with energy-efficient features and permanently lower utility bills then one of those features that cost 2% to 3% less' (NAHB, 2013, p. 2). Likewise, a study of consumer demand for stricter building codes in Miami, Florida, showed that consumers were willing to pay a premium of an average of 1.6% for houses built to more stringent codes, and in excess of 10% for properties that were most vulnerable to damage (Jacobsen & Kotchen, 2010).

However, while there is evidence that homebuyers are willing to pay *something* for a more resilient and energy-efficient home, most of the cost analysis has centred on the consumers' willingness to pay for the cost of the changes, not on the impact of the investment in up-to-date codes on the financial model of the home-builders (Walls, Palmer, & Gerarden, 2013; Jacobsen & Kotchen, 2010; NAHB, 2013) In reality, however, the absolute willingness of consumers to pay for the increase in building costs of up-to-date codes is less important than the timing of the investment, the profit margin for the improvements and the difference between the cost of the improvements to the builder and the amount consumers are willing to pay.

First and foremost, the premium consumers are willing to pay may be less than the cost of compliance with the more stringent codes. Randy Dumm and co-authors (2009) provide a robust overview of the scholarship on consumers' unwillingness to value fully more stringent codes, although their literature survey is not related to climate change or energy efficiency specifically.

Some reasons that consumers may not fully value more efficient and resilient codes include:

- the externalized costs, like damage from climate change, are not internalized by consumers or builders

- consumers are unaware of the impact of building codes on characteristics of value, like energy efficiency

- in less disaster-prone regions, consumers underestimate the value of resilient structures

- low energy prices lower the perceived financial benefit of energy efficiency and make payback periods longer

- people significantly discount the value of uncertain benefits

- substitute mechanisms for achieving the same or similar goals are available, for example, substituting insurance for greater resilience

- government practices, like the ready availability of public money for rebuilding after a disaster, reduces the value of mitigation measures

In addition, consumers' willingness to pay an increased price that is equal to or greater than the cost of the improvements is not a relevant figure. Rather, the consumer must be willing to pay the cost of the improvements, *plus the 17–20% gross margin of the builder* (Home Innovation Research Labs, 2015).

Even if homebuyers are willing to pay the increased price, more resilient codes mandate a minimum basket of features that a builder must offer. These features may not be the most profitable basket of features that a homebuilder can offer. In other words, a buyer may be willing to pay US$1000 more for more efficient windows or US$1000 more for granite work surfaces in the kitchen. If the builder stands to make a greater profit on the granite work surfaces rather than the windows, they will prefer to offer these and oppose code changes that require installation of the windows.

It is undoubtedly the case that the homeowner or building owner that realizes the benefits of more resilient structures and greater energy efficiency savings. It is society overall that benefits from reduced rebuilding costs, lower emissions and amelioration of climate change. But it is the builder who must bear the first costs of any increase in construction costs. Therefore, overcoming political resistance to updated codes requires changing the financial incentives of both the homebuyer and the homebuilder. A policy decision must be made about who should bear the costs and how to incentivize the change.

Currently, taxpayers and insurance companies bear the cost of a less resilient building stock, not individual homebuyers. Governments (funded by taxpayers) pay billions of dollars in disaster recovery relief. Insurance companies (funded by their policy owners) also bear the cost of damage to property and people.

When climate change impact is taken into consideration, the marginal increase in costs for more resilient codes is well worth the investment for society. The goal of code policy, then, is either to internalize the cost externalities currently being borne by society at large or to create more cost-effective ways of publicly funding the development of a more resilient building stock.

By improving the resilience of building codes, building practices and technology will become more commonplace and less expensive. This improves the profit margin for the builder, either by improving the profitability of the products themselves or by lowering building costs overall, allowing the builder to offer more profitable elements as well. In the meanwhile, however, incentives could be provided to defray the increased costs and overcome builder opposition.

Another option is to shift more of the societal costs onto the building owner, potentially in the form of increased insurance premiums or by changing levels of government funding for jurisdictions that do not adopt up-to-date codes and ensure robust enforcement.

The time when consumers demand more robust codes is directly in the wake of a natural disaster (Spence, 2004). This is also the time when the federal government tends to provide resources to the states for rebuilding. More robust building code adoption and enforcement could be a contingency for receiving post-disaster aid.

Acknowledging that withholding disaster aid can be politically challenging, the government can use this funding leverage proactively as well. The federal government can (and sometimes does) offer additional funding for adoption, implementation and enforcement of up-to-date codes. However, this funding is typically minimal. By contrast, the larger budget allocations for weatherization assistance programmes and state energy programme block grants could also be contingent upon code adoption and enforcement.

There is precedent for tying federal funds to building code adoption. In 2010, the DOE determined that the 2009 IECC represented such a significant increase in building energy efficiency that it should be adopted throughout the country. It decided to tie American Reinvestment and Recovery Act (ARRA or 'stimulus') funding to the adoption and enforcement of the 2009 IECC. Governors from all 50 states signed commitments to adopt the 2009 IECC and achieve 90% compliance with it by 2017 (Burt, 2010). While this 90% target will probably never be achieved, it has already increased adoption and compliance with energy codes significantly.

The federal government could also send a strong price signal to the market by incorporating the real cost of risks from climate change into premiums for flood insurance available through the National Flood Insurance Program (NFIP). In 2012, the Biggert–Waters Act attempted to take steps towards aligning the price of flood insurance premiums with risk.

Unfortunately, realpolitik intervened. Senator Mary Landrieu (D-La.) was facing a tough re-election, and many NFIP reforms were reversed or delayed through the Homeowner Flood Insurance Affordability Act of 2014 (Knowles, 2014). Nonetheless, the increases that were permitted are already showing signs of changing consumer preferences (Klibanoff, 2015).

However, there are some limitations with using disaster recovery funding and the other pools of funding discussed above as leverage:

- it is not evenly divided among states

- it withholds funds from the neediest households

- it could discourage activity that is intended to be encouraged

- the connection between the funding and building codes is attenuated

By contrast, the federal government has a financial tool that is used nationwide and relates specifically to the risks from climate change: federally backed mortgages.

As of 2009, Fannie Mae and Freddie Mac owned or guaranteed 'roughly half of all outstanding mortgages in the United States […] and they financed three-quarters of new mortgages originated that year' (Congressional Budget Office, 2010).

Fannie Mae and Freddie Mac set underwriting requirements for these mortgages, which already include a contingency that houses comply with the then-current building codes. Indeed, the Department of Housing and Urban Development (HUD) currently requires adherence to the 2009 IECC for HUD-assisted housing.7 C.F.R. 91 (Beardsley, 2015). Up-to-date codes and a robust enforcement infrastructure could be a contingency for the federal government to guarantee mortgages in a specified jurisdiction. The BCGES ratings could be used as an evaluation tool to ease administration of this requirement. Even the threat of this sort of policy change would be extremely motivating for politicians.

The public sector is not the only, and perhaps not the most efficient, driver for more robust building codes. Insurance and mortgage companies can take a larger role in demanding adoption and enforcement of up-to-date code. For example, the insurance industry could provide discounts on premiums for proactive mitigation efforts like adoption of up-to-date codes and effective enforcement on the part of the individual building owner and the community.

Whether through the public or private sector, if property or flood insurance becomes very expensive or unavailable for structures built to lower code standards or in communities that do not enforce building codes well, building owners would receive a strong price signal encouraging greater consideration of the risk of damage from climate change into their decision making and preferences.

In summary, residential homebuilders oppose more stringent building codes based on first costs and lack of consumer demand. Some of the builders' claims regarding exorbitant increases in first-costs for resilient and energy efficient codes are questionable. However, to the extent that there are increased costs for more resilient codes, these costs are outweighed by the overall benefits to society of up-to-date codes. The public and private sector can use incentives for builders and changes to insurance, mortgages, property taxes and disaster relief to better align individual and societal preferences.

Policing

In the words of Sutter, 'overall, the provision of quality certification by the local public sector [through building code enforcement] cannot be characterized as exemplary' (Sutter, 2008, p. 9). If general code enforcement needs improvement, energy codes are particularly neglected in terms of enforcement (Misuriello, Penney, Eldridge, & Foster, 2010). Building code officials tend to believe that energy code enforcement is not important to protecting the health, safety and welfare of the public. Therefore, they spend the least amount of time if any on reviewing compliance with the energy codes. In addition, many code officials have not been sufficiently trained to evaluate compliance with the new, more stringent energy efficiency requirements (Misuriello et al., 2010).

However, the code official is only one professional responsible for code compliance. Architects, engineers and builders are all responsible for ensuring compliance with the building codes. This is particularly true of the architects and engineers who stamp construction documents, averring that the plans comply with applicable law. Unfortunately, data show that technical knowledge regarding energy code requirements is insufficient (Kaplan, 2012). In many states, builders do not have any training or licensing requirements (Gibson, 2010). Therefore, there is little incentive for them to seek out building code training.

Code training exists for energy codes, and is available through state and federally funded venues. The DOE is currently investing heavily in improving energy code compliance and training for construction industry professionals, providing millions of dollars in grant funding to develop training and tools to increase energy code compliance (US DOE, 2014). Some training is being developed in resiliency as well. Virginia, for example, has specialized training for code officials in the resilience aspects of the codes (IBHS, 2015a).

Climate-related code training should be mandated for code officials, and expanded to architects, engineers, contractors and other construction professionals.

Contractor training is particularly neglected. Architects and engineers generally must be licensed and have continuing education requirements under state law. By contrast, in many states there is no requirement that contractors be licensed, and still fewer states that require continuing education (NCARB, 2015). Implementation of licensing and continuing education requirements would provide incentive for contractors to receive additional training. Like the architects, engineers and code officials, this training should include education in the energy codes and the resiliency aspects of building codes.

Funding is a perennial challenge when it comes to code enforcement and education programming. To overcome this challenge, several states are supplementing federal funding by allowing utilities to fund energy code-related activities as part of the utilities' mandated energy efficiency programming. California has done so

for decades. More recently, Ohio, Arizona, Rhode Island, Massachusetts and several other states have implemented similar programmes (Sarah Stellberg, 2012).

Utility-funded energy code programmes have taken different forms, including funding for training of design professionals, builders and code officials, funding for compliance collaboratives and incentives to cover first costs associated with construction to higher efficiency levels (Sarah Stellberg, 2012). These code programmes have been highly cost-effective ways to increase energy efficiency.

The Clean Power Plan (CPP), the new regulations limiting carbon emissions from existing power plants, provides another potential avenue to expand funding for building code programming. Pursuant to the Environmental Protection Agency (USEPA, 2015) final regulations released in August 2015, states can use energy code adoption and compliance programming as a component of their plans to comply with the emissions reduction requirements of the CPP. This provides new impetus for states to invest in code-related programming, and opens up new funding opportunities for these efforts.

Conclusions

Achieving climate resilience through building codes will continue to meet with political resistance until and unless there is activity across the 'Three P's' – participation, pricing and policing. Fortunately, these areas provide many opportunities to shift the political atmosphere over time. Some organizations have already started to become more engaged at the state and local levels in building code adoption and policy. Insurance companies and the NFIP are starting to introduce pricing that incorporates the growing risk from climate change. New programmes, like the Virginia programme for training code officials on the resilience provisions of the model codes, are showing leadership in integrating code enforcement with the larger effort to create a more resilient built environment. Through these three avenues, the potential for climate resilience from building codes will become the reality on the ground.

Acknowledgements

Many thanks to Professor Cary Coglianese, Director of the University of Pennsylvania Program on Regulation, and the author's research assistant Elijah Kroos.

Disclosure statement

Shari Shapiro has performed consulting work for the International Code Council and the National Fire Protection Association.

References

American Institute of Architects (AIA). (2012). *Code of ethics & professional conduct*. Washington DC: AIA.

American Society of Heating, Refrigerating and Air-Conditioning Engineers (ASHRAE). (2009). *ASHRAE position document on climate change*. Atlanta: ASHRAE.

American Society of Heating, Refrigerating and Air-Conditioning Engineers (ASHRAE). (2013, January 30). ASHRAE Code of Ethics.

Beardsley, E. (2015, May 14). *US moves forward with raising the bar for energy efficiency in affordable housing*. Washington DC: US Green Building Council. Retrieved December 17, 2015, from http://www.usgbc.org/articles/us-moves-forward-raising-bar-energy-efficiency-affordable-housing

Building Owners and Managers Association (BOMA). (2015). *Energy efficiency and green building codes*. BOMA. Retrieved 12 14, 2015, from http://www.boma.org/industry-issues/building-codes/Pages/Energy-Efficiency-and-Green-Building-Codes.aspx

Burt, L. (2010, January 5). *Is your governor keeping his/her promises about efficiency?* Natural Resources Defense Council Staff Blog. Retrieved December 17, 2015, from http://switchboard.nrdc.org/blogs/lburt/is_your_governor_keeping_hishe.html

Clean Air Council v. Department of Labor and Industry of the Commonwealth of Pennsylvania and Uniform Construction Code Review and, No. 1041 C.D. (2015). (Commonwealth Court of Pennsylvania 9 October, 2015).

Code Breaker. (2015, April 17). *The News & Observer*.

Congressional Budget Office. (2010). *Fannie Mae, Freddie Mac, and the federal role in the secondary mortgage market*. Washington DC: Congressional Budget Office.

Delaware Valley Green Building Council. (2014). *PA green building questionnaire: Gubernatorial Candidates' responses*. Delaware Valley Green Building Council. Retrieved December 18, 2015, from: http://dvgbc.org/policyadvocacy/pagovcandidates-responses

Dewar, B. (2015). *The NAHB 'pricing out' farce-spinning statistics*. The Fire Sprinkler Times, 1. Retrieved from http://newsletter.nfsawi.org/2014/12/nahb-pricing-out-farce-spinning-statistics/

Dornfeld, S. (2012, April 5). *Minnesota builders warn energy-code changes would raise price of homes*. Retrieved from www.minnpost.co: http://www.minnpost.com/cityscape/2012/04/minnesota-builders-warn-energy-code-changes-would-raise-price-homes

Dumm, R., Sirmans, S., & Smersh, G. (2009). *The capitalization of stricter building codes in Miami, Florida house prices*. Tallahassee, FL: Florida Catastrophic Risk Management Center.

Emrath, P. (2009, February 11). How long buyers remain in their homes. *NAHB.org*.

Fair, M. (2015, June 29). *Pa. building code review process illegal, enviro group says*. Law 360. Retrieved December 17, 2015, from http://www.law360.com/articles/673446/pa-building-code-review-process-illegal-enviro-group-says

Federal Emergency Management Agency (FEMA). (2006). *Hurricane Katrina in the Gulf Coast: Mitigation assessment team report*. Washington DC: FEMA.

Galbraith, K. (2009, May 28). Building code battles heats up in Midwest. *nytimes.com*.

Gibson, S. (2010, December 6). *Are we really better off with building codes?* http://www.greenbuildingadvisor.com/blogs/dept/qa-spotlight/are-we-really-better-building-codes#ixzz3ug2Gprat Retrieved December 17, 2015, from Green Building Advisor: http://www.greenbuildingadvisor.com/blogs/dept/qa-spotlight/are-we-really-better-building-codes

Hassler, U. and Kohler, N. (2014). Resilience in the built environment. *Building Research & Information*, 42(2), 119–129. doi:10.1080/09613218.2014.873593

Home Innovation Research Labs. (2015). *Estimated cost of the 2015 IRC code changes*. Upper Marlboro, MD: Home Innovation Research Labs.

Hurricane Sandy Rebuilding Task Force. (2013). *Hurricane Sandy rebuilding strategy report*. Washington DC: Hurricane Sandy Rebuilding Task Force.

Institute for Market Transformation. (2015, March). *SAVE act*. Washington DC: Institute for Market Transformation. Retrieved December 14, 2015, from http://www.imt.org

Insurance Institute for Business & Home Safety (IBHS). (2015a). *Katrina: 10 years later*. Tampa, FL: IBHS.

Insurance Institute for Business & Home Safety (IBHS). (2015b). *Rating the states: 2015*. Tampa, FL: IBHS.

Insurance Institute for Business & Home Safety (IBHS). *The benefits of statewide building codes*. Tampa, FL: IBHS.

International Code Council (ICC). (2014). *Reducing flood losses through the international codes*. ICC.

Jacobsen, G. D., & Kotchen, M. J. (2010). *Are building codes effective at saving energy? Evidence from residential building data in Florida*. Cambridge, MA: National Bureau of Economic Research.

Kaplan, M. (2012). *New York energy code compliance study*. Albany, NY: New York State Energy Research and Development Authority.

Keystone Municipal Services Inc. (2010). *Rittenhouse club final report*. Norristown: Keystone Municipal Services Inc.

Klibanoff, E. (2015, September 10). *Wake-up call or washout: Flood insurance reform threatens Pa. river towns*. University Park, PA: WPSU.

Knowles, S. G. (2014, March 23). *Flood zone foolishness*. Slate. Retrieved December 17, 2015, from http://www.slate.com/articles/health_and_science/science/2014/03/biggert_waters_and_nfip_flood_insurance_should_be_strengthened.html

Kotchen, M. J. (2011, Fourth Quarter). Energy efficiency codes: Plan B for climate change? *Milken Institute Review*, 58–69.

Kromm, C., & Sturgis, S. (2013, June 6). North Carolina's tug-of-war. *The American Prospect*.

Lacapra, V. (2015, August 12). *Proposed energy efficiency standards for new homes spark controversy in St. Louis County*. St. Louis Public Radio. Retrieved from http://news.stlpublicradio.org/post/proposed-energy-efficiency-standards-new-homes-spark-controversy-st-louis-county

Levinson, A. (2015). *How much energy do building codes really save? Evidence from California*. NBER Working Paper No. 20797. Cambridge, MA: National Bureau of Economic Research.

Mallach, A. (2013, March 26). *5 things cities and CDCs don't get about code enforcement*. Retrieved from www.rooflines.org: http://www.rooflines.org/3143/5_things_cities_and_cdcs_dont_get_about_code_enforcement/

McGrath, G. (2007, July 6). Flood Insurance Rates May Surge. *Star News*.

Mendon, V., Lucas, R., & Goel, G. (2013). *Cost-effectiveness analysis of the 2009 and 2012 IECC residential provisions – technical support document*. Richland, WA: Pacific Northwest National Laboratory.

Misuriello, H., Penney, S., Eldridge, M., & Foster, B. (2010). *Lessons learned from building energy code compliance and enforcement evaluation studies*. Washington DC: American Council for an Energy-Efficient Economy.

NAHB. (2014, October 25). Builders show their political clout.

NAHB Research Center. (2012a). *2012 IECC cost effectiveness analysis*. NAHB Research Center.

NAHB Research Center. (2012b). *Commercially viable energy efficiency solution package*. US Department of Energy.

National Association of Home Builders (NAHB). (2013). *What home buyers really want*. BuilderBooks.

NCARB. (2015). *Continuing education requirements by state*. NCARB: Continuing Education Requirements by State Retrieved December 17, 2015, from.

Pennsylvania Review and Advisory Council (RAC). (2015, February 25).

Risky Business Project. (2014). *Risky business: The economic risks of climate change in the United States*. Risky Business Project.

Sarah Stellberg, A. C. (2012). *Role for utilities in enhancing building energy code compliance.* Washington DC: ACEEE.

Siniavskaia, N. (2014). State and metro area house prices: The 'priced out' effect. *Housingeconomics.com.*

Sorrell, S. (2007). *The rebound effect: An assessment of the evidence for economy-wide energy savings from improved energy efficiency.* London, UK: UK Energy Research Centre.

Spence, R. (2004). Risk and regulation: Can improved government action reduce the impacts of natural disasters? *Building Research & Information, 32*(5), 391–402. doi:10.1080/0961321042000221043

Spicuzza, M. (2012, September 25). Walker reverses course on removing electrical safety codes. *Wisconsin State Journal.* Retrieved from http://host.madison.com/news/local/govt-and-politics/walker-reverses-course-on-removing-electrical-safety-codes/article_6916957c-066b-11e2-9ad8-001a4bcf887a.html

Sutter, D. (2008). *Quality assurance by the public sector: An analysis of building code enforcement.* Fairfax, VA: Mercatus Center, George Mason University.

Swift, R. (2010, May 16). Two sides raise pitch of sprinkler debate. *The Times-Tribune.* Retrieved December 17, 2015, from http://thetimes-tribune.com

Top Lobbyist of 2014. (2014, October 22). *The Hill.*

Union of Concerned Scientists. (2015). *The hidden cost of fossil fuels.* Cambridge, MA: Union of Concerned Scientists. Retrieved December 17, 2015, from http://www.ucsusa.org/clean_energy/our-energy-choices/coal-and-other-fossil-fuels/the-hidden-cost-of-fossil.html#.VnQluoujfFJ

Union of Concerned Scientists. (n.d.). Is global warming fueling increased wildfire risks? Retrieved from http://www.ucsusa.org/global_warming/science_and_impacts/impacts/global-warming-and-wildfire.html#.VgVu78tViko

US Department of Energy. (2015, April). *Achieving energy savings and emission reductions from building energy codes.*

US Environmental Protection Agency. (2015). *Clean power plan.* Washington DC: USEPA.

US Government Accountability Office. (2015). *Hurricane Sandy: An investment strategy could help federal government enhance national resilience to future disasters.* Washington DC: United States Government Accountability Office.

US Department of Energy. (2014, August 14). *Department of Energy invests $6 million to increase building energy code compliance rates.* Washington, DC: US Department of Energy/Energy Efficiency & Renewable Energy. Retrieved December 17, 2015, from https://www.energycodes.gov/department-energy-invests-6-million-increase-building-energy-code-compliance-rates

Vaughan, E., & Turner, J. (2013). *The value and impact of building codes.* Washington, DC: Environmental and Energy Study Institute.

Volokh, A. (2014). The new private-regulation skepticism: Due process, non-delegation, and antitrust challenges. *Harvard Journal of Law and Public Policy, 37*(3), 931–1006.

Walls, M., Palmer, K., & Gerarden, T. (2013). *Is energy efficiency capitalized into home prices?* Washington DC: Resources for the Future.

Transforming building regulatory systems to address climate change

David A. Eisenberg

The challenge of climate change for building regulation is explored based on long-term, personal observations of US building regulatory systems that are intended to safeguard the public. Systemic problems and patterns have allowed significant, large-scale hazards (*e.g.*, those related to climate change) to be excluded from consideration in regulatory systems. Regulatory systems are typically not comprehensive, designed, integrated systems. Instead, they are silos of regulatory responsibility with gaps in authority. The system lacks formal processes to address emergent hazards. There is an absence of formal processes to assess and balance risks across hazard types, locations, timeframes and scales. The current regulatory goal of preventing or limiting known harm is compared with the positive outcome goals of the Living Building Challenge (LBC). The more comprehensive scope of the LBC and its goals surpass the existing code, but ironically, LBC projects often struggle to gain regulatory approval due to its use of innovative approaches. Potential avenues to create more comprehensive and effective regulatory systems are suggested. It is proposed that the purpose of the regulatory role is expanded from policing the arbitrary boundary between what is legal and illegal to one that includes enabling the most regenerative and positive outcomes.

Our greatest responsibility is to be good ancestors.
Dr Jonas Salk
(quoted in Fahey and Randall, 1998, p. 332)

Introduction

An examination of the regulatory challenges inherent in addressing climate change will reveal the need to rethink the goals and mechanisms of regulatory response and responsibility. The reality that hazards of significant scope and seriousness can escape regulatory attention and response for so long raises the question of why this occurs. This commentary will briefly outline a set of problematic patterns in the regulatory realm that have enabled other serious, though potentially less consequential, hazards to emerge without effective response. Those patterns are explored below, as well as potential ways to transform the current regulatory processes for the built environment.

Observations from over two decades working to incorporate issues of sustainability into US building regulatory systems, as well as engaging in code-related discussions in Europe (Belgium, Denmark, France, Germany, England, Scotland and Wales), Australia, New Zealand, Canada, Israel, Jordan and Saudi Arabia, provide insights into how significant emergent hazards are systematically excluded from consideration within current regulatory structures. Whether they result from changes in circumstances or the environment, new technologies, or new understandings about practices once deemed safe, the existing systems of codes, standards and other public policies in the US have not effectively recognized or addressed hazards occurring at larger scales, over longer timeframes, occurring away from the building site, or of types not previously recognized or addressed.

Historically, modern building and land-use regulations in the US and many other developed countries paid little attention to human health beyond basic sanitation, adequate ventilation and light, or the impacts of industrial processes on people and the environment.

These regulatory systems were responses to disasters and serious failures, created through *ad hoc* processes aiming to avoid or minimize those problems.

Over time these systems developed into separate regulatory and functional silos, often without comprehensive, unifying goals or principles applying to the whole system and each of its parts. Designations of safety or acceptable risk given to mainstream practices by these regulatory systems can depend on which categories of risk are included or excluded from each particular scope of regulatory concern. That scope has often been determined in an even less integrated or well-informed process: the acts of legislative and political bodies establish and limit regulatory authority over activities thought to need regulatory control. The political influence of industries and private-interest groups on these processes is one of the most substantial barriers to addressing the changes needed for more effective regulation at every level of government.

In order to adequately address the hazards tied to climate change, a transition will be necessary to comprehensive systems-based regulatory structures. These would include, but extend well beyond, the goal of avoiding worst outcomes in specific areas of focus. There has been a growing recognition of the need for action at a larger scale to address climate change for both mitigation and adaptation. The concept of resilience is increasingly recognized by researchers and practitioners as an appropriate framework for thinking about how to deal with the increasing uncertainties tied to a changing climate and could be a useful addition to regulatory thinking (Hassler & Kohler, 2014).

What focus?

The modern history of regulation of the built environment is filled with well-intentioned codes, standards and policies followed by the discovery of unintended consequences and unrecognized hazards, and the eventual response to those newly identified problems.

Fire Safety
Structural Integrity
Means of Egress
Light
Ventilation
Heat
Water & Wastewater
Electrical & Gas
Energy Efficiency

Figure 1 Risk viewed through the microscope of codes. These are the categories of hazard of regulatory focus typically addressed by building codes

These responses often create their own unintended outcomes and the process repeats. To a degree this is unavoidable, since it is not possible to foresee all the connections and possibilities resulting from any action. However, such unintended outcomes are amplified in fragmented regulatory systems not intentionally designed as comprehensive and integrated systems with clearly articulated system goals and principles, and formal processes to identify and integrate emerging risks and hazards.

The issues stem from a form of myopia. Focusing on a problem to be solved can be exclusionary and does not automatically lead to the exploration of causal linkages or the possible chain of consequences of action or inaction. Rather, the tendency is to focus on and remain in the details. The relevance of this to building codes is illustrated in Figures 1–3. Looking at buildings through codes is like looking through a microscope. Important information about things such as site conditions, the design, characteristics of materials, structural integrity, fire safety, means of egress can all be viewed through that lens.

Stepping back to observe the larger set of impacts of a built project, other kinds of hazards and risks become visible. Though an individual project may contribute a small amount to a given hazard, the cumulative impact of all projects creates the larger hazard. As long as those remote, distributed, cumulative and longer-time-frame impacts are hidden from view, there is no way to understand what may actually be at risk or how those impacts might be addressed (Figure 3).

The broader basis for acceptable, safe practices can only result from considering the full risk profile and the wider set of impacts that arise. This capability is missing in most regulatory systems today. A process that can reveal the overall risks and promote actions to balance them is necessary in order to address climate change and the related emerging crises.

This larger context for understanding risk and responsibility corresponds with a systems-based approach to addressing problems (Meadows, 1999). Hawken, Lovins, and Lovins (1999, p. 117) state that 'Optimizing components [of a system] in isolation tends to pessimize the whole system'. Most regulatory reform efforts to date have focused on optimizing the subsystems, without acknowledging that the nature of actual risks is not limited to the arbitrary boundaries of professional disciplines, regulatory silos, timeframes, scales or site-based impacts. One might say that optimizing parts of the built environment in isolation has been pessimizing the global climate system.

Today's regulatory systems are the cumulative result of thousands of essentially *ad hoc* efforts to address identified hazards and risks, taken at all levels of

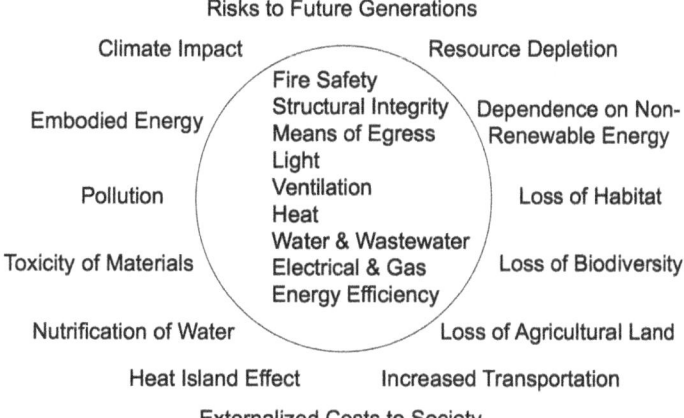

Figure 2 The bigger picture of risk. The actual risk profile is much larger, including remote and longer timeframe of impacts, most of which are invisible when viewed through the lens of building codes

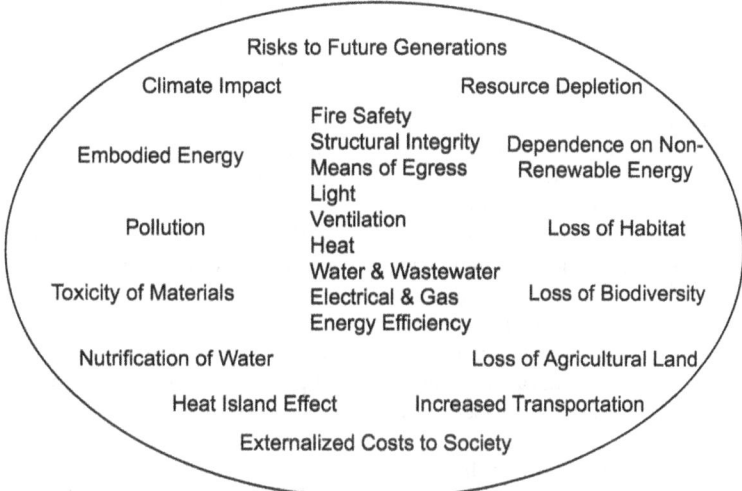

Figure 3 Addressing these risks is not about either/or, it is about a balanced approach. With the full set of impacts and hazards in view, it becomes possible to consider them all together, across all regulatory boundaries and address them in a process that balances risks and seeks the best overall risk performance

government and by agencies, organizations and groups of varied expertise, capability, authority, geographical location and focus. The products of those varied processes often become local, regional or national codes, standards and policies. Despite periodic review and improvement processes for building codes and standards, the overarching goals, principles and overall effectiveness are rarely examined. No formal, high-level process exists to identify gaps through which serious hazards have escaped regulatory attention. Even when major changes or reorganizations of regulatory entities occur, as with the consolidation of the three regional building code organizations (and their codes) in the US into the International Code Council, completed around the year 2000,[1] or with the shifts to performance- or outcome-based codes in some

countries, the focus rarely goes beyond changes within the existing scope and the details already included. One might ask: where in the goals, principles and contents of existing regulatory systems is a continuous and explicit representation of the rights and welfare of future generations? Few gaps in regulatory responsibility could have greater significance or potentially graver consequences.

Regulatory patterns: disregarding large-scale unregulated hazards

Regulatory systems are reactive. Regulations and regulatory entities are typically created in response to serious disasters, failures or problems after they

43

occur, not before. They respond to known, persistent and widespread problems serious enough to demand official response. Thus they are backward-facing by nature, with limited anticipatory capacity or authority. Emergent hazards and new understandings about practices previously viewed as harmless are often resisted or overlooked, sometimes long after they have become serious and widely recognized.

Most regulations are based on minimum requirements establishing thresholds for what is deemed to be allowable. Though logical and necessary, the result is a set of intricate, somewhat arbitrary, and often inflexible boundaries between illegal and barely acceptable practices. Minimums, unfortunately, tend to become the accepted standards of practice. They also shift the regulatory focus to the most particular level of hazard and away from assessing which hazards may present cascading risks to larger populations, rather than affecting only individual buildings and their occupants.

The widespread assumption by regulators is that the content already in codes and standards represents safe practice. Moreover, it is also assumed these codes and standards serve as the basis for establishing the criteria by which alternatives must be judged. A consequence of this thinking is that it limits or precludes the hazards which can be considered when evaluating those alternatives. While this seems plausible, it ignores the reality that a current classification of safety or acceptability may have been based on excluding certain categories of harm from consideration. These assumptions also impede the ability to assess the hazards and risks across categories of harm, or at different scales, by removing those other hazards from consideration.

In the current US building code development processes, several significant issues are deemed to be outside the scope of building codes, *e.g.*, the toxicity of materials throughout their lifecycle and their human and environmental health impacts. Attempts to introduce some of these issues through proposed code changes have been rejected by the building codes community based on assertions that these hazards are beyond their scope of concern, and other, more appropriate regulatory entities exist to address them. However, no formal process exists to ensure that another entity exists with the authority and capacity to perform that regulatory role, or to assess whether code provisions and referenced standards result in the use of materials that pose serious hazards, yet go unregulated.

The success of regulatory systems in addressing the specific hazards they were designed to address reinforces confidence in their adequacy. With no formal processes to consider potential risks from significantly changing circumstances such as those associated with the mitigation of and adaptation to climate change, *ad hoc* processes remain the only way to address them. Even with serious emerging hazards, it takes considerable time and effort to begin to get these issues the attention they deserve. This includes issues like the decreasing availability, reliability or quality of essential resources (energy, water, materials or the utility systems that buildings rely upon for safe operation), or the inadequacy of historical climate, flood, seismic or other data on which hazard assessments and requirements are based.

Finally, effective enforcement is essential for regulations to provide their intended outcomes and benefits. Many issues impede optimal enforcement, including staffing and training, and the unreliable funding mechanisms that support building, planning and other regulatory agencies. Enforcement of energy codes is key for addressing climate impacts. In the US, enforcement of energy codes has been an ongoing issue, in part because these codes were largely developed and viewed as outside the traditional remit of building codes on life safety. As a consequence, many building officials view them as beyond their core responsibilities. As energy codes became more complex and stringent, resentment was more widely expressed that they were time-consuming, complicated and seen as interfering with the their central mission of life safety. This was a failure to develop awareness about the full range of benefits and impacts of highly energy-efficient buildings, including climate impacts. This failure provides an important lesson about the care required to seek alignment and buy-in at the start of new efforts.

Potential pathways forward

As observed, current regulatory 'systems' for the built environment are rarely fully designed, comprehensive, integrated systems. What would a fit-for-purpose regulatory system look like? It would be grounded in systems thinking, and be based on positive system goals and principles. It would include formal processes to identify, assess and balance hazards across silos of regulatory authority and responsibility, including emergent hazards and new knowledge about previously unrecognized impacts of accepted practice. It would treat innovative approaches, materials and designs aiming to reduce systemic risks as worthy of full and immediate consideration. The shift in thinking about the purpose and role of the regulatory system could be characterized as becoming community and public resources enabling the best and most regenerative built projects, which includes the essential traditional policing function to prevent the worst outcomes, but not as the lone focus of their mission.

It is not possible or reasonable to require all built projects to meet the highest performance standards. It is possible and reasonable to look for ways to discourage practices with the greatest large-scale impacts and encourage better practices that reduce them. Once there is awareness of the harm created by conventionally 'safe' projects and the risks emanating from the entire enterprise of creating built projects, it is a shorter step to grasping the benefits of allowing projects that do not follow the old minimum standards and practices.

To date, most efforts to address these larger impacts have been voluntary – market-based – rather than mandatory government regulations. They include green building programmes, rating and certification systems, incentive programmes, and a few reach or stretch codes. Sometimes they incorporate green provisions into mainstream codes and standards. These systems share the same basic goal as regulations: to reduce or eliminate harmful outcomes. The International Living Future Institute's (ILFI)[2] Living Building Challenge (LBC) certification system differs from nearly all others by aiming to shift development, design and building practice to net-zero negative impact or better (positive impact, net-positive) across the spectrum of impacts of built projects: not just energy or water, but carbon footprint, land impact, materials, toxicity and more.

LBC projects embrace a wider, higher set of performance criteria than anything required by current regulation. Therefore, one might assume that they would move more easily through regulatory review and approval processes. Experience has shown this is not the case. To achieve those higher, deeply integrated goals, project designers are pulled toward innovative designs, materials, equipment and methods of construction outside typical practice. Although the designs exceed what is required by regulation, the regulators often demand greater evidence of the validity of proposed performance. These projects use integrated project delivery systems, with integrative design processes that include early design charrettes. One strategy to address proactively the regulatory challenges is to invite representatives of the regulatory agencies to participate in design charrettes. The regulators then learn what the designers are trying to accomplish, how and why, and the designers have the benefit of the regulators' concerns and ideas about how best to address them early in the process, creating collaborative relationships rather than adversarial ones.

Unlike regulatory systems that require obtaining approvals before a building is occupied, LBC projects must operate for a year before they can certify, making the goals measurable and outcome based. The ILFI has sponsored research[3] into regulatory challenges,[4] case studies of solutions, and collaborative efforts with local governments supporting such projects, including the extensive report *Code, Regulatory and Systemic Barriers Affecting Living Building Projects* (Eisenberg & Persram, 2009) One possible improvement to the LBC system in dealing with regulatory issues would be to incorporate by reference the International Code Council's *Performance Code for Buildings and Facilities*[5] as the building code of choice for LBC projects. If the requirements for certification even informally included code compliance, it would make clear that the goal is for the highest levels of both performance and safety. The use of this code by designers creating the most regenerative projects would help improve the performance code and could influence and lead the trajectory of the rest of the codes.

The Rocky Mountain Land Use Institute (RMLUI) at the University of Denver Sturm College of Law created the Sustainable Community Development Code Framework (2009)[6] that compiled a broad spectrum of precedents, ordinances, best practices and other resources into a usable framework to help communities incorporate more sustainable land-use policies that align and better support sustainable building and development practices. The framework could be expanded to include building-related best regulatory practices as well. Coupled with the idea of reimagining the purpose of building and development regulation to support the best community and overall outcomes, one could envision a comprehensive regulatory resource covering the spectrum of regulations affecting built projects with basics (the minimum standards and requirements), and good, better, and best practices and links to precedents, incentives and relevant resources. Such a resource could expand regulatory possibilities and aspirations; reframing the notion of a regulatory arena to be more inclusive and constructive than the present narrow boundary between what is legal and illegal.

Conclusions

The imperative to address the impacts of climate change is a public safety issue for both mitigation and adaptation. What appears reasonably certain is that climate change is already happening and it is accelerating. The long lifespan of buildings and infrastructure means that everything designed and built today will exist in different conditions than exist now. Given the long timeframes involved in changes to codes, standards, regulations and policies, the urgency of acting now is clear.

A process is needed to identify, evaluate and consider hazards across categories, timeframes, locations of impact and impacted populations. The regulatory realm has an important role in the provision and encouragement of alternatives to mainstream practice that can provide an improved overall risk-performance profile.

Initiating a process of change will require clear and bold leadership – to transform these systems and to share progress (including setbacks and unintended outcomes that arise). The regulatory systems that emerge must be able to adapt to changing realities, and respond to risks larger than any current systems were designed to address. This is not a call to abandon what has been learned, but rather a call to integrate it into the larger context of risk and responsibility needed today. There are leadership roles for those inside regulatory agencies at all levels. The design community (architects, engineers, consultants, interior designers, planners) can play a more direct and active role in changing codes and standards and influencing lawmakers and state and local officials to provide leadership and support for regulatory changes. The design community has a distinct advantage of a 'middle out' role (Janda & Parag, 2013) as they are situated between policy/regulation and clients/occupants and therefore have the possibility to transmit ideas and insights in both directions. Forward-thinking developers, landowners and builders have much to contribute to the process as well. There is also a need for more citizen involvement to help elected officials understand that support exists for addressing these issues in proactive and aggressive ways.

Acknowledgements

The author would like to thank his wife, Patricia Eisenberg, for her assistance in editing the manuscript.

Disclosure statement

The author has no known conflicts of interest. In the interest of full disclosure, he is chief executive of a non-profit 501-(c)(3) public interest organization that produced the International Living Future Institute's (ILFI) referenced report *Code, Regulatory and Systemic Barriers Affecting Living Building Projects* for the Cascadia Green Building Council. The author has no existing or expected financial relationship with ILFI or the LBC. The views presented are his own.

References

Eisenberg, D., & Persram, S. (2009). *Code, regulatory and systemic barriers affecting living building projects: Cascadia region Green Building Council*. Retrieved from http://living-future.org/sites/default/files/photos/09-0729%20code%20paper%20Eisenberg%281%29.pdf

Fahey, L., & Randall, R. M. (Eds.). (1998). *Learning from the Future: Competitive foresight scenarios*. New York: Wiley.

Hassler, U., & Kohler, N. (2014). The ideal of resilient systems and questions of continuity. *Building Research & Information*, 42(2), 158–167. doi:10.1080/09613218.2014.858927

Hawken, P., Lovins, A., & Lovins, H. (1999). *Natural capitalism: Creating the next industrial revolution*. Boston: Little Brown.

Janda, K. B., & Parag, Y. (2013). A middle-out approach for improving energy performance in buildings. *Building Research & Information*, 41(1), 39–50. doi:10.1080/09613218.2013.743396

Meadows, D. (1999). *Leverage points: Places to intervene in a system*. Hartland, VT: The Sustainability Institute. Retrieved from: http://www.donellameadows.org/wp-content/userfiles/Leverage_Points.pdf

Endnotes

[1] About ICC - History, see http://www.iccsafe.org/about-icc/overview/about-international-code-council/.

[2] International Living Future Institute, see http://living-future.org/.

[3] International Living Future Institute Ideas In Action - Research, see http://living-future.org/ilfi/ideas-action/.

[4] International Living Future Institute Research - Building Codes, see http://living-future.org/ilfi/ideas-action/research/building-codes/.

[5] International Code Council's Performance Code for Buildings and Facilities, see http://shop.iccsafe.org/codes/2015-international-codes-and-references/2015-iccr-performance-code-for-buildings-and-facilities.html/.

[6] Rocky Mountain Land Use Institute - Sustainable Community Development Code Framework, see http://www.law.du.edu/index.php/rmlui/rmlui-practice/code-framework/model-code/.

RESEARCH PAPER

The impact of regulations on overheating risk in dwellings

Mark Mulville and Spyridon Stravoravdis

Many new and emerging regulations and standards for buildings focus on climate change mitigation through energy and carbon reduction. In cool climates, such reductions are achieved by optimizing the building for heat retention. It is increasingly recognized, however, that some degree of climate change is now inevitable. New and existing buildings need to consider this to ensure resilience and an ability to adapt over time. In this context, the current approach to regulation that largely remains focused on the 'point of handover' may not be fit for purpose. This paper focuses on a 'typical' dwelling designed to a range of standards, representing current or emerging approaches to minimizing energy use, using a range of construction methods, where a number of adaptations are available to occupants. It considers, through the use of building performance simulation, how each configuration is likely to perform thermally over time given current climate change predictions. It is demonstrated that the current approach to assessing overheating risk in dwellings, coupled with the regulatory focus on reducing energy consumption, could result in significant levels of overheating. This overheating could, in the near future, present a risk to health and result in the need for significant interventions.

Introduction

The UK Committee on Climate Change (2014) noted that one-fifth of homes in England could already experience overheating (in a mild summer) and that this percentage is likely to increase in a warming climate and therefore it called for new building standards to be developed to address the overheating risk in dwellings. Parry, Canziani, Palutikof, van der Linden, and Hanson (2007) reported that the 2003 heatwave in Europe accounted for in the region of 35 000 excess deaths. This increased mortality was related to high internal temperatures over an extended time period (Wright, Young, & Natarajan, 2005) and affected the most vulnerable in society, such as the elderly, infants and those with underlying health problems (Johnson et al., 2005). If such events are to become more common (which is predicted, with a similar event affecting southern Europe in 2007), this could have significant consequences for health, housing design (Wright et al., 2005) and arguably the regulatory framework in the built environment.

The UK building regulations primary focus is one of health, safety and well-being. The growing realization, though, that CO_2 emissions associated with the built environment can have a significant negative impact on the environment and contribute to climate change (in 2009 buildings were responsible for 43% of all CO_2 emissions in the UK; Department for Communities and Local Government (DCLG), 2015) has resulted in the requirements of national and regional legislation (such as the UK Climate Change Act and The European Energy Performance of Buildings Directive) being interpreted through building regulations. In recent years building regulations in the UK (UK Part L revisions 2010 and 2013) (Figure 1), European Union (European Energy Efficiency Directive) and, indeed, globally (Dubai Green Building Regulations and Specifications, Green Building Code of Peoples Republic of China) have increasingly sought to set more stringent energy performance targets. In cool or temperate climates such as mid-latitude and northern Europe, characterized by extended winter heating seasons, these regulations aim to mitigate the worst effects of climate change by reducing CO_2 emissions and optimizing buildings for heat retention. However, as noted by McLeod, Hopfe, and Kwan (2013), it is

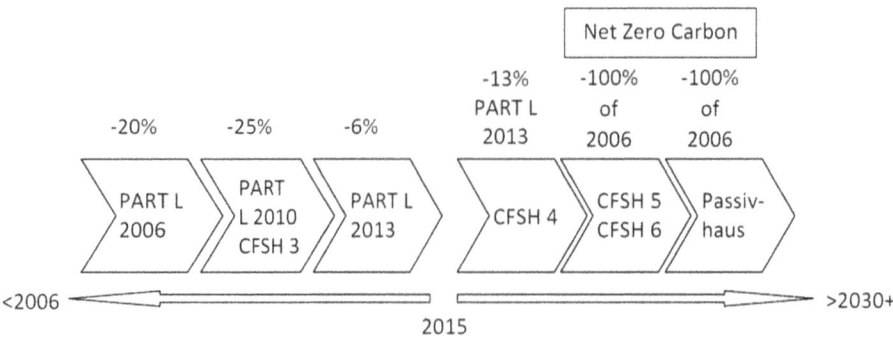

Figure 1 Timeline of energy-related regulations and emerging standards (UK) including energy-reduction targets

being increasingly recognized that higher global emissions scenarios are becoming more likely, increasing the risk of greater levels of warming. As a result, it has been suggested that dwellings in cool climates may in future suffer from overheating, and indeed there is evidence that this may already be happening (Dengel & Swainson, 2012). de Wilde and Coley (2012), in a review of the implications of climate change for buildings, point out that the rationale and requirements of the building regulations in relation to overheating are largely based on historic data and may not therefore enable a realistic assessment of the potential future overheating risk.

Taking this into consideration, this paper explores the idea that the current focus of increased levels of insulation and other heat loss-reduction measures for new or recently constructed dwellings may, at least during warm periods, be counterproductive. As noted by Eames, Kershaw, and Coley (2011), overheating could have severe health implications and as such could result in uninhabitable buildings that are technically obsolete due to the 'locked-in' (de Wilde & Tian, 2011) impacts of climate change. The paper, supported by detailed probabilistic predictions, demonstrates the level of overheating that could be experienced depending on the standard or regulation the dwelling is constructed to and suggests how an alternative, more robust approach to overheating risk assessment could be developed.

Background
Evidence of overheating
Until recently the general assumption in relation to domestic overheating has been that older dwellings were more vulnerable, with increased levels of insulation protecting new dwellings from such issues (Dengel & Swainson, 2012). However, there is now increasing evidence that new and low-energy dwellings may suffer significant problems in future and indeed that many such dwellings may already experience high levels of overheating (Dengel & Swainson,

2012; Lomas & Kane, 2013; Mavrogianni, Taylor, Davies, Thoua, & Kolm-Murray, 2015; Peacock, Jenkins, & Kane, 2010; Rodrigues, Gillott, & Teltow, 2013). It has been suggested that lightweight, air-tight dwellings with little access to cross-ventilation (such as single-aspect flats) may be at a particularly high level of risk of overheating (Dengel & Swainson, 2012).

What is clear is that overheating presents a significant risk to dwellings and occupant health and that this could impact a wide range of buildings and building types. Using probabilistic assessment, Jenkins et al. (2014) found that by the 2030s up to 76% of flats and 29% of detached dwellings could be at risk of overheating. Furthermore, Rodrigues et al. (2013), in a study considering the performance of a low-energy steel frame house design under current and future weather scenarios, found that the building could be uncomfortably warm for 30% of the year.

As noted by de Wilde and Coley (2012), there appears to be at least some debate in relation to highly insulated buildings, with Crawley (2008) noting greater resilience and Wang, Chen, and Ren (2010) noting less resilience to the impacts of climate change. Gupta and Gregg (2012) cite the Passiv-on project which found that the high levels of insulation in Passive House buildings in southern Europe worked to keep the building cool during warm weather. However, McLeod et al. (2013) cites a number of studies (Larsen & Jensen, 2011; Mlecnik et al., 2012; Schnieders, 2005) in relation to Passive House dwellings in several European locations where it is noted that there may be a risk of overheating in the current climate unless a number of alterations including active cooling were implemented. McLeod et al. (2013) do, however, point out that Passive House dwellings may offer marginally more protection from overheating than other highly insulated options. (Findings for the 2050s heavy weight Passive House are greater than 25°C for 6.6% of the year, whereas a heavy weight 'well insulated' building is greater than 25°C for 8.2% of the year.) The core reasons for these differences are unclear, although Gupta and Gregg (2012), when considering a range of 'typical' dwelling types in Oxford, found insulation's position to be of importance. As such insulation may be a 'double edged sword' (Hacker, Holmes, Belcher, & Davies, 2005) with increased insulation reducing winter heat loss but increasing overheating risk especially in air-tight buildings where it is difficult to dissipate internally generated gains. This is further supported by Orme and Palmer (as cited in Dengel & Swainson, 2012) who noted that increasing the level of insulation in dwellings resulted in higher levels of overheating.

Peacock et al. (2010) note the potential for increased internal gains through the proliferation of electronic devices in the home which could lead to higher internal temperatures. They go on to note that despite this climatic effects are still likely to be more influential than internal gains and for the UK such internal gains are likely to remain useful throughout the heating season. Building on this, it can be argued that measures designed to mitigate future climate change must also take account of the impact these measures may have on overheating risk. The research presented in this paper attempts to take this into consideration. There is evidence that behavioural and technical adaptations may at least in some part be able to address overheating risk (Coley, Kershaw, & Eames, 2012); however, as noted by Hills (2012) for low-income households any requirement for technical intervention through artificial cooling may result in summer fuel poverty. This is particularly important for countries like the UK as installed cooling capacity remains low (Hulme, Beaumont, & Summers, 2011).

Overheating adaptation
A number of authors have explored potential solutions to overheating in domestic buildings (*e.g.*, Gupta & Gregg, 2012; van Hoff, Blocken, Henson & Timmermans, 2015; Porritt, Cropper, Shao, & Goodier, 2012). These studies found that adaptation measures such as shutters and fixed shading, reductions in solar/fabric gains, increased surface albedo and the use of thermal mass may help reduce future overheating, although the magnitude of the influence of thermal mass has been questioned (Kendrick, Odgen, Wang, & Baiche, 2012). Gupta and Gregg (2012) found that user-controlled shading, surface albedo and thermal mass could help reduce overheating risk, although it was also found that no 'passive' measures alone could completely remove the risk and that behavioural and active measures may also be needed. Jones, Mulville, and Brooks (2013) set out an approach to adaptation planning related to the potential impact of climate change where the lifecycle of the building is taken into consideration. It suggests that where risks are identified at the design stage, an adaptation plan can be developed, and where such adaptations

in future may otherwise prove prohibitively expensive, preparatory or enabling works can be carried out during the construction phase to enable the future adaptation. As noted by Gupta and Gregg (2012), mitigation measures to reduce the contribution of the built environment to climate change and adaptation measures to allow the building to adapt to climate change that does occur should not negatively impact on the performance of the building. Combining the adaptation planning approach suggested by Jones et al. (2013) with the measures that have been shown to reduce overheating risk may have the potential to ensure the buildings can perform over time. In such a scenario adaptation measures are enabled at the design stage but only implemented when needed, therefore not having a negative impact on current performance (such as increasing winter heat load due to increased solar shading).

Literature review
UK regulatory framework
In the UK, building regulations do not stipulate a single maximum temperature in the workplace or the home. For domestic buildings, checks are required at the design stage, comparing mean summer internal temperature with a threshold temperature (using the Standard Assessment Procedure – SAP) to assess the risk of overheating (Zero Carbon Hub, 2015). Further guidance that does refer to specific temperature is provided by the Chartered Institution of Building Services Engineers (CIBSE) (2015), but this is beyond the requirements of the regulations. The UK Workplace (Health, Safety and Welfare) regulations (Health and Safety Executive, 2013) specify minimum but not maximum temperatures in the workplace; the same standards do not exist for private residential buildings, but minimum and maximum temperatures may apply to other residential-type buildings (see, for example, Northern Irelands Residential Care Homes minimum standards; Department of Health, Social Services and Public Safety, 2011).

By way of comparison, for new dwellings, the Danish Code for Indoor Thermal Climate (DS 474) sets a maximum of 100 hours not in excess of 26°C and not more than 25 hours above 27°C, calculated using a simple software tool (BE10) (Kunkel, Kontonasiou, Arcipowska, Mariottini, & Atanasiu, 2015) similar to the UK SAP. The Swedish Building Code sets minimum room temperatures and maximum surface temperatures, but not maximum room air temperatures (Kunkel et al., 2015). The Republic of Ireland follows a similar approach to that of the UK with the Dwelling Energy Assessment Procedure (DEAP) used to carry out an overheating risk assessment (Sustainable Energy Authority of Ireland, 2013), although this assessment is optional and not mandated by regulation. The Norwegian building regulations have implemented requirements for the consideration of solar gains in non-domestic buildings, but not domestic buildings (Schild, 2009). As can be seen from the above comparisons, set maximum summer temperatures for dwellings in cool climates are still uncommon. Where such standards do exist, the assessment method used tends to be simplistic in nature and, as will be discussed below, may not be fit for purpose.

Metrics

There remains considerable discussion with regard to the most appropriate metrics for defining overheating (de Dear, Kim, Candido, & Deuble, 2015; Hacker et al., 2005; Nicol & Spires, 2013; Nicol, Hacker, Spires, & Davies, 2009; de Wilde & Tian, 2011) with the merits of single-temperature exceedance or a traditional approach versus adaptive comfort standards (Nicol & Spires, 2013), considerations of potential health and mortality impacts (Dengel & Swainson, 2012; Jenkins et al., 2014), along with considerations of the relevant importance of night-time temperatures (Dengel & Swainson 2012; Peacock et al., 2010), forming the main points of discussion.

The main criticism of the single-temperature exceedance method (which sets an allowable exceedance of a percentage of occupied hours above a given temperature) is that there is evidence of a correlation between acceptable internal temperatures and external temperature (Nicol & Humphreys, 2010), and as such the comfort range is in reality a moving target that a single-temperature criterion cannot take account of. Furthermore, a single-temperature criterion does not give an indication of the severity of overheating experienced, while the adaptive comfort approach considers both severity and length of exposure (Nicol & Spires, 2013). Length of exposure is of particular importance in relation to potential health impacts with, as previously noted, many of the premature deaths associated with the 2003 European heatwave being related to the extended period of high temperatures (Wright et al., 2005). The adaptive thermal comfort approach as detailed by Nicol and Humphreys (2010) is built on field studies that note the ability of occupants to adapt to climatic changes, whereas the temperature exceedance approach, it could be argued assumes a more passive approach taken by occupants. However, as noted by Beizaee, Lomas, and Firth (2013), EN15251, which forms the basis of the adaptive comfort approach for 'free running' buildings, was developed primarily for commercial buildings and the adaptations available for occupants of dwellings are significantly different to those of an office. Furthermore, Mavrogianni et al. (2015) question how under such conditions vulnerable and potentially immobile people may be able to adapt. It has been noted (Roaf, Brotas, & Nicol, 2015) that the single-temperature

exceedance method (in comparison with the adaptive comfort method) would appear to overestimate the overheating risk and that, in reality, the risk may be lower than many studies in this field have predicted. With the potential health risks associated with overheating such assertions need to be approached with caution and, as suggested by Mavrogianni et al. (2015), it may be wise to combine the adaptive comfort approach with a static overheating criterion.

What is clear from the literature is that high temperatures in bedrooms overnight are a significant risk (Naughton et al., 2002, as cited in Peacock et al., 2010) and that temperatures above 24°C may begin to impair sleep and have other associated health impacts (Dengel & Swainson, 2012). With the lack of a clear definition of overheating in dwellings, studies in this area are as a result subject to a degree of uncertainty (Dengel & Swainson, 2012).

Overheating assessment methods

In addition to lack of clarity regarding the appropriate metrics used to define overheating, there is evidence that the current methods of assessing overheating risk in dwellings may not be reliable (Jenkins, Ingram, Simpson, & Patidar, 2013). It has been noted that much of the data used to analyse current overheating risk come from the past (de Wilde & Coley, 2012) and, therefore, do not make allowance for the potential impacts of a changing climate. The overheating prediction method currently used in the SAP (Department of Energy and Climate Change (DECC), 2014) assigns a level of risk to potential overheating, where higher levels of overheating are detected the assessor has the option to use window opening to alleviate these problems (window opening may help to reduce overheating by increasing the ventilation rate, which will reduce the ambient room temperature where outside air is cooler and can also help aid personal cooling through the evaporation of moisture from the skin). Although this may be suitable in rural areas, in areas subject to high levels of pollution and noise (Mavrogianni et al., 2015) this may be unrealistic and in both cases (urban and rural) such an approach may present a security risk (Mavrogianni, Wilkinson, Davies, Biddulph, & Oikonomou, 2012). This is supported by the findings of Skinner and Grimwood (2005) who note that 54% of dwellings in the UK during the daytime and 67% during night-time are exposed to noise levels in excess of World Health Organisation (WHO) guidelines. In addition, there are strong safety-related arguments for the inclusion of restrictors on upper-floor windows (McLeod et al., 2013) which reduce the potential for increased ventilation. As such, the reliance on window opening may be over-optimistic and, furthermore, as suggested by several authors (*e.g.*, Peacock et al., 2010; Roaf et al., 2015) there may in future be an upper limit to the effectiveness of window opening in reducing overheating.

The Passive House Standard, as developed by the 'Passivhaus Institut' (Cotterell & Dadeby, 2012) (see http://www.passiv.de), uses a separate method of overheating risk assessment administered through the Passive House Planning Pack (PHPP). Although, like the other standards discussed here, the data contained in PHPP are from the past, the exceedance limitation of 25°C for a maximum of 10% of the year (Feist, 2013) assessed through PHPP would appear to be more robust than the SAP assessment method (below 23.5°C) as it is not wholly based on a reliance on window opening.

Building performance simulation

In most cases of overheating risk assessment at the design stage, there is some form of building performance prediction, usually through building performance simulation (such as the compliance assessments used in SAP and PHPP or dynamic simulation more commonly used in research such as that of Coley et al. (2012)). Although widely used, there are some significant uncertainties that must be taken into account when considering building simulation-based performance assessments. As noted by de Wilde and Tian (2011), this includes modelling, numerical and specification uncertainty along with scenario uncertainties related to occupant behaviour and, when considering performance over time, predicted climatic conditions and likely renovation scenarios. Future weather files are a key component of any climate change impact study using building simulation. In the UK, the data used to generate such weather files (as used in building simulation studies) originate from the UK Climate Change Projections (UKCP) (Jenkins et al., 2009). It has been argued that due to the inherent difficulty in predicting future weather scenarios a probabilistic approach should be taken (de Wilde & Tian, 2011), and in that sense the UKCP09 offers a significant advantage over the previous UKCP02 as it offers a wide range of probabilistic scenarios. This has the knock-on impact of introducing additional complication to such prediction due to the required processing. With the goal of addressing this, the PROMETHEUS project at the University of Exeter developed a range of weather data files based on the UKCP09 predictions which represent a range of probabilities (10th, 50th and 90th percentiles for low, medium and high scenarios respectively) for three timescales (2030s, 2050s and 2080s) producing both test reference years (TRYs) to represent a 'normal' year and design summer years (DSYs) to represent a 'near extreme' year. The methods used to develop the files are detailed by Eames et al. (2011). Several studies considering domestic overheating have used these weather data files (e.g., Coley et al., 2012; Gupta & Gregg, 2012; Mavrogianni et al., 2012),

and although the methods used to develop the files reduce the computing needed, some considerations remain in relation to how the files are implemented in climate change impact studies. Gupta and Gregg (2012) suggest that climate impact studies carried out using worst case scenarios (90th percentile) would ensure resilient designs. Coley et al. (2012), however, suggest that a more median approach (50th percentile) with allowance only for 'hard' adaptations to resolve any overheating, as this would allow designers to avoid potentially unnecessary and costly adaptations. Jenkins et al. (2013) suggest that both the DSYs, which are intended to test the building for overheating, and TRYs, used to represent more normal conditions for energy estimates, be used so that both 'near extreme' and higher probability scenarios can be represented. This in turn allows for at least some of the probabilistic capabilities of UKCP09 to be realized, although, as noted by Coley et al. (2012), for some UK cities DSYs may show a cooler climate than TRYs and use of DSYs therefore should be approached with caution.

Methods

For this research, five building standards were chosen to represent a range of construction specifications, requiring increasing levels of energy savings. This included the regulatory minimums of Part L 2006 and 2010, the voluntary standards of the Code for Sustainable Homes (CfSH) levels 4 and 5, and the Passive House Standard. Figure 1 depicts how the energy consumption predictions for each standard noted relate to each other. As can be seen, they follow the push towards 'net zero carbon' dwellings in the UK (National House Building Council Foundation, 2009), and although this has recently been put on hold (HM Treasury, 2015) it is likely to remain a long-term goal.

Once the relevant standards were established, a 'typical' UK dwelling was developed based on a dwelling stock analysis of England and the wider UK. This indicated that since 2001, 66% of dwellings constructed were houses (DCLG, 2014a; National House Building Council, 2014); approximately one-third of these were semi-detached, with an average of three bedrooms and an average floor area of 91.7 m² (DCLG, 2014b). Of the total dwelling stock in England and Wales (24.6 million, DCLG, 2014c; Statistics for Wales, 2015), 30% or 7.38 million are semi-detached houses (Office for National Statistics (ONS), 2011). Although flats/apartments and particularly single-aspect flats have been shown to be at high risk of overheating (Mavrogianni et al., 2015), and several studies have explored these issues (Dengel & Swainson, 2012), detailed exploration of the impact on houses is less prevalent, and given the number of dwellings involved (7.38 million), consideration of the potential impacts on these units is merited. Following this, a three-bedroom semi-detached property aligning with these criteria, with a glazing ratio of 22.5% and layout based on Roberts-Hughes's (2011) 'case for space' guidelines, was chosen to represent the 'typical' dwelling.

The dwellings did not include solar shading, other than in places where the regulation or standard in question required it. It is accepted that details of design will have a significant impact on overheating risk and it is a limitation of this work that the 'typical' case presented here cannot be fully representative of the stock of semi-detached dwellings. The criteria used aim to ensure that the 'typical' case reflects the likely approach to delivering such dwellings built to the standards and regulations under consideration in this paper.

The fabric specifications of the dwellings were developed in accordance with the relevant compliance assessment for the given standard: SAP (DECC, 2014) for Part L, Passive House Planning Pack (Feist, 2013) for Passive House and the DCLG (2011) recommendations for the Code for Sustainable Homes for CfSH levels 4 and 5. For each specification three variants were established, representing three levels of thermal mass, and three different construction methods: low mass approximately $100 \, \text{kJ/m}^2 \, \text{K}$, medium mass about $250 \, \text{kJ/m}^2 \, \text{K}$ and high mass about $450 \, \text{kJ/m}^2 \, \text{K}$, based on SAP 2012 (DECC, 2014). Predicted internal loads were then established in accordance with CIBSE Guide A (CIBSE, 2015).

Details of the specifications used can be found in Table 1.

This 'typical' building was then modelled in 3D using the Ecotect software (Marsh, 1996), before being exported to the Heat Transfer in Buildings 2 (HTB2) software (Lewis & Alexander, 1990) for thermal performance simulation and analysis. Exporting to HTB2 was necessary as this software allows for more in-depth analysis and also simulates the building dynamically as opposed to the admittance or steady-state method used in Ecotect.

Once the various versions of the buildings had been developed, models were run using current and future probabilistic reference years based around the UKCP09 weather generator as developed by the PROMETHEUS project at the University of Exeter and detailed by Eames et al. (2011). For the purposes of this study the 50th percentile predictions of the median scenario where used for both TRYs (representing the more likely prediction) and DSYs (representing near extreme predictions). It is accepted that using a range of predictions (10th, 50th and 90th percentiles for low, medium and high scenarios respectively) would give a wider indication of probability; however, as Coley et al. (2012) note, using the 50th percentile provides a more median result, allowing for consideration of overheating risk while avoiding designing to the worst-case scenario. Such an approach also reduces the risk of interventions made now to avoid future overheating risk, increasing current energy use.

Table 1 Model input parameters

Element	Units	Part L 2006	Part L 2010	CFSH 4	CFSH 5/6	Passive House
Roof U-value	W/m² K	0.18	0.15	0.13	0.1	0.1
Wall U-value	W/m² K	0.28	0.23	0.18	0.15	0.15
Floor U-value	W/m² K	0.2	0.18	0.15	0.1	0.1
Window U-value	W/m² K	1.8	1.4	1.4	1.1	0.8
Window G-value	0–1	0.72	0.64	0.64	0.60	0.54
Air permeability	m³/h/m² at 50 pa	10	5	3	1	0.7
Ventilation rate	As noted	27 l/s	27 l/s	27 l/s	27 l/s	30 m³/h/person
Thermal bridging	W/m² K	0.15	0.08	0.06	0.04	0.04
Window position ventilation rate[a]		Slightly open: one air change per hour (ACH)				
		Half open: four ACHs				
		Fully open: eight ACHs				

Notes: [a]Based on SAP 2012 (DECC, 2014) and Technical Memorandum 36 (Hacker et al., 2005).
For the 'typical' building to comply with Passive House overheating criteria a 600 mm shading device was included on the south facade (or west for an alternative orientation).
Passive House air permeability of 0.7 relates to 0.6 ACH at 50 pa for a building with a volume of 219.65 m³.
Mass was varied by choosing between raised timber floors and slab on grade, timber frame construction, cavity wall construction and blockwork with external insulation and between low- and high-mass party walls and internal partitions.

As building occupants are likely to adapt to warmer climates, four 'anytime' window-opening scenarios (closed, slightly open, half open, fully open) and two night-time purge-ventilation scenarios (half or fully open) were included in the simulations, with associated ventilation rates for each (Table 1). The findings of the 'night purge' window positions are only included for the high thermal mass scenarios where they had the greatest impact. In each case the software used enabled the window-opening position to be triggered by internal temperatures of 23.9°C. This aligned with the work of Jenkins, Patidar, Banfill, and Gibson (2011) who suggested that 23.9°C was the temperature at which occupants are likely to begin to adapt.

Along with building standard, thermal mass and ventilation rate (window position), orientation can be expected to have an impact on overheating risk. For this study two possible orientations were considered (north–south and east–west). These were chosen to reflect potential extremes of solar gain (high gains for south facing, but easier to control versus the east–west scenario). It is a limitation of the study that a wider range of orientations could not be considered, which arguably would be more representative of the as-built stock.

Overall five building standards, built using three thermal mass variations, in two orientations with up to six possible window positions, were simulated (Table 2), along with some additional simulations for comparison between London and Edinburgh (a full set of scenarios/models were run for London with a selection for Edinburgh for comparative purposes), which resulted in 1788 individual model runs.

As discussed previously, there is some debate regarding the most appropriate metrics to be used for predicting overheating in free-running dwellings. To allow for

further comparison between the two most common approaches (single-temperature exceedance and the adaptive comfort method), both were included in this study. Overheating criteria 1–3 are based on adaptive comfort standards as detailed by Nicol and Spires (2013), while criteria 4 and 5 are based on the more traditional single-temperature exceedance criteria; the specifics of each criterion can be found in Table 3. For each of the adaptive comfort criteria, the threshold used is a function of the relationship between the running mean outdoor temperature and the indoor temperature. As the software used was not capable of calculating a running mean of temperature (directly at least), the data generated were exported and manually analysed to create a running mean, thus allowing for a consideration of the adaptive comfort measures. As noted by Nicol and Spires (2013), exceedance of any two of these criteria is considered to represent overheating.

Results

Figure 2 displays the findings of the building simulations carried out for London (with a comparative for Edinburgh included in the 2080s) combining the building standard, thermal mass, window position and climatic scenario for a dwelling with either a north–south or an east–west aspect. Although as noted by Coley et al. (2012), for some UK cities DSY (for testing near extreme) may show a cooler climate than TRY (for a more normal climate). This was not found to be an issue in this scenario as TRY was consistently cooler than DSY. As a result of this, the findings presented in this section are based on DSYs only.

A number of initial observations can be drawn from the analysis, such as an increase in overheating over time, as can be seen in Table 4. Taking the CfSH level 4, medium thermal mass, north–south orientation with slightly

Table 2 Sample modelling plan (Part L 2010 – model images are included for illustration)

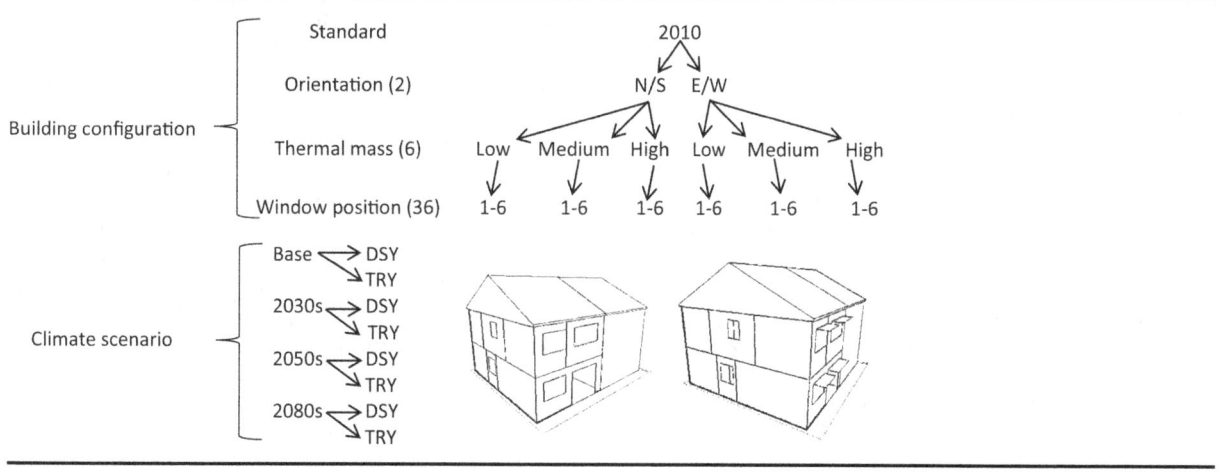

Table 3 Traditional and adaptive comfort overheating criteria

Adaptive comfort method[a]	
Criterion 1	Limits the how often the internal temperature exceeds the comfort range (by 1 K for 3% of occupied hours) during the summer months (May–September inclusive)
Criterion 2	Considers the severity of overheating in any one day based on how much the space overheats (by how many degrees the space exceeds the prescribed temperature) and for how long, again during the summer months
Criterion 3	Sets an absolute maximum temperature that reflects the point at which normal adaptations may be insufficient to ensure comfort
Traditional method[b]	
Criterion 4	Sets a limit of 1% of occupied hours above 25°C, which relates to the temperature at which discomfort may begin
Criterion 5	Sets a limit of 1% of occupied hours above 28°C, which relates to an unacceptably high temperature

Sources: [a]Nicol and Spires (2013).
[b]Hacker et al. (2005).

open windows possible and measured against exceedance of two of the three adaptive comfort criteria (criteria 1–3) no overheating occurs in the base case; however, for the 2030s exceedance occurs 7.8% of the time, 11.1% for the 2050s and 27.4% for the 2080s. Interestingly, when measured against the more traditional steady-state criteria (criterion 4), predicted overheating increases significantly (base 25.2%, 2030s 54.8%, 2050s 64.8% and 2080s 75%) highlighting, as noted by Roaf et al. (2015), that the traditional criteria predict higher levels of overheating than adaptive comfort standards. This research did not explore the merits of the adaptive comfort versus traditional steady-state overheating criteria in detail; for clarity the remainder of this study focuses on the adaptive comfort standards.

Secondly, buildings with higher thermal mass are at less risk of overheating than more light-weight buildings when the construction standard built to is kept constant (for instance, built to CfSH level 4, but the thermal mass varied) (Table 5). Taking Part L 2010 as an example, in the 2030s higher thermal mass can help to reduce incidents of overheating by up to 15.0%, by 15.6% in the 2050s and by 19.6% in the 2080s (all based on slightly open windows when temperatures are appropriate to take advantage of thermal mass). Furthermore, thermal mass becomes more important as insulation levels increase, while at higher levels of insulation (CfSH level 5 and Passive House) the benefit of thermal mass reduces between the 2050s and 2080s as the climate warms. This is based on the difference in the percentage of overheating observed between the low and high thermal mass versions across the building standards, as noted in Table 5.

This reduction in the benefit of thermal mass is likely a result of the decrease in diurnal temperature ranges over time and, as can be seen in Figure 3, the general reduction in the difference between internal and external temperatures. This has the impact of reducing the ability of ventilation and thermal mass to store or transfer energy to the cooler outside air. Figure 3 also demonstrates that along with the potential for increased overheating over time, temperatures in excess of 24°C in bedrooms at night-time may be regularly experienced from the 2030s onwards, which presents a particular cause for concern.

When considering the performance of the buildings across the range of standards (Part L 2006 to the Passive House Standard), during peak temperatures (taken as August) it was found that as the levels of insulation increase and external temperatures increase over time, increasing levels of fabric gains are observed overnight (Figure 4). This overnight gain maybe a contributory factor in high night-time bedroom temperatures. Figure 4 is based on mean daily temperature ranges for August, showing a comparison between the 2050s and 2080s for dwellings built to the Part L 2006 and CfSH level 5 building standards.

As previously noted, the adaptive comfort criteria include an assessment of the severity and duration of the overheating (criterion 2 in this study). As can be seen in Figure 2, in the low and medium thermal mass scenarios, as higher levels of insulation are introduced (CfSH level 4 and 5 and Passive House) from the 2030s onwards this criterion is regularly exceeded unless windows are in the half-open position. Such window opening as previously noted may be impractical due to noise, pollution or security issues. Consistent exceedance of this criterion is an additional cause for concern as persistent exposure to elevated temperatures can result in serious health impacts, particularly when they occur at night-time (Dengel & Swainson, 2012). This criterion does not consider longer-term exposure to higher temperatures (beyond the length of time within a single day), which although of importance was beyond the scope of this paper.

Using the adaptive comfort criteria, a cross-analysis between the five building standards (Table 5) found that, for the slightly open window position in the

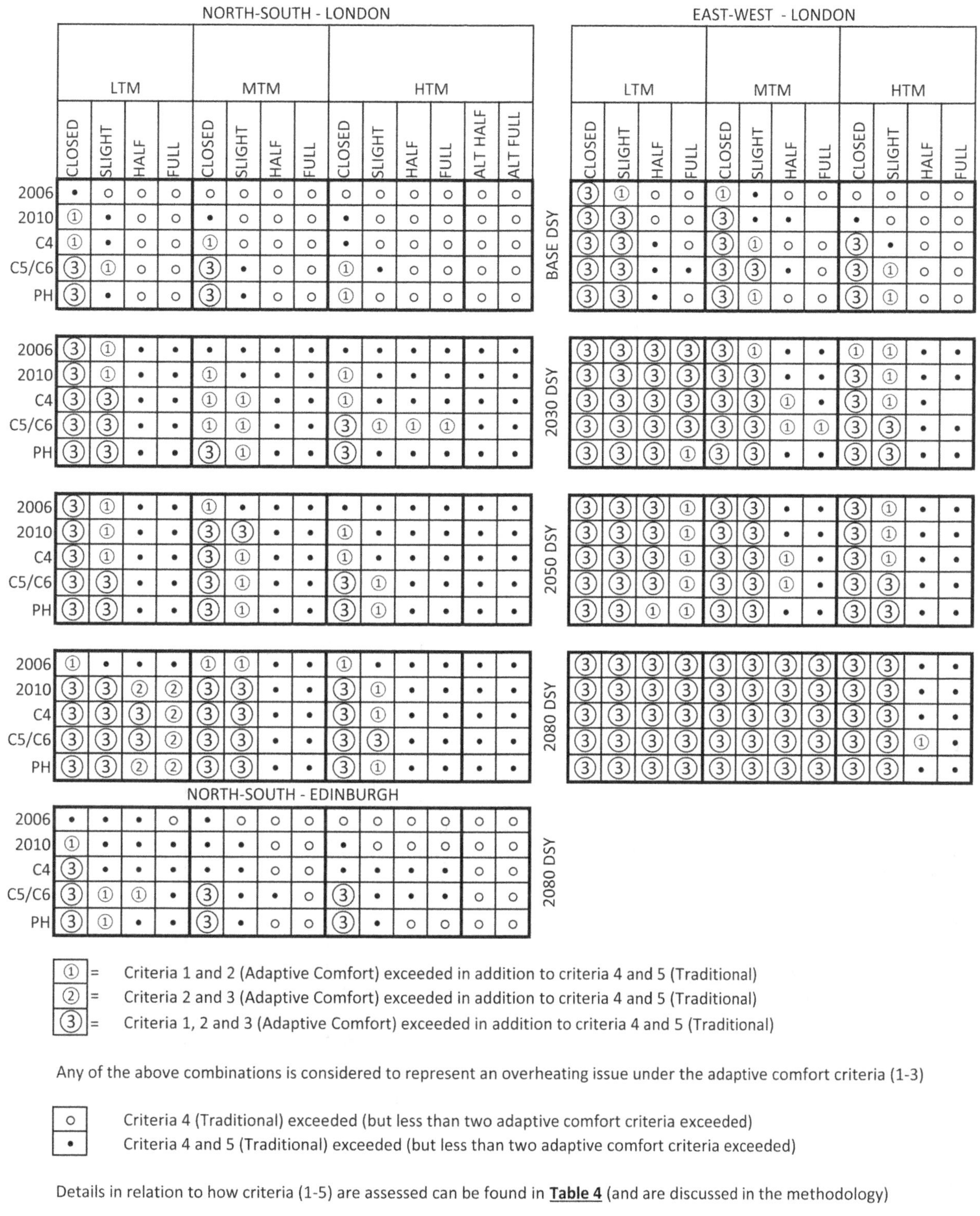
NORTH-SOUTH - LONDON
EAST-WEST - LONDON
LTM
MTM
HTM
CLOSED
SLIGHT
HALF
FULL
ALT HALF
ALT FULL
2006
2010
C4
C5/C6
PH
BASE DSY
2030 DSY
2050 DSY
2080 DSY
NORTH-SOUTH - EDINBURGH
① = Criteria 1 and 2 (Adaptive Comfort) exceeded in addition to criteria 4 and 5 (Traditional)
② = Criteria 2 and 3 (Adaptive Comfort) exceeded in addition to criteria 4 and 5 (Traditional)
③ = Criteria 1, 2 and 3 (Adaptive Comfort) exceeded in addition to criteria 4 and 5 (Traditional)
Any of the above combinations is considered to represent an overheating issue under the adaptive comfort criteria (1-3)
○ Criteria 4 (Traditional) exceeded (but less than two adaptive comfort criteria exceeded)
• Criteria 4 and 5 (Traditional) exceeded (but less than two adaptive comfort criteria exceeded)
Details in relation to how criteria (1-5) are assessed can be found in Table 4 (and are discussed in the methodology)

Figure 2 Overheating risk assessment (semi-detached house, cross-ventilation possible)

2030s (near future) and low thermal mass construction, the 2006 building overheats 5.9% of the time, the 2010 building overheats 15% of the time, CfSH level four 18.2% of the time, CfSH level five 35.9% of the time and the Passive House building 31.3% of the time, all measured during the summer months (May–September inclusive).

Building on this, and extrapolating the data in Table 5, it can be said that in the above scenario building

Table 4 CfSH 4 overheating risk as a percentage, during summer months (May–September)

| | Low thermal mass | | | | Medium thermal mass | | | | High thermal mass | | | |
| | Adaptive | | Traditional[a] | | Adaptive | | Traditional[a] | | Adaptive | | Traditional[a] | |
Overheating criteria / Window position	Slightly open	Half open	Slightly open	Half open	Slightly open	Half open	Slightly open	Half open	Slightly open	Half open	Slightly open	Half open
Base	1.3	0	31.1	14.8	0	0	25.2	12.2	0	0	24.5	11
2030s	18.3	5.8	55.8	34.4	7.8	1.9	54.8	32.8	2.6	0	53.8	32.7
2050s	20.2	9.1	66.6	41.7	11.1	2.6	64.8	39.6	4.5	0	63.8	40.5
2080s	42.4	11.7	77.5	55.1	27.4	3.2	75	52.8	19.6	0	74.7	53.3

Note: [a]Traditional overheating criteria relates to criterion 4 only.

Table 5 Overheating risk as a percentage of occupied hours during summer months (slightly open wins – north–south orientation)

| | Part L 2006 | | | Part L 2010 | | | CfSH 4 | | | CfSH 5 | | | Passive House | | |
Thermal mass	Low	High	Difference	Low	High	Difference	Low	High	Difference	Low	High	Difference	Low	High	Difference
Base	0	0	0	0	0	0	1.3	0	1.3	11.7	0	11.7	7.8	0	7.8
2030s	5.9	0	5.9	15	0	15	18.2	2.6	15.6	35.9	15	20.9	31.3	11.7	19.6
2050s	11.1	2.6	8.5	15.6	0	15.6	20.2	4.5	15.7	41.8	21.5	20.3	36	15	21
2080s	13	5.2	7.5	32	12.4	19.6	42.4	19.6	22.8	61.4	45	16.4	58.1	41	17.1

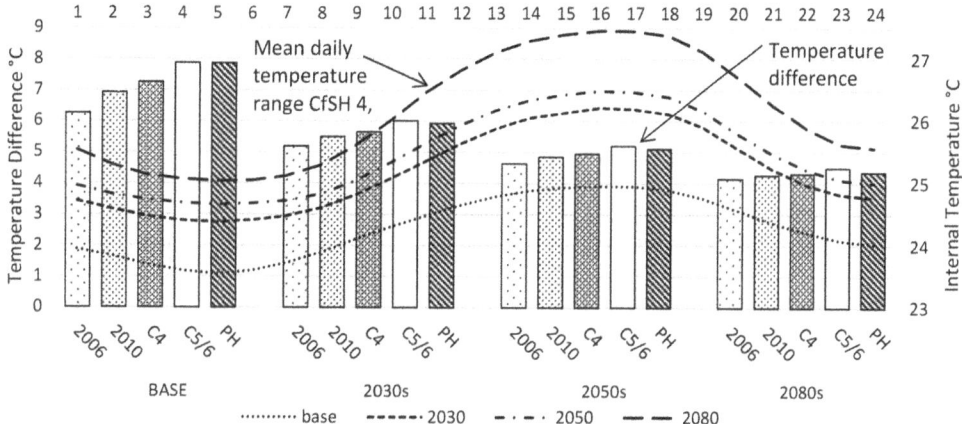

Figure 3 Internal to external temperature difference (high thermal mass)

standard (assuming thermal mass is kept constant) can increase potential incidents of overheating by up to 11.7% in the base case (11.7–0), for the 2030s this can increase to 30% (35.9–5.9), 30.7% in the 2050s (41.8–11.1) and for the 2080s 48.4% (61.4–13). It is worth noting that although the fabric specifications of CfSH level 5 and the Passive House Standard are similar (when in 'free running' mode), there is evidence that the overheating assessment criteria in the Passive House Standard are more robust. Reductions in incidents of overheating of 6.6% were observed for the 2030s, 6.8% for the 2050s and 7.2% for the 2080s for the Passive House Standard in comparison with the CfSH level 5, all based on a medium thermal mass scenario with slightly open windows (the data are stated but not presented in a separate table). As noted below, however, orientation presents a significant overheating risk for the Passive House Standard, which, if shading is not carefully considered, may reduce or reverse the benefit.

As can be seen from Figure 2, changing the orientation of the building from north–south to east–west has a noticeable impact on the incidents of overheating across all standards, when measured against the adaptive comfort criteria. This highlights the importance of solar shading and the impact it can have on overheating (east–west orientations are more difficult to shade than north–south orientations due to the low sun angle). For the east–west orientation, in line with the findings from the north–south orientation, further increases in incidents of overheating are experienced as the building standard increases (greater levels of insulation and reduced infiltration), and the building is optimized for heat retention.

For comparative purposes, thermal performance simulations were also carried for Edinburgh in the 2080s (Figure 2). Unsurprisingly, this resulted in significantly reduced levels of predicted overheating, with, in all scenarios, only the low thermal mass version of CfSH

level 4 and all thermal mass versions of CfSH level 5 and the Passive House Standard presenting significant overheating issues. By comparison, for London all standards exhibited at least some incidents of significant overheating, with the low thermal mass version (excluding the 2006 version) experiencing this across all window-opening positions.

Implications

The preceding results suggest that the current standards used in relation to overheating risk in dwellings in cool or temperate climates may no longer be fit for purpose. The lack of regulation in relation to maximum allowable temperatures and the opportunity for designers to utilize potentially unrealistic adaptations at the design stage, such as window opening, results in unreliable predictions. Coupled with this, the use of historic climatic data means that the current approach cannot make predictions about likely overheating risk beyond the point of handover, thus ignoring the lifecycle of the building and potential impact of climate change. As noted by Meikle and Connaughton (2006), new and existing housing is predicted to have to last for an extended period, and in many cases this may be beyond the end of the century. In this context, an alternative approach is required if the regulatory framework is to ensure the delivery of comfortable and healthy dwellings that offer resilience to predicted climate change.

In addition to the issues raised with the approach to assessing overheating risk in dwellings, the drive to reduce energy use in new dwellings through current and emerging regulations and building standards (thus optimizing the building for heat retention) has the potential to increase the overheating risk. The high proportion of CO_2 emissions that have been linked to the built environment (43% of all CO_2 emissions in the UK; DCLG 2015), and the link between this and

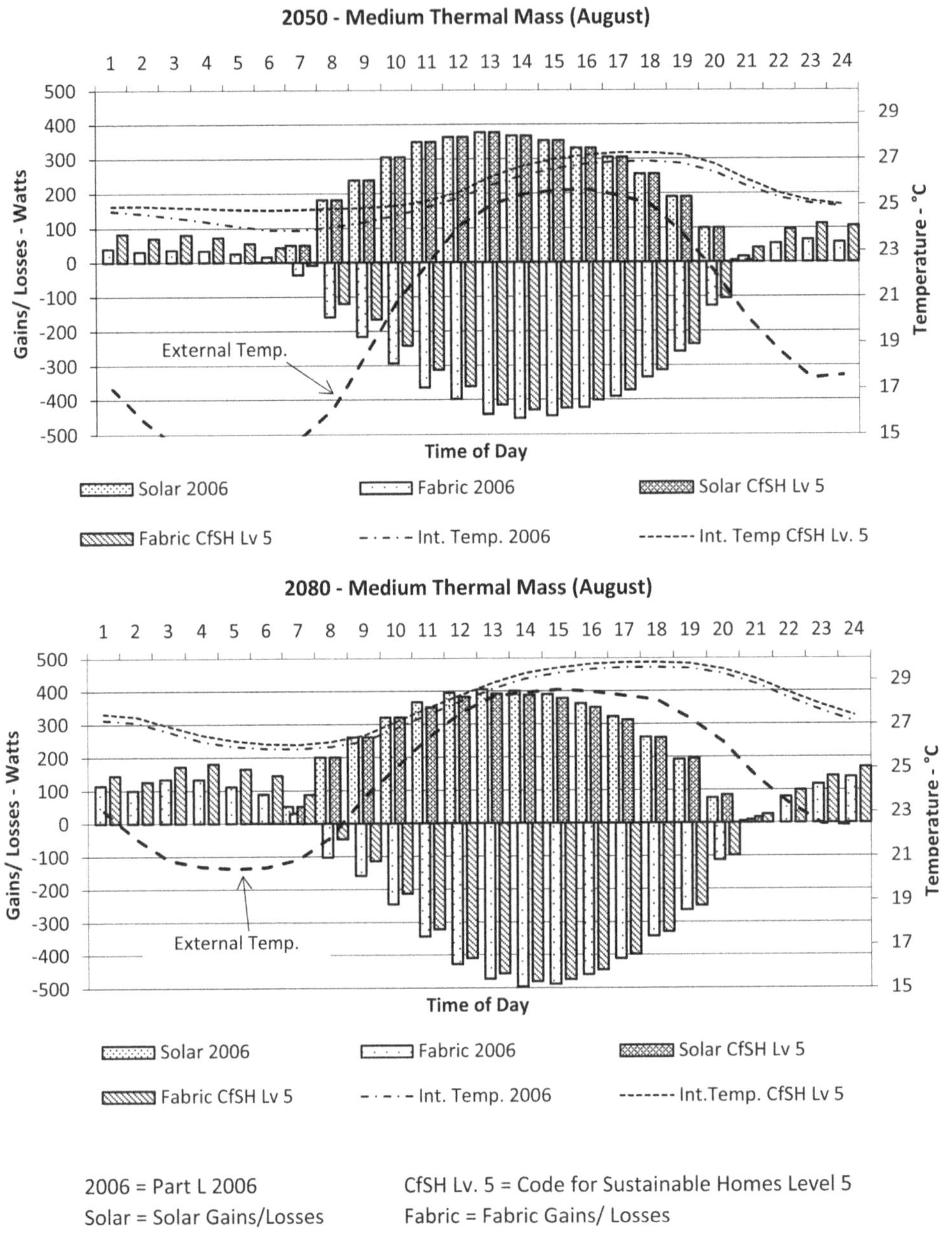

2050 - Medium Thermal Mass (August)

2080 - Medium Thermal Mass (August)

▨ Solar 2006	▭ Fabric 2006	▨ Solar CfSH Lv 5
▨ Fabric CfSH Lv 5	– · – · Int. Temp. 2006	– – – – Int. Temp CfSH Lv. 5

▨ Solar 2006	▭ Fabric 2006	▨ Solar CfSH Lv 5
▨ Fabric CfSH Lv 5	– · – · Int. Temp. 2006	– – – – Int.Temp. CfSH Lv 5

2006 = Part L 2006	CfSH Lv. 5 = Code for Sustainable Homes Level 5
Solar = Solar Gains/Losses	Fabric = Fabric Gains/ Losses

Figure 4   Gains and losses (2050 and 2080 comparing Part L 2006 and CfSH level 5)

anthropogenic climate change, creates a particular challenge for regulation in the built environment, where increased short- or near-term energy savings must be balanced against increased long-term overheating risk.

The approach outlined in this paper and the findings as presented in Figure 2 could be used as the basis of an alternative risk-based overheating assessment method that takes account of predicted climate change. This would require a careful definition of how exposure to high temperatures for a specific period of time constitutes an unacceptable overheating risk, which would provide clarity and allow methods to be developed and refined accordingly (Holmes, Phillips, & Wilson, 2016).

In addition to this, such an approach could also incorporate climate change adaptation planning at the design stage, as outlined by Jones et al. (2013), thus ensuring that short- or near-term energy efficiency is not compromised for longer-term comfort (by, for instance, adding shading now that may reduce future overheating risk but also increase current heating loads).

Conclusions

Although the current drive within the regulatory framework of the built environment to reduce heat loss (in cool climates) is not without its merits, as levels of insulation increase and infiltration decreases, there is increasing risk of summer-time overheating linked to climate change, particularly in urban areas. Furthermore, there is a risk of extended periods of overheating and unacceptably high temperatures in bedrooms overnight. The current standards in the UK used for predicting overheating risk may not be fit for purpose as they are based on historic data which take no account of potential climate change and make unrealistic assumptions in terms of occupant adaptations. Higher levels of thermal mass and the potential for increased ventilation rates offer benefits, however these may reduce over time as internal to external temperature differences decrease. Although this investigation is based on modelling UK domestic dwellings and the predicted changes to the UK climate, other countries with similar climates may experience similar risks.

Predicting overheating in buildings, given the range of variables involved, is complex and subject to a degree of uncertainty. Instead of a single temperature or hours of exceedance metric requiring complex building simulation to predict overheating, a risk-based scale may be more appropriate. This risk-based assessment, embedded in regulations and standards, could take account of the duration of the high temperatures experienced and the predicted impact of climate change (taking account not only of increased temperatures but also of the reduced ability of thermal mass and ventilation to minimize overheating). The approach to climate change risk assessment in dwellings as detailed in this paper could be expanded upon to provide such a design-stage risk assessment for a range of dwelling types. This approach presents a clear role for standards and regulations in defining anticipated scenarios in relation to overheating risk linked to climate change with the aim of ensuring resilience in both the predicted 'normal' future climate and during 'extreme' events.

This research must be considered in the context of a number of limitations. Future predictions of overheating risk based on building simulation are subject to uncertainty and rely on a number of assumptions in relation to variables that cannot be easily predicted (such as internal gains, occupancy patters and occupant adaptations). The paper has tried to reduce this uncertainty through the use of 50th percentile climate predictions (using weather data files produced from the PROMETHEUS project), offering a range of possible adaptations (window opening) and taking conservative predications in relation to internal gains, in order to avoid more extreme and potentially less realistic predictions. The research presented is based on a single building type, namely a semi-detached dwelling where cross-ventilation is possible, but consideration of a wider range of buildings using the same methodology would be of benefit. Further granularity could be added to the predictions by considering a range of

shading configurations. For all standards considered the buildings are assumed to be in 'free running' mode, as such this ignores the potential for a dwelling with a mechanical ventilation system to implement purge ventilation, although, as noted, the benefits of increased ventilation may reduce over time.

Further research is needed to consider how occupants of dwellings, particularly vulnerable occupants, may adapt to overheating. This could, for example, include further research in relation to window-opening patterns in urban areas where noise, pollution or security may place restrictions on such adaptations.

Disclosure statement

No potential conflict of interest was reported by the authors.

References

Beizaee, A., Lomas, K., & Firth, S. (2013). National survey of summertime temperatures and overheating risk in English homes. *Building and Environment*, *65*, 1–17. doi:10.1016/j.buildenv.2013.03.011

Chartered Institution of Building Services Engineers. (2015). *CIBSE guide A: Environmental design*. London: Chartered Institution of Building Services Engineers.

Coley, D., Kershaw, T., & Eames, M. (2012). A comparison of structural and behavioural adaptations to future proofing buildings against higher temperatures. *Building and Environment*, *55*, 159–166. doi:10.1016/j.buildenv.2011.12.011

The Committee on Climate Change. (2014). *Managing climate risks to well-being and the economy: Adaptation Sub-Committee progress report 2014*. Retrieved from Committee on Climate Change: https://www.theccc.org.uk/wp-content/uploads/2014/07/Final_ASC-2014_web-version.pdf

Cotterell, J., & Dadeby, A. (2012). *The Passivhaus handbook: A practical guide to constructing and retrofitting buildings for ultra-low energy performance*. Totnes: Green Books.

Crawley, D. (2008). Estimating the impacts of climate change and urbanization on building performance. *Journal of Building Performance and Simulation*, *1*, 91–115. doi:10.1080/19401490802182079

de Dear, R., Kim, J., Candido, C., & Deuble, M. (2015). Adaptive thermal comfort in Australian school classrooms. *Building Research & Information*, *43*, 383–398. doi:10.1080/09613218.2015.991627

Dengel, A., & Swainson, M. (2012). *Overheating in new homes: A review of the evidence*. Retrieved from the Zero Carbon Hub: http://www.zerocarbonhub.org/sites/default/files/resources/reports/Overheating_in_New_Homes-A_review_of_the_evidence_NF46.pdf

Department of Communities and Local Government. (2014a). *House building: September quarter 2014*. Retrieved from gov.uk: https://www.gov.uk/government/uploads/system/uploads/attachment_data/file/382244/House_Building_Release_-_Sept_Qtr_2014_-_revised.pdf

Department of Communities and Local Government. (2014b). *English housing survey headline report*. Retrieved from Gov.uk: https://www.gov.uk/government/uploads/system/uploads/attachment_data/file/406740/English_Housing_Survey_Headline_Report_2013-14.pdf

Department of Communities and Local Government. (2014c). *Dwellings stock estimates 2013, England*. Retrieved from gov.uk: https://www.gov.uk/government/uploads/system/

uploads/attachment_data/file/285001/Dwelling_Stock_Estimates_2013_England.pdf

Department of Energy and Climate Change. (2014). *The government's standard assessment procedure for energy rating of dwellings.* Retrieved from the Building Research Establishment: http://www.bre.co.uk/filelibrary/SAP/2012/SAP-2012_9-92.pdf

Department for Communities and Local Government. (2015). *2010–2015 government policy: Energy efficiency in buildings.* Retrieved from Gov.uk: https://www.gov.uk/government/publications/2010-to-2015-government-policy-energy-efficiency-in-buildings/2010-to-2015-government-policy-energy-efficiency-in-buildings

Department of Health, Social Services and Public Safety. (2011). *Residential care homes: Minimum standards.* Retrieved from rqia.org.uk: http://www.rqia.org.uk/cms_resources/care_standards_-_residential_care_homes_August_2011.pdf

Departments of Communities and Local Government. (2011). *Cost of building to the code for sustainable homes.* London: Departments of Communities and Local Government.

Eames, M., Kershaw, T., & Coley, D. (2011). On the creation of future probabilistic design weather years from UKCP09. *Building Services Engineering Research and Technology, 32,* 127–142. doi:10.1177/0143624410379934

Feist, W. (2013). *The Passive House Planning Package* (PHPP). Darmstadt: Passivhaus Institute.

Gupta, R., & Gregg, M. (2012). Using UK climate change predictions to adapt existing English homes for a warming climate. *Building and Environment, 55,* 20–42. doi:10.1016/j.buildenv.2012.01.014

Hacker, J., Holmes, M., Belcher, S., & Davies, G. (2005). *Climate change and the indoor environment: Impacts and adaptation.* London: The Charted Institution of Building Services Engineers.

Health and Safety Executive. (2013). *Workplace (health, safety and welfare) regulations 1992.* Retrieved from Hse.gov.uk: http://www.hse.gov.uk/pUbns/priced/l24.pdf

HM Treasury. (2015). *Fixing the foundation: Creating a more prosperous nation.* Retrieved from Gov.uk: https://www.gov.uk/government/uploads/system/uploads/attachment_data/file/443897/Productivity_Plan_print.pdf

Hills, G. (2012). *Getting the measure of fuel poverty, final report of the fuel poverty review* (CASE report 72). Retrieved from London School of Economics and Political Science: http://sticerd.lse.ac.uk/dps/case/cr/CASEreport72.pdf

van Hoff, T., Blocken, B., Henson, J., & Timmermans, H. (2015). On the predicted effectiveness of climate change adaptation measures for residential buildings. *Building and Environment, 82,* 300–316. doi:10.1016/j.buildenv.2014.08.027

Holmes, S. H., Phillips, T., & Wilson, A. (2016). Overheating and passive hability: Indoor health and heat indices. *Building Research & Information, 44*(1), 1–19.

Hulme, J., Beaumont, A., & Summers, C. (2011). *Energy follow-up survey 2011, report 9: Domestic appliances, cooking & cooling equipment* (BRE report number 288143). Retrieved from Gov.uk: https://www.gov.uk/government/uploads/system/uploads/attachment_data/file/274778/9_Domestic_appliances__cooking_and_cooling_equipment.pdf

Jenkins, D., Ingram, V., Simpson, S., & Patidar, S. (2013). Methods for assessing domestic overheating for future building regulation compliance. *Energy Policy, 56,* 684–692. doi:10.1016/j.enpol.2013.01.030

Jenkins, D., Patidar, S., Banfill, P. F., & Gibson, G. (2011). Probabilistic climate projections with dynamic building simulation: Predicting overheating in dwellings. *Energy and Buildings, 43,* 1723–1731. doi:10.1016/j.enbuild.2011.03.016

Jenkins, G. J., Murphy, J. M., Sexton, D. M., Lowe, J. A., Jones, P., & Kilsby, C. G. (2009). *UK climate projections: Briefing report.* Retrieved from metoffice.gov.uk: http://ukclimateprojections.metoffice.gov.uk/media.jsp?mediaid=87868&filetype=pdf

Jenkins, K., Hall, J., Glenis, V., Kilsby, C., McCarthy, M., Goodess, C., ... Smith, D. (2014). Probabilistic spatial risk assessment of heat impacts and adaptations for London. *Climate Change, 124,* 105–117. doi:10.1007/s10584-014-1105-4

Johnson, H., Kovats, R. S., McGregor, G., Stedman, J., Gibbs, M., & Walton, H. (2005). *The impact of the 2003 heat wave on daily mortality in England and Wales and the use of rapid weekly mortality estimates.* Retrieved from Eurosurveillance.org: http://www.eurosurveillance.org/ViewArticle.aspx?ArticleId=558

Jones, K., Mulville, M., & Brooks, A. (2013). FM, risk and climate change adaptation. In K. Alexander (Ed.), *EuroFM journal: International Journal of Facilities Management: Conference papers 12th EuroFM research symposium* (pp. 120–128). Naarden: EuroFM Publ.

Kendrick, C., Odgen, R., Wang, X., & Baiche, B. (2012). Thermal mass in new build UK housing: A comparison of structural systems in future weather scenarios. *Energy and Buildings, 48,* 40–49. doi:10.1016/j.enbuild.2012.01.009

Kunkel, S., Kontonasiou, E., Arcipowska, A., Mariottini, F., & Atanasiu, B. (2015). *Indoor air quality, thermal comfort and daylight: Analysis of residential building regulations in eight EU member states.* Retrieved from Bpie.eu: http://bpie.eu/uploads/lib/document/attachment/120/BPIE_IndoorAirQualityThermalComfortDaylight_2015.pdf

Larsen, T., & Jensen, R. (2011, November). *Comparison of measured and calculated values for the indoor environment in one of the first Danish Passive houses.* Paper presented at building simulation 2011: Conference of international building performance simulation association (IBPSA), Sydney, Australia. Retrieved from https://www.ibpsa.org

Lewis, P., & Alexander, D. (1990). HTB2: A flexible model for dynamic building simulation. *Building and Environment, 25,* 7–16. doi:10.1016/0360-1323(90)90035-P

Lomas, K. J., & Kane, T. (2013). Summertime temperatures and thermal comfort in UK homes. *Building Research & Information, 41*(3), 259–280.

Marsh, A. (1996, December). *Performance modelling and conceptual design.* Paper presented at building simulation 1996: Conference of international building performance simulation association (IBPSA), Sydney, Australia. Retrieved from https://www.ibpsa.org

Mavrogianni, A., Taylor, J., Davies, M., Thoua, C., & Kolm-Murray, J. (2015). Urban social housing resilience to excess summer heat. *Building Research & Information, 43,* 316–333. doi:10.1080/09613218.2015.991515

Mavrogianni, A., Wilkinson, P., Davies, M., Biddulph, P., & Oikonomou, E. (2012). Building characteristics as determinants of propensity to high indoor summer temperature on London dwellings. *Building and Environment, 55,* 117–130. doi:10.1016/j.buildenv.2011.12.003

McLeod, R. S., Hopfe, C. J., & Kwan, A. (2013). An investigation into the future performance and overheating risks in Passivhaus dwellings. *Buildings and Environment, 70,* 189–209. doi:10.1016/j.enpol.2010.08.027

Meikle, J., & Connaughton, J. (2006). How long should housing last? Some implications of the age and probable life of housing in England. *Construction Management and Economics, 12,* 315–321. doi:10.1080/01446199400000041

Mlecnik, E., Schütze, T., Jansen, S., de Vries, G., Visscher, H., & van Hal, A. (2012). End-user experiences in nearly zero-energy houses. *Energy and Buildings, 49,* 471–478. doi:10.1016/j.enbuild.2012.02.045

National House Building Council. (2014). *Housing supply update* (Issue 68). Retrieved from National House Building Council: http://www.nhbc.co.uk/ExternalAffairs/documents/HousingNewsletters/filedownload,57892,en.pdf

National House Building Council Foundation. (2009). *Zero carbon homes – An introductory guide for house builders.* Retrieved from the Zero Carbon Hub: http://www.zerocarbonhub.org/sites/default/files/resources/reports/

Zero_Carbon_Homes_Introductory_Guide_for_House_Builders_%28NF14%29.pdf

Naughton, M., Henderson, A., Mirabelli, M., Kaiser, R., Wilhelm, J., Kieszak, S., ... Rubin, C. (2002). Heat-related mortality during a 1999 heat wave in Chicago. *American Journal of Preventative Medicine, 22*, 221–227. doi:10.1016/S0749-3797(02)00421-X

Nicol, F., Hacker, J., Spires, B., & Davies, H. (2009). Suggestion for new approach to overheating diagnostics. *Building Research & Information, 37*(4), 348–357.

Nicol, F., & Humphreys, M. (2010). Derivation of the adaptive equations for thermal comfort in free-running buildings in European standard EN15251. *Building and Environment, 45*, 11–17. doi:10.1016/j.buildenv.2008.12.013

Nicol, F., & Spires, B. (2013). *Climate change and the indoor environment: Impacts and adaptations.* London: The Charted Institution of Building Services Engineers.

Office for National Statistics (ONS). (2011). *Social trends: Housing.* Retrieved from Ons.gov.uk: http://www.ons.gov.uk/ons/rel/social-trends-rd/social-trends/social-trends-41/housing-chapter.pdf

Parry, M., Canziani, O., Palutikof, J. P., van der Linden, P., & Hanson, C. (Eds.). (2007). *Contribution of working group II to the fourth assessment report of the intergovernmental panel on climate change, 2007.* Cambridge and New York: Cambridge University Press.

Peacock, A., Jenkins, D., & Kane, D. (2010). Investigating the potential of overheating in UK dwellings as a consequence of extant climate change. *Energy Policy, 38*, 3277–3288. doi:10.1016/j.enpol.2010.01.021

Porritt, S., Cropper, P., Shao, L., & Goodier, C. (2012). Ranking of interventions to reduce dwelling overheating during heat waves. *Energy and Buildings, 55*, 16–27. doi:10.1016/j.enbuild.2012.01.043

Roaf, S., Brotas, L., & Nicol, F. (2015). Counting the cost of comfort. *Building Research & Information, 43*, 269–273. doi:10.1080/09613218.2014.998948

Roberts-Hughes, R. (2011). *The case for space: The size of England's new homes.* Retrieved from Architecture.com: https://www.architecture.com/Files/RIBAHoldings/PolicyAndInternationalRelations/HomeWise/CaseforSpace.pdf

Rodrigues, L., Gillott, M., & Teltow, D. (2013). Summer overheating potential in a low-energy steel frame house in future climate scenarios. *Sustainable Cities and Societies, 7*, 1–15. doi:10.1016/j.scs.2012.03.004

Schild, P. G. (2009). *Norway: Impact, compliance and control of EPBD legislation.* Retrieved from Buildup.eu: http://www.buildup.eu/publications/8491

Schnieders, J. A. (2005). *A first guess passive home in southern France.* Passive-On Project report. Retrieved from Une maison passive à Nice: http://www.maison-passive-nice.fr/documents/FirstGuess_Marseille.pdf

Skinner, C., & Grimwood, C. (2005). The UK noise climate 1990–2001: Population exposure and attitudes to environmental noise. *Applied Acoustics, 66*, 231–243. doi:10.1016/j.apacoust.2004.07.009

Statistics for Wales. (2015). *Dwelling stock estimates for Wales, 2013–2014.* Retrieved from Gov.wales: http://gov.wales/docs/statistics/2015/150429-dwelling-stock-estimates-2013-14-en.pdf

Sustainable Energy Authority of Ireland. (2013). *Introduction to DEAP for professionals.* Retrieved from Seai.ie: http://www.seai.ie/Your_Building/BER/BER_Assessors/Technical/DEAP/Introduction_to_DEAP_for_Professionals.pdf

Wang, X., Chen, D., & Ren, Z. (2010). Assessment of climate change impact on residential building heating and cooling energy requirement in Australia. *Building and Environment, 45*, 1663–1682. doi:10.1016/j.buildenv.2010.01.022

de Wilde, P., & Coley, D. (2012). The implications of a changing climate for buildings. *Building and Environment, 55*, 1–7. doi:10.1016/j.buildenv.2012.03.014

de Wilde, P., & Tian, W. (2011). Towards probabilistic performance metrics for climate change impact studies. *Energy and Buildings, 43*, 3013–3018. doi:10.1016/j.enbuild.2011.07.014

Wright, A., Young, A., & Natarajan, S. (2005). Dwelling temperatures and comfort during the August 2003 heat wave. *Building Services Engineering Research & Technology, 26*, 285–300. doi:10.1191/0143624405 bt136oa

The Zero Carbon Hub. (2015). *Assessing overheating risk: Evidence review.* Retrieved from Zerocarbonhub.org: http://www.zerocarbonhub.org/sites/default/files/resources/reports/ZCH-OverheatingEvidenceReview-Methodologies.pdf

RESEARCH PAPER

Framework for selecting occupancy-focused energy interventions in buildings

Aslihan Karatas, Allisandra Stoiko and Carol C. Menassa

Energy policy tools aimed at achieving energy savings for buildings have often yielded less than the predicted savings. One reason for these results is that the design of these tools often neglects the actions and behaviours of the building occupants, and focuses more on the cost and ease of implementation. This research highlights the importance of identifying the diverse characteristics of occupants that significantly contribute to environmental problems, and the factors that make sustainable behavioural patterns attractive. A multilevel intervention strategy tailored to various occupants' characteristics to produce and maintain large-scale energy savings for buildings over time is proposed. To achieve this, the framework adopts a motivation/opportunity/ability (MOA) approach from the consumer and social marketing fields, where intervention strategies can be regarded as advertisements enticing the building occupants to adopt certain energy-use characteristics. The conceptual framework involves: measuring occupants' pre- and post-intervention exposure MOA level and energy-use profiles; clustering occupants based on identified characteristics; and accordingly choose energy-efficiency intervention strategies. The results of a case study of an actual building highlight the capabilities of the proposed framework in the selection of effective intervention strategies to achieve the required energy reductions.

Introduction

In the US, the existing building sector accounts for 40% of the total energy consumption by the built environment (Environmental Information Administration (EIA), 2014). The United Nations Environment Programme (UNEP) (2014) reported that the existing building sector represents an excellent opportunity to achieve large-scale energy-use reductions in a cost-effective manner through efficiency and conservation strategies. The objective is to alleviate economic, environmental and social problems associated with diminishing natural resources and global warming. Several studies argue that if large-scale energy reductions are to be maintained in the existing buildings sector, decision-makers need to analyze the determinants of individual occupant behaviours and the impact of such behaviours on energy consumption in buildings (Azar & Menassa, 2014; Carrico &

Riemer, 2011; Schweiker & Shukuya, 2010; Von Borgstede, Andersson, & Johnsson, 2013). They also pointed out that the analysis is of critical importance in the design of effective energy policy tools.

To achieve this, energy policy tools can be designed either (1) to affect directly individual behaviours through regulation tools and inducement tools (*i.e.*, incentives and/or sanctions); and/or (2) to affect directly the environment in which individual behaviours are manifested through knowledge tools that indirectly affect the behaviour (*i.e.*, knowledge based and/or persuasion) (Hand, 2012). Regulatory tools, often referred to as government command-and-control systems, are intended to change behaviour by forcing people to obey the law or other regulations without providing a promise of a positive incentive. Inducement tools aim to motivate an individual through the promise of a reward or a penalty to behave in a certain way without the level of government coercion inherent in regulation. On the other hand, knowledge tools are intended to change the individual behaviour based on the provisions of information. These methods strive to achieve desired outcome through continuous monitoring of how individual or group behaviours change (Hand, 2012).

Despite the fact that policy analysts often suggest knowledge tools as a starting point for policy intervention because of their non-coercive nature (Hand, 2012), energy policy tools are mostly designed as either regulatory tools (*e.g.*, occupancy-focused interventions that implement technological solutions such as heating, ventilation and air-conditioning (HVAC) scheduling and resets according to outside conditions) (Akridge, 1998; Jollands et al., 2010; Ruzelli, 2010; Wong et al., 2003); or inducement tools (*e.g.*, occupancy-focused interventions that implement reward/ penalty systems such as varying the energy costs with on- and off-peak consumption) (Braun, 1990; Kouveletsou et al., 2012; Ruzelli, 2010).

The design of energy policy tools as such often leads to three main challenges for decision-makers when applied in a context of energy reduction in the built environment. First, the compliance of energy policy tools with the regulations and incentives mostly resulted in inefficiencies due to the higher costs of implementation of appliance standards, building energy codes, financial incentives and public sector energy leadership programmes (Azar & Menassa, 2014; Hand, 2012; Levine et al., 2007; Lopes, Antunes, & Martins, 2012). Second, energy policy-makers assume that even though occupants have different behavioural characteristics, they will respond in the same way to the intervention and converge in their behaviour (*e.g.*, a fixed set-point room temperature) (Azar & Menassa, 2012; Mahdavi & Pröglhöf, 2009). Third, Hand (2012) highlighted that the lack of information or capacity (ability) is mostly the primary barrier that can be overcome by knowledge tools that relay the appropriate information to the occupants. In that case, it is unnecessary to incentivize or sanction the target in order to obtain the desired behaviour.

The limitations mentioned above highlight the need for robust policy measures based on the comprehensive understanding of the factors that will impact people to act in an environmentally friendly manner (Von Borgstede et al., 2013). Therefore, when designing policy measures aimed at reducing energy use in buildings, it is important to identify diverse energy-use characteristics of occupants that significantly contribute to environmental problems, and to identify the factors (*e.g.*, occupancy-focused interventions) that make sustainable behavioural patterns attractive.

To address and overcome the aforementioned challenges in the development of large-scale energy-policy tools, this study focuses on development of a conceptual framework that proposes multilevel building energy-use intervention strategies. This is achieved by examining the fundamental differences among occupancy-focused intervention strategies, and identifying which interventions are more effective based on the occupants' characteristics and energy-use profiles.

Objectives

The aim of this paper is to address the identified research gap in the development of energy policy tools (*i.e.*, their higher costs for building occupants and managers, and ignoring the impact of occupants' impact on buildings energy consumption), and present a conceptual framework for assisting policy-makers in designing effective energy policy pools. This framework proposes a multilevel building energy-use intervention strategy capable of systematically evaluating the effects of occupancy-focused interventions on occupants' behaviour, and also of designing energy policy tools tailored to various behavioural and energy-use characteristics of occupants. The main goal is to design a flexible framework that can be used to measure occupancy characteristics and to propose interventions in both commercial and residential building settings. To accomplish this, the proposed framework is developed to answer the following questions:

- *How can occupants' energy-use profile be measured and classified before and after implementation of energy policy tools?*
  The answer to this question helps determine the energy-use profiles (i.e., occupants' situational characteristics pre- and post-exposure to any intervention) to evaluate the benefits and effectiveness of different energy policy tools before actual implementation on a large stock of buildings (e.g., in a residential community, city or campus environment).

- *What type of building energy-use intervention strategies can be selected by decision-makers to achieve the required energy reductions at lower costs based on the occupants' energy-use characteristics?*
  This supports decision-makers in formulating effective energy policy tools to achieve cost-effective occupancy-focused intervention programmes that increase the attractiveness of pro-environmental behaviour, and encourage a more sustainable behaviour pattern.

- *How can the proposed framework be implemented in real-life examples to achieve large-scale energy reductions by delivering interventions to the occupants through the most-effective energy policy tools?*

The results of a case study of a real building demonstrate the capabilities of the proposed framework in evaluating the intervention strategies and designing effective energy policy tools based on the occupants' energy-use characteristics.

Occupancy-focused energy-efficiency interventions

To achieve large-scale energy reductions in buildings, four broad energy management tools are identified (Figure 1): knowledge translation (Abrahamse, Steg, Vlek, & Rothengatter, 2005; Backer, 1991; Bzdel, Wither, & Graham, 2004; Estabrooks, Thompson, Lovely, & Hofmeyer, 2006; Kaufman & Rousseeuw, 2009; Norcross, Koocher, & Garofalo, 2006; Spring et al., 2008; Wensing, Bosch, & Grol, 2010); persuasion (Grier & Bryant, 2005; Kotler, Roberto, & Lee, 2002; Peattie & Peattie, 2009), law and public policy (Hedlund, 2000; Houston & Richardson, 2007), and technological interventions (Environmental and Energy Study Institute, 2006). A detailed description of each intervention type, their limitations and relation to energy policy is now provided.

Knowledge-based interventions

Knowledge-based interventions consist of presenting informative messages to invoke voluntary behaviour change. It attempts to raise awareness about the benefits of changing a behaviour without presenting an explicit reward or penalty. Through outlets like posters, articles, videos and brochures, researchers have attempted to convey information to influence consumers in contexts such as encouraging vegetable consumption, stair use and sunscreen use (Armstrong, Idriss, & Kim, 2011; Blamey, Mutrie, & Tom, 1995; Sahota et al., 2001).

In an energy-use context, Abrahamse et al. (2005) evaluated 38 studies that reported on interventions in household environments, and categorized these studies into antecedent strategies (*i.e.*, information, modelling) or consequence strategies (*i.e.*, feedback). A further review of the existing literature resulted in dividing these types of studies into four main categories, some of which were not included in Abrahamse et al.'s review (*e.g.*, interactive programmes and peer-pressure comparisons). The first category of studies used information-distribution outlets (*e.g.*, posters, videos, brochures) to influence occupants (Agha-Hossein et al., 2014; Hayes & Cone, 1977; Hutton & McNeill, 1981; Marans & Edelstein, 2010; Midden, Meter, Weenig, & Zieverink, 1983; Zografakis, Menegaki, & Tsagarakis, 2008). Results from these studies show that the distribution of energy-consumption facts and reduction guidelines does not appear to influence occupants effectively, and provided mixed results on its own.

The second category considered interactive programmes with more personalized approaches to

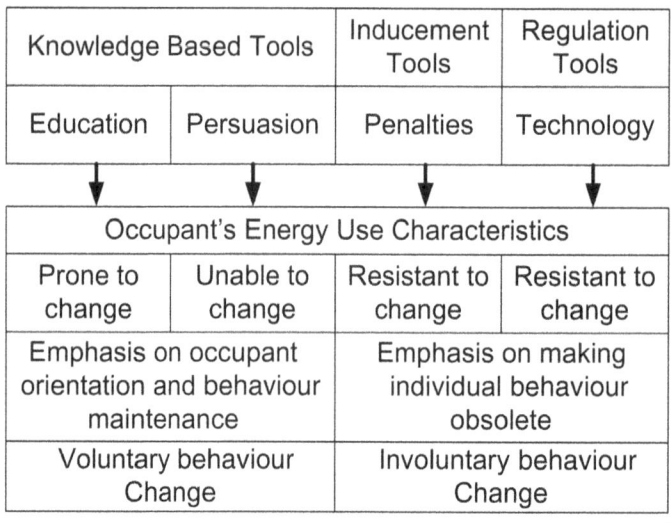

Figure 1 Building energy-use intervention strategies

information delivery. For example, McMakin, Malone, and Lundgren (2002) created site-specific video programmes to motivate military families to conserve energy. Combined with other awareness material, the programme resulted in a 3% reduction in gas use and 7% reduction in energy use after one year. However, Geller (1981) found that motivated individuals attending a three-hour workshop on energy conservation did not tend to apply skills either immediately or six weeks later. This implies that the short-term delivery of information may not be effective, and an extended programme may yield better results.

The third category of studies considered feedback methodology that provides consumers with personalized evaluation and a means to monitor progress (Dolan & Metcalfe, 2013; Van Houwelingen & Van Raaij, 1989). These studies generally concluded that even if a feedback method is an effective knowledge-based approach for building energy reduction, it may become less effective in changing behaviours over an extended period of time (e.g., two years).

Finally, the last category focused on peer comparison of monthly or quarterly energy use (Allcott, 2011; Ayres, Raseman, & Shih, 2013; Siero, Bakker, Dekker, & Van den Burg, 1996). Allcott (2011) determined that energy use decreased approximately 2% with these reports, and that energy use began to regress between quarterly reports. Ayres *et al.* (2013) discussed studies that experienced energy savings of 1.2–2.1%. Peschiera, Taylor, and Siegel (2010) concluded that peer comparison is more effective in decreasing energy use than energy consumption information alone. The participants' energy use dropped by 16% after the first peer comparison period and by 32% after the second peer comparison, both relative to baseline levels. These studies generally concluded that peer comparison is a successful means to reduce energy use, but suffered from significant rebound once the intervention is halted.

Persuasion interventions (social marketing)

Persuasion involves providing rewards to encourage a favourable behaviour. Researchers studied persuasion interventions to incite voluntary changes through incentives like money and fast food (Burn & Oskamp, 1986; John et al., 2011). In the context of energy use, several studies have been conducted to study the effect of incentives, such as monetary incentives and pledging, for reducing energy use. These are divided into two main categories.

The first category of studies considered monetary incentives, both small and large, to decrease energy use (Alahmad, Wheeler, Schwer, Eiden, & Brumbaugh, 2012; Darby, 2006; Handgraaf, Van Lidth

de Jeude, & Appelt, 2013; Katzev & Johnson, 1984; McClelland & Cook, 1980; Whitsett, Justus, Steiner, & Duffy, 2013). Abrahamse et al. (2005) included this type of intervention as part of the consequence strategies where occupants expect a reward (*i.e.*, consequence) as a result of adopting a certain behaviour. Examples of this type of intervention in buildings include Katsev and Johnson (1984) who provided occupants with US$3.00–10.00 for reducing energy use during two-week periods. They measured no significant difference in energy use between control groups during the study or follow-up. Winett, Kagel, Battalio, and Winkler (1978) provided households with variably high (US$0.30/1% reduction) and low rewards (US$0.013/1% reduction) for decreased kilowatt-hours of electricity compared with the average. Initially, both high- and low-reward groups saved energy, with the high-reward group decreasing their energy consumption by 16% more than the low-reward group. Upon receiving a reward, however, there was a larger increase in energy use for the high-reward group than the low-reward group. This larger relapse indicates that the high-reward groups may become less intrinsically motivated than the low-reward groups. While monetary incentives appear to be an effective persuasion method to reduce energy use, low-to-medium monetary rewards could lead to more sustained motivation for the occupants.

The other category of studies considered pledging campaigns as incentives to encourage sustainable energy conservation behaviour (Boyce & Geller, 2001; Burn & Oskamp, 1986; Whitsett et al., 2013). Abrahamse et al. (2005) included this type of intervention as part of the antecedent strategies under the commitment and goal-setting categories. Schick and Goodwin (2011) reported that in a voluntary 'pledge to save', participants decreased energy by a factor of three when compared with non-pledgers.

Law and public policy interventions (penalties)

Penalty interventions consist of negative consequences, often legal, that discourage an unfavourable behaviour. They provide a set of rules that result in sanctions and penalties for non-compliance. Several studies have focused on the influence of penalties in the context of energy conservation.

For example, multiple studies considered dynamic building control by altering energy costs with on- and off-peak consumption (Braun, 1990; Heberlein & Warriner, 1983; Kouveletsou et al., 2012). These studies observed that increased price of on-peak electricity reduced consumption of energy during that period. Heberlein and Warriner (1983) determined that consumers with higher price ratios (8:1) used less on-peak electricity than those with lower price ratios (2:1). Therefore, it appears that increasing energy cost during on-peak periods may be very effective in reducing energy use, particularly by increasing the difference in the ratio between on- and off-peak costs.

Technology interventions

Technology interventions consist of tools and systems that automatically solve problems without continued human influence. It is often dictated by laws and other regulations that require non-voluntary changes in behaviour for compliance. Research suggests that careful building design, including the incorporation of building automation systems, can dramatically reduce energy use (Akridge, 1998; Erickson, Achleitner, & Cerpa, 2013; Erickson & Cerpa, 2010; Mathews, Botha, Arndt, & Malan, 2001; Ruzelli, 2010; Wong et al., 2003). Examples of such solutions include: replacing electrical components (Environmental and Energy Study Institute, 2006), changing from non-condensing to condensing boilers (Harvey, 2009), use of compact or LED light bulbs, and using modern energy-efficient appliances (Environmental and Energy Study Institute, 2006).

Limitations of the existing approaches

The existing occupancy-focused energy interventions (*i.e.*, knowledge based, persuasion, penalties and technology) have several limitations that contributed to their limited success and maintenance of the desired behaviour over longer periods of time. First, these studies support what Dietz, Gardner, Gilligan, Stern, and Vandenbergh (2009, p. 18453) emphasized:

> single policy tools have been notably ineffective in reducing household energy consumption. Mass media appeals and informational programmes can change attitudes and increase knowledge, but they normally fail to change behaviour.

Thus, there is an urgent need for an effective multilevel intervention strategy targeted towards the extremely diverse occupancy characteristics to produce and maintain energy-use reduction in buildings over time. For example, occupants with extreme energy-use behaviours will react differently to interventions as compared with others with more moderate energy-use patterns. Education might be sufficient to reduce the energy use of the latter, but might need to be supplemented with interventions from higher levels shown in Figure 1 if extreme energy-use patterns are observed among the building occupants (Azar & Menassa, 2015). Second, most of the interventions resulted in temporary change in energy-use patterns. No specific studies investigated what occupancy characteristics are required for interventions to succeed and sustain energy-use behaviour over time. Finally, these studies assume that occupants react in the same way to the intervention strategies. None of these studies addresses the different occupancy characteristics and social relationships that play a key role to induce energy-change patterns and effectively achieve intervention objective.

On the other hand, the main limitation of existing technological solutions is their highly uncertain energy-reduction benefits (Chidiac, Catania, Morofsky, & Foo, 2010; DOE, 2011; Entrop, Brouwers, & Reinders, 2010; Schneider & Rode, 2010); relatively high initial costs (Nemry et al., 2010; Yudelson, 2010); reluctant stakeholder commitment, especially while taxes and energy prices are still low (Bernstein & Russo, 2009); and lack of information about the existing building systems making integration of new technologies a daunting task (Juan, Gao, & Wang, 2010). The other main limitation is the focus on integrating these solutions with the building systems, while ignoring the important human impact factors have often times contributed to less-than-desirable energy-reduction benefits from these approaches.

Relationship to energy policy

From a policy-making perspective, relaying the appropriate information to the target is mostly the major barrier to obtain the desired behaviour (Hand, 2012). Therefore, knowledge-based interventions can be delivered through knowledge energy policy tools that enable or encourage a voluntary behaviour change by the occupants with minimum economic and environmental costs (Figure 1). Persuasion interventions can also be delivered through knowledge energy-policy tools to change occupants' energy-use behaviour based on the given information and in the desired manner. This will also provide voluntary behaviour change for the occupants to produce and maintain energy-use reduction in buildings over time (Figure 1).

If extreme energy-use patterns are observed among the building occupants, lower-level interventions such as knowledge based and persuasion might not be sufficient to reduce the energy use effectively. Therefore, there is a need to supplement knowledge-based methods with interventions from higher levels (*i.e.*, penalty, technology), which results in higher economic and environmental costs. Penalty interventions can be delivered through inducement energy policy tools, since they mainly aim to motivate individuals through the promise of a reward or a penalty to change their behaviour in the desired manner. Finally, technology interventions can be delivered through regulatory energy policy tools, since they are intended to regulate the individuals' desired behaviour through legislative command-and-control systems (Hand, 2012). Both cases involve an involuntary (forced) change in the behaviour of occupants, at increased economic costs.

Conceptual framework for targeted occupancy-focused interventions

When designing energy-policy measures to reduce building energy use, it is important to identify occupants' behaviours that significantly contribute to environmental problems, and then identify the factors that make sustainable behavioural patterns attractive (Von Borgstede et al., 2013). Therefore, the framework in this study proposes a multilevel intervention strategy targeted towards the diverse human characteristics to sustain energy-use reduction in different types of buildings, including commercial and residential. To achieve this, the presented framework for designing effective energy-policy tools includes five main stages (Figure 2):

- measuring the occupants' pre-exposure motivation/opportunity/ability (MOA) level

- clustering occupants' pre-exposure MOA level, and identifying their energy-use profiles (i.e., occupants' situational characteristics prior to any intervention)

- measuring occupants' post-intervention exposure MOA levels

- clustering post-exposure MOA levels, and identifying energy-use profiles (i.e., occupants' situational characteristics after any intervention) to determine the effectiveness of the intervention strategies

- identifying the link between occupants MOA levels and the multilevel energy-efficiency intervention strategies, and accordingly energy-policy tool types (i.e., knowledge tools, inducement tools, and regulation tools)

Measuring the impact of occupancy characteristics on building energy use

To identify the role of occupants' characteristics and their impact on the interventions, the proposed framework adopts an analogy from the MOA model in the consumer and social marketing field, where intervention strategies can be regarded as advertisements enticing the building occupants to adopt certain energy-use characteristics. The MOA model in consumer and social marketing is typically used to identify the consumers' information-processing level, and the main influential factors in encouraging consumers to adopt desired behaviour (Buurma, 2001; Celsi & Olson, 1988; Machleit, Madden, & Allen, 1990; MacInnis, Moorman, & Jaworski, 1991; Moorman, 1990; Polonsky, Binney, & Hall, 2004; Rothschild, 1999). For example, MacInnis et al. (1991) propose an MOA model to explain the certain outcomes influenced by the extent of brand information processing

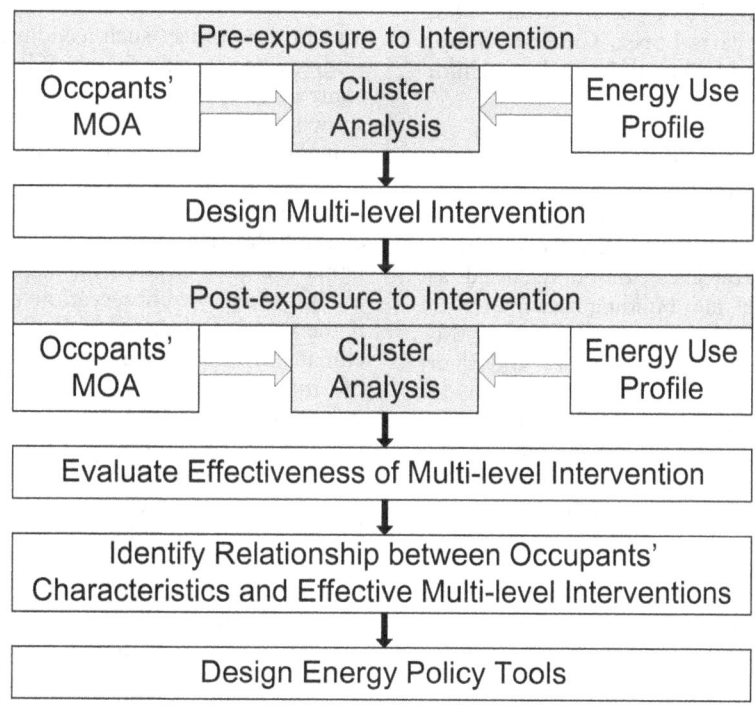

Figure 2 Framework for designing effective energy policy tools

from ads. Based on this model, consumers' motivation, ability and opportunity levels have a great impact on the level of processing brand information during advertisements or after exposure to the advertisements. Hastak, Mazis, and Morris (2001) emphasize the important mediation role that MOA plays in determining the communication effectiveness of advertisements in consumer research. Consumers loyal to a particular product usually have high MOA levels that help facilitate the adoption of this product. Bigné, Hernández, Ruiz, and Andreu (2010) applied an MOA model to explore the key drivers of online airline ticket purchase intentions, and also identified which perceived channel benefits are more effective for consumers in using the internet to purchase airline tickets. Results from this study highlighted that the MOA model is a useful methodology to predict consumers' ticket-purchasing intentions and explains 55% of variations in adopting the desired behaviour.

In a consumer and social marketing field context, motivation is a goal-directed arousal to engage consumers in the desired behaviour to process brand information in the advertisement (Govindaraju, Hadining, & Chandra, 2013; Richins & Bloch, 1986; Steg & Vlek, 2009; Zaichkowsky, 1985); opportunity is defined as executional factors (e.g., exposure time to ads) that are not in the control of consumers to enable the desired actions (Bigné et al., 2010; Hallahan, 2001; MacInnis et al., 1991; Rothschild, 1999); and ability is defined as consumers' perception of their capacity to access the brand information, and interpret this information to create new knowledge structures (Bigné et al., 2010; Celsi & Olson, 1988; MacInnis et al., 1991; Parra-Lopez, Gutierrez-Tano, Diaz-Armas, & Bulchand-Gidumal, 2012; Rothschild, 1999).

In light of the aforementioned approach, this study establishes an analogy between the MOA characteristics of people to process brand information in their environment and MOA levels where occupancy-focused intervention strategies can be regarded as advertisements enticing the building occupants to adopt certain energy-use characteristics. By using this analogy, multilevel intervention strategies and their related energy-policy tools can be developed to reduce energy use in buildings based on identified MOA characteristics of the building occupants. Therefore, a set of metrics were identified based on the MOA energy-use characteristics of occupants, and organized as shown in Tables 1–3.

Measuring the motivation level of occupants

The motivation (M) level of an occupant can be measured as the perceived link between an occupant's needs, goals and values (self-related knowledge), and the level of involvement with the energy-use information (e.g., level, impact and consequences) (Table 1). M is a key factor if lower levels of intervention (e.g., knowledge based) are to be successful. A high motivation level can be identified by occupants' desire to adjust room temperature (e.g., setting the office heating point to a lower temperature when unoccupied hours), adjust the lighting system (e.g., turning off the office lights when not in use), and adopt energy-conservation strategies (e.g., bringing an extra jacket instead of setting the heating point to higher degrees). Lack of motivation implies that occupants will resist any change and will require higher levels of intervention (e.g., penalties) and energy policy tools (i.e., inducement and regulatory tools) to improve their M level and get involved in energy conservation. The motivation level of each individual occupant will vary depending on whether he/she is queried in a residential environment where they might be motivated to conserve energy to reduce their monthly bills, or commercial setting where the motivation level might reduce significantly due to lack of awareness of one's own impact among a group of other employees in the building.

Measuring the opportunity level of occupants

The opportunity (O) level of an occupant represents the availability and accessibility of energy-reduction strategies to the occupant in his/her environment (MacInnis et al., 1991; Rothschild, 1999). However, there is a conditional relationship among the opportunity (O), motivation (M) and ability (A) levels of occupants. For example, highly motivated occupants are not able to present an energy-conservation behaviour in their offices such as adjusting the temperature setting when leaving the office if they do not have any control on their room's thermostat. Studies have shown that people have more tendency to make permanent changes in their energy-use behaviours when they have the available resources and if the new behaviours are easy and convenient to perform (McMakin et al., 2002). Therefore, the opportunity level of occupants can be measured by factors such as the availability of energy-conservation control systems (e.g., if the occupant has any individual control on indoor climate in their office), the availability and amount of information (e.g., having energy-conservation information available for occupants on bulletin boards), the format of the information (e.g., organized by an energy-reduction target like heating set point and the impact of changing that on energy use, or by measures to avoid changing the heating set point like wearing layers), the modality (e.g., information presented in workshops gives occupants a short period of time to process versus having the information available on bulletin boards affording occupants continuous opportunities to attend to and comprehend), and how often the information is available (e.g., weekly e-mails) (Table 2). As in the case of

Table 1 Metrics for measuring occupancy motivation level for energy conservation

Metrics of construct	Pre-exposure measures	Post-exposure measures
Measures of motivation (M) Concern level about their personal energy consumption Looking for ways to conserve energy at their offices Important factors to conserve energy	Assess *self-awareness* about importance of energy-use consumption Measure *desire to receive* energy-use knowledge Evaluate knowledge on conserving energy in a building Detect *norms of avoiding* energy-use knowledge (*e.g.*, not interested in attending workshops or receiving e-mails)	Assess *self-awareness* about importance of energy-use knowledge Measure the *desire to process (continue to receive)* energy-use knowledge Evaluate *knowledge (continue to improve)* on conserving energy in a building Detect *norms of avoiding* energy-use knowledge (*e.g.*, still not interested in attending workshops or receiving e-mails)

Table 2 Metrics for measuring occupancy opportunity level for energy conservation

Metrics of construct	Pre-exposure measures	Post-exposure measures
Measures of opportunity (O) Availability: 　Control systems 　Office environment Information: 　Exposure to information 　Peer pressure	*Determine availability:* Availability of energy-conservation control system (*e.g.*, indoor lighting control) Office condition level (*e.g.*, physical characteristics of offices) *Determine number of times:* Attend awareness seminars Read information on general advertisement boards (self-reported) Read e-mails (ask for a response with a blank e-mail) Discuss with peers (self-reported)	*Measure information recall by recording:* Number of arguments about the impact of energy use on the global environment Number of arguments about the impact of energy use on the building footprint Number of arguments related to the benefits of conservation methods

Table 3 Metrics for measuring occupancy ability level for energy conservation

Metrics of construct	Pre-exposure measures	Post-exposure measures
Measures of ability (A) Perceived energy consumption level Energy-use knowledge (external stimuli): 　Impact 　Consequences 　Conservation strategies	Measure *subjective knowledge* of energy use relative to the average person: 　Perceived effectiveness of the intervention strategy to reduce impacts/consequences Measure *actual knowledge* (*i.e.*, factual information): 　Terminology 　Possible impacts/consequences 　Criteria to evaluate impacts/consequences Measure the *extent* conservation strategies are used (*e.g.*, estimate the number of times lights are consciously turned off at the end of the day)	Measure *improvement in actual knowledge* (*i.e.*, factual information): Terminology Possible impacts/consequences Criteria for evaluating impacts/consequences Perceived effectiveness of conservation strategies to reduce impacts/consequences

motivation, it is expected that the opportunity level of the residential building occupants is high when one measures access and control over lighting, HVAC and plug loads. On the other hand, in a commercial building where occupants have limited control over the thermostat or lighting setting, their opportunity level will be low. The intervention in this case will focus on other aspects such as ensuring equipment (*e.g.*, computers and printers) are setback or turned off prior to leaving the office.

Measuring the ability level of occupants
The ability (A) level of occupants can be measured by evaluating their prior knowledge about energy use, its impact and consequences, as well as knowledge

69

about possible conservation strategies (Table 3). The probability that occupants will process energy-reduction information from intervention strategies is directly related to this pre-existing knowledge. The ability level of occupants can be measured by the perceived energy-consumption knowledge level, and the actual level of knowledge on energy consumption facts. Abrahamse and Steg (2009) concluded that occupants with higher perceived energy-consumption knowledge levels (*e.g.*, 'I know how I can save energy') and better knowledge on energy consumption facts (*e.g.*, 'the depletion of fossil fuels is a problem') could more likely save energy. Additionally, a set of studies have shown that people need to have sufficient ability (*e.g.*, self-efficacy) before they can actively care enough to take environmentally responsible actions that benefit others (Geller, 1995).

The identified metrics are presented in Tables 1–3. The motivation level of an occupant can be measured by investigating occupants' concern level about their personal energy consumption, how often they are looking for ways to conserve energy, and which factors are important for them to conserve energy. The opportunity level of occupants can be measured by investigating: the availability of energy-conservation control systems (*e.g.*, if the occupant has any individual control on the indoor climate of their office), indoor environment satisfaction level (*e.g.*, the quality of natural lighting overall in the office), and exposure to information and peer pressure about environmental concerns (*e.g.*, having energy-conservation information available for occupants on bulletin boards, discuss environmental-conservation strategies with colleagues). The ability level of an occupant can be measured by identifying occupants' perceived energy consumption knowledge level, and their level of knowledge on energy consumption facts. These identified metrics enable decision-makers to measure occupants' pre-exposure MOA level and energy-use profiles (*i.e.*, occupants' situational characteristics prior to any intervention) and post-exposure MOA levels (*i.e.*, occupants'

situational characteristics after any intervention) to determine the effectiveness of an intervention level to use for a given energy-reduction strategy (*e.g.*, encouraging occupants to turn the office lights off when not in use).

Clustering occupants' MOA levels and energy-use profiles

In this stage of the framework development, occupants' pre-existing MOA levels are classified into main target clusters relative to their potential to process energy-use characteristics. Azar and Menassa (2012) identified the different occupants' energy-use characteristics as 'high energy consumers' that represents occupants consuming more energy than other occupants in the same building; 'medium energy consumers' that represents occupants making minimal efforts towards energy savings form the category; and 'low energy consumers' that represents occupants that use energy efficiently.

Based on these identified categories, this study proposes to compare occupants' energy-use characteristics on an equal basis (*i.e.*, occupants in the same building). Therefore, occupants' MOA levels are identified in three categories (Figure 3): 'prone to react' (*i.e.*, occupants who are willing to adopt energy-reduction strategies immediately), 'unable to react' (*i.e.*, occupants who are willing to adopt energy-reduction strategies immediately but who do not have the necessary knowledge and tools to do that), and 'resistant to react' (*i.e.*, occupants who are unwilling to adopt energy-reduction strategies regardless of whether or not they have the necessary knowledge and tools). As shown in Figure 3, each motivation, opportunity and ability level of an occupant is measured using ranges from very high to low. For example, if an occupant has motivation, opportunity and ability levels in the range of high to very high, this occupant is categorized as 'prone to react'. If an occupant has a motivation level of low/medium to high and an ability level of low to very high but an opportunity level is low to medium, then this occupant is categorized as 'unable

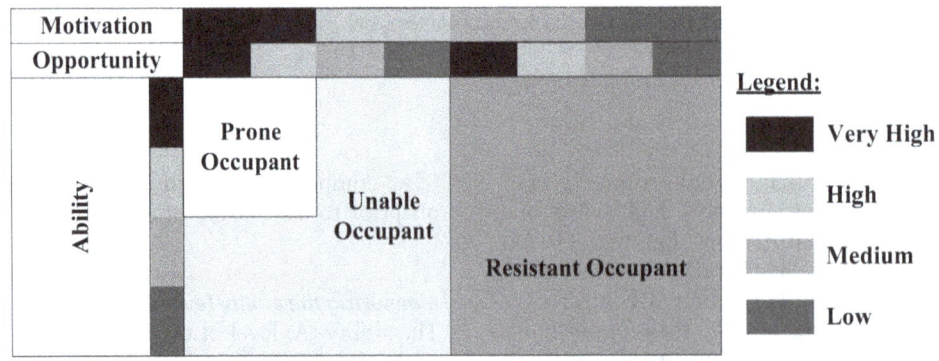

Figure 3 Categories of building occupants based on their motivation, opportunity and ability levels

to change'. If an occupant has a low to very high opportunity and ability levels, but his/her motivation level is between low to medium, this occupant is categorized as 'resistant to react'.

After identifying occupants' pre-exposure energy-use characteristics, the effectiveness of intervention strategies is determined by measuring occupants' post-exposure MOA level. Actual energy-use data before and after application of the intervention to a group of occupants can be compared with develop effectiveness distributions for the four levels of intervention strategies relative to the overall pre-existing MOA level (i.e., prone, unable or resistant) in the building. It is also important to evaluate the collected data with extraneous factors that may have an impact on the energy consumption of occupants. These factors can be listed as: weather conditions (e.g., temperature, cloudy/sunny), age of the building, schedules of occupants and demographic characteristics of occupants (e.g., age, gender). The effectiveness of each intervention strategy is expected to be highest when it corresponds to a certain MOA level (e.g., persuasion is most effective when overall MOA level is in the medium range).

Identifying multilevel building energy-use intervention strategies

This stage of the framework development focuses on linking occupants' MOA levels to the multilevel energy-efficiency intervention strategies, and accordingly energy policy tool types (i.e., knowledge, inducement and regulation tools) to provide decision-makers with some guidelines to help in the selection of the intervention approaches for a particular building. These guidelines are developed and shown in Table 4, and they take into account: benefits and solutions presented to the target (decision-maker), expected benefits and costs also to the target, expected occupant reaction and time to achieve benefits. These characteristics have been identified in previous research on interventions to resolve public health and social behaviour issues like obesity and smoking (Brug, Oenema, & Ferreira, 2005; Dann, 2010; Dolan, Hallsworth, Halpern, Dominic, & Vlaev, 2011; Grier & Bryant, 2005; Hoek & Gendall 2006).

First, the occupant MOA levels (e.g., 'prone to react' that refers to the occupant who has a self-interest to adopt energy-reduction strategies without additional reinforcement) are presented in Table 4. These MOA levels are then linked to the energy policy tools and their related four energy-efficiency intervention strategies (i.e., knowledge based, persuasion, penalties and technology). Subsequently, these intervention strategies are classified into: benefits and solutions presented to the target (e.g., adding choices with comparative advantage and favourable cost–benefit relationships), the expected benefits/costs (e.g.,

sanctions, penalties and legal consequences to behave in a certain way without the level of government coercion inherent in regulations for non-compliance energy-use behaviour), the expected reaction of the occupants (i.e., the targets) to each intervention (e.g., uncoerced free-choice behaviour that refers to the change of energy-use behaviour voluntarily), and the time to achieve the expected benefits (e.g., direct and timely exchange for desired energy-use behaviour that refers to direct reinforcement by the government command-and-control systems).

Accordingly, Table 4 provides a guideline for policy-makers to design cost-effective and efficient energy policy tools to deliver multilevel building energy-use intervention strategies based on identified energy-use characteristics of the occupants. For example, if the majority of occupants in a building are identified as 'prone' to react to conserve energy, then decision-makers should focus on designing knowledge energy policy tools to deliver knowledge-based interventions for the occupants. These interventions are based on teaching and creating an awareness about the benefits of energy-reduction strategies. They only provide suggestions for the occupants and comply with voluntarily behaviour change to adopt energy-reduction strategies.

Case study

Data were collected from employees of an energy-efficiency consulting company that occupies a single floor in an 85 000 ft^2 building located in Madison, Wisconsin. The offices are equipped with an intelligent building automation system (BAS) with a centralized monitoring and control of the indoor environment to maintain the operational performance of the facility, and comfort of building occupants. This BAS system provides occupants with different levels of occupancy control over building environment. For example, not all occupants surveyed had the capability to control their environment through individual thermostats while occupants in single offices had that capability. In both cases the opportunity level of the occupants is expected to vary.

To identify occupants' MOA level pre-exposure to occupancy-focused interventions, an online survey was distributed to the occupants in the case study building. This survey includes 39 questions and focuses on evaluating occupants' control level on energy systems (e.g., indoor lighting control), office environment conditions, energy-conservation motivation level and energy-conservation knowledge level. Previous studies (e.g., in the medical field, on a ticket purchasing website) found out that motivation is directly associated with most behaviours (Bigné et al., 2010; Moorman, 1990). However, opportunity and

Table 4 Characteristics of intervention strategies from a building energy-conservation perspective

Occupant MOA: characteristics	Energy policy tools	Intervention	Benefits and solutions presented to the target	Expected benefits/costs	Occupant reaction	Time to achieve benefits
MOA = *prone to react* Merely uninformed occupant Strong self-motivation No additional reinforcement necessary	Knowledge tools	*Knowledge based*: attempts to teach and create an awareness about the benefits of a particular behaviour. *Objective*: influence knowledge, attitudes and beliefs	*Suggests* benefits but does not deliver them explicitly *Does not add new choices* (uses existing choices) Target is required to initiate quest/solution to achieve a benefit	No explicit reward/penalty	Uncoerced free-choice behaviour *Voluntary compliance*	Promise of future potential payback Unable to reinforce directly
MOA = *Unable to slightly resistant to react* Strong self-motivation Insufficient opportunity	Inducement tools	*Persuasion*: offers reinforcing incentives/consequences *Objective*: invite voluntary exchange	*Offers* benefits they want *Adds choices* with comparative advantage and favourable cost-benefit relationships Target receives solutions through well-known distribution channels	Positive reward/ punishment delivered when the exchange transaction is completed	Uncoerced free-choice behaviour *Voluntary exchange* (self-monitoring – self-sanctioning)	Direct and timely exchange for the desired behaviour Direct reinforcement Expects free-market exchange
MOA = *resistant to react* Existing low-level motivation and ability cannot be overcome with additional rewards through exchange	Regulation tools	*Penalties*: prescribes a body of rules of action/conduct *Objective*: coerce and threaten to achieve non-voluntary compliance	*Forces* benefit by providing external motivation in the absence of an internal one *Adds proffered choices* Target bound by legal force	Sanctions, penalties and legal consequences for non-compliance	Coerced behaviour *Non-voluntary compliance*	Direct and timely exchange for desired behaviour Direct reinforcement
		Technology: control behaviour change and referred to governmental authority and legitimacy *Objective*: coerce and threaten to achieve non-voluntary compliance		Law or other costly regulations without requiring a promise of a positive incentive		

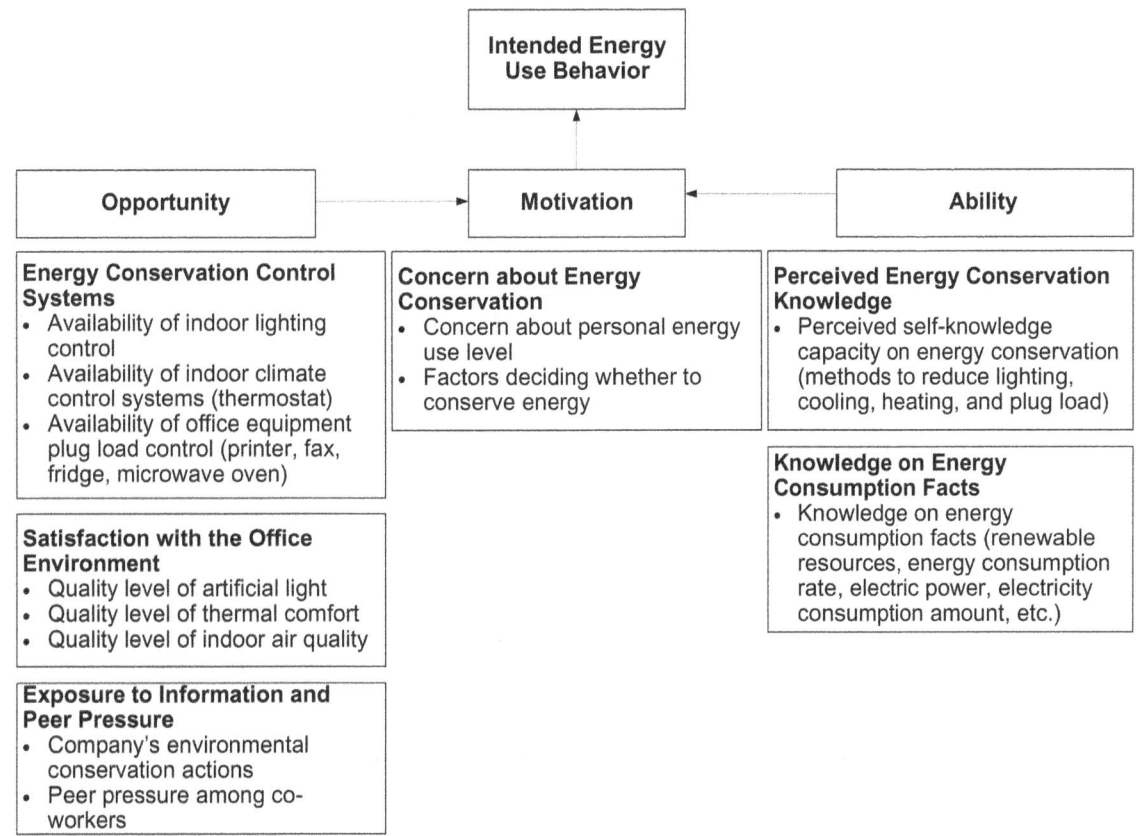

Figure 4   Measures for MOA level of occupants' energy-use behaviour

ability affect behaviours only when motivation is present and therefore moderate the impact of motivation on behaviours. Thus, the survey is developed by stating a set of research hypotheses and their relevant measures to investigate occupants' energy-use characteristics through assessing their MOA level on adopting energy-saving behaviours. These hypotheses are designed based on the extended context of MOA levels of occupants in energy-use characteristics, and incorporated with a set of measures that is identified based on a comprehensive literature review (Figure 4).

The company surveyed had a total of 39 employees. Of those, 19 responded to the online survey with a response rate of 49%. Occupants' energy-use characteristics pre-exposure to occupancy-focused interventions were identified using k-means clustering analysis performed in MATLAB (2014). Framework implementation was accomplished in four steps. First, all survey questions are categorized to measure motivation (M), opportunity (O) or ability (A) using logic presented in Figure 4; M, O and A for each respondent are determined as the average of the responses to the corresponding questions; and use k-means to cluster the occupants into three main clusters. The resulting clusters each had the number of occupants shown in Figure 5(a) with the centroids (i.e., the average M, O and A of the respondents in each

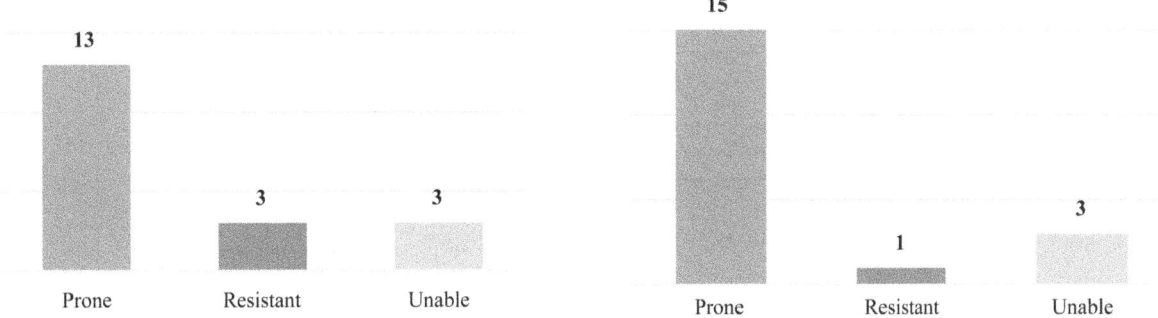

Figure 5   Occupants' energy-use characteristics: (a) before and (b) after intervention

cluster respectively) as follows: for resistant to react (M = 0.12, O = 0.33, A = 0.27), for the unable to react (M = 0.73, O = 0.20, A = 0.87), and for the prone to react (M = 0.89, O = 0.80, A = 0.84). These results can be interpreted as follows:

- Thirteen occupants (*i.e.*, 68% of total respondents) with 'prone to react' behaviour that refers to the occupant who has flexible behaviour characteristics and is willing to adopt energy-reduction strategies without additional reinforcement. These occupants have very high motivation, opportunity and ability levels when measured on a scale of 0–1. This result is expected given the company profile that the respondents work for. For example, an occupant in this group has self-motivation to adjust the plug loads when not in use to conserve energy, is satisfied with the thermal comfort and lighting quality levels at his/her office environment, and has a high level of perceived self-knowledge capacity on energy conservation.

- Three occupants (*i.e.*, 16% of total respondents) with 'unable to react' behaviour that refers to the occupants who have self-interest to adopt energy-reduction strategies voluntarily but do not have the necessary ability and/or occupancy control level to do that (*e.g.*, no/low control on thermostat settings in his/her office room). This is reflected in the low average opportunity level that occupants in this category had. These occupants have very high motivation level and ability levels. Even if occupants in this group have similar motivation and ability levels compared with occupants with 'prone to react' behaviour, their lower opportunity level makes them 'unable to react' to change their energy-conservation behaviour. For example, an occupant in this group is not able to turn off the lights when not in use due the lack of control on lighting, even if she/he has the self-motivation to adjust the lights when not in use, and a high level of knowledge capacity on energy-conservation

strategies about lights. In this particular case, there is not much behavioural interventions that can be done to change the profile of these respondents to prone to react.

- Three occupants (*i.e.*, 16% of total respondents) with 'resistant to react' behaviour that refers to the occupants who are reluctant to adopt energy-reduction strategies. Occupants in this category had a very low motivation level with moderate to low opportunity and ability levels. For example, occupants in this category are identified as unwilling to turn off their computers when not in use, while exhibiting strong knowledge about the methods to reduce plug loads in their office space.

Due to the higher percentage of 'prone to react' occupants in this case study, the conceptual framework presented in this study proposed that energy savings can be achieved effectively by delivering knowledge-based interventions to the occupants. Accordingly, the proposed intervention strategy for this case study is evaluated using three criteria: *attributes* (*e.g.*, offers the occupant an alternative to changing heating set point), *consequences* (*e.g.*, I will have to bring an extra jacket; I will be rewarded for turning the lights off with a gift card to my favourite store), and *time and cost of implementing* the intervention (*e.g.*, knowledge based is the cheapest as it does not require the exchange of rewards or penalties for achieving the required change and can be implemented immediately) (Figure 6).

Moreover, occupants' behaviour change after delivering the education interventions is also investigated to highlight the impact of intervention on occupants' behaviour. Azar and Menassa (2015) used simulation analysis and estimated the occupancy conversion rate from extreme (*i.e.*, resistant) to low (*i.e.*, prone) users at 10% when discrete interventions are used (*e.g.*, educational campaigns). This will result in an increase in the number of 'prone to react' occupants. Accordingly,

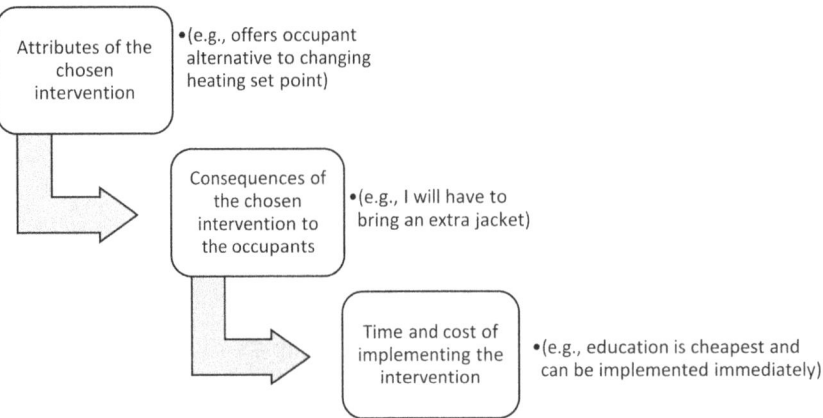

Figure 6 Criteria for evaluating intervention strategies

this study assumed a 10% occupancy energy-use profile conversion rate to analyze the energy-use behaviour change after exposure to education intervention. The results from this analysis are presented in Figure 5(b). In this case, two 'resistant to react' occupants are converted to 'prone to react' occupants by increasing their motivation level through education interventions. Additional interventions (*e.g.*, rewards like gift cards) can be used to entice the remaining resistant occupant to adopt more conservative energy-use behaviour.

Discussion

In energy policy-making, common approaches, such as assuming that all occupants have the same behaviour pattern (*e.g.*, a fixed set-point room temperature), and presuming that people are rational agents making considered decisions based on the available information and resources, have resulted in unintended consequences of individual behaviours and diminished the impacts and effectiveness of the interventions (Prendergrast, Foley, Menne, & Isaac, 2008; Schweiker & Shukuya, 2010). To overcome these negative consequences, it is important to evaluate systematically the effects of interventions and their impacts' on occupants' behaviour in buildings (Abrahamse et al., 2005; Steg & Vlek, 2009). Therefore, this study proposed a conceptual framework that provides a linkage between energy policy tools and their related multilevel building energy-use intervention strategies to evaluate systematically and change the environmental behaviour of occupants, and accordingly achieve large-scale energy reductions. Using this framework, occupancy-related actions (*e.g.*, after-hours equipment use; occupied and unoccupied hours; cooling and heating temperature set points) can be examined, and well-tuned interventions to change the relevant occupants' behaviours can be provided efficiently and cost-effectively.

Moreover, the set of criteria presented in Figure 6 can be used for all four intervention strategies (*i.e.*, education, persuasion, penalties and technology) to guide decision-makers in the selection of intervention strategies by providing them with research-based evidence of their effectiveness to achieve the required energy reductions given the occupants' MOA characteristics. Accordingly, decision-makers can determine (1) how occupants' energy-use profile in a building can be measured and clustered pre- and post-implementation of energy policy tools; and (2) how to design cost-effective energy policy tools to deliver multilevel energy-efficiency occupancy-focused interventions that promote actions for improving sustainable behaviour pattern. The conceptual framework presented in this study demonstrates a scalable, adaptable and expandable approach able to support decision-makers (*e.g.*, energy planners, local administrators) in identifying the most effective energy policies and strategies.

The case study results illustrate that buildings indeed have different occupancy energy characteristics that need to be taken into account while designing an intervention strategy. However, the fact that this proposed framework was tested on a small sample of occupants provides limited opportunities to extract general results, especially when it comes to measuring the effectiveness of the intervention strategies. Future research should focus on validating the application and benefits of using this approach. First, data need to be collected from a wide range of building types (*e.g.*, commercial and residential), sizes and functions. Second, the occupancy clustering into the three groups in the case study might need to be expanded to a larger number of clusters, especially in buildings with high occupancy rates. Finally, the evaluation of the interventions should be based on real tests and experiments to validate the results obtained from simulation.

Conclusions

This study presents a conceptual framework capable of analyzing the relationship between occupants' energy-use characteristics and the effectiveness of occupancy-focused intervention strategies to design cost-effective and efficient energy policy tools to improve energy savings potential from occupancy interventions in both commercial and residential buildings (*e.g.*, university campus, community and city). First, to address the question of measuring the occupants' energy-use profile, the study proposed an approach based on the MOA model. In this approach, a set of survey questions are used to measure each individual occupant's motivation, opportunity and ability levels. The results are further analyzed using a *k*-means clustering algorithm to classify the occupants into different categories (*e.g.*, in this paper occupants are classified into three main categories). Second, to address the question of mapping the energy-use interventions to the occupancy categories in the building, a comprehensive literature review of existing interventions is conducted. These interventions are classified into four main categories distinguished by the benefits and solutions presented to the target, the expected benefits/costs, occupant reaction and time to achieve the benefit (Table 4). In addition, the intervention strategies are mapped onto the three occupancy types that correspond to those obtained from the clustering analysis. Finally, the implementation of the proposed framework is illustrated through a case study example. The results of this example highlighted that even when the building size is small, occupants have varying energy-use characteristics that need to be considered in designing any intervention approach.

This new proposed framework will assist researchers and policy-makers in:

- systematically evaluating the effects of occupancy-focused interventions on occupants' behaviour providing feedback on the performance of specific policies and measures; and

- designing and implementing more effective energy policy tools specifically targeted to diverse human energy-use characteristics. This will help to formulate and deliver occupancy-focused intervention programmes in an efficient, cost-effective and appropriate manner.

This framework will also help encourage pro-environmental behaviour for occupants by providing for a more sustainable behaviour pattern.

Currently the authors are testing the implementation of the proposed framework in a large case study of buildings. Data from the occupants and their energy consumption are being collected from several buildings. The effectiveness of intervention strategies on different occupants' characteristics will then be investigated to validate the proposed framework on a large-scale building stock. The authors' future study will also consider the effect of the interactions between occupants by understanding their underlying social network topology (SNT), which defines the relationships between occupants and the strength of their influence on each other based on the frequency of interaction and interdependence. Understanding SNT will help decision-makers to analyze how energy knowledge diffuses between occupants, and how this diffusion contributes to enhance the impact of the chosen intervention strategy on improving the post-exposure MOA levels and the building's energy consumption.

Funding

The authors acknowledge the financial support for this research received from the US National Science Foundation (NSF) [grant numbers CBET 1407908 and 1349921]. Any opinions and findings in this paper are those of the authors and do not necessarily represent those of NSF.

References

Abrahamse, W., & Steg, L. (2009). How do socio-demographic and psychological factors relate to households' direct and indirect energy use and savings? *Journal of Economic Psychology, 30*, 711–720.

Abrahamse, W., Steg, L., Vlek, C., & Rothengatter, T. (2005). A review of intervention studies aimed at household energy conservation. *Journal of Environmental Psychology, 25*, 273–291.

Agha-Hossein, M., Tetlow, R., Hadi, M., El-Jouzi, S., Elmualim, A., Ellis, J., & Williams, M. (2014). Providing persuasive feedback through interactive posters to motivate energy-saving behaviours. *Intelligent Buildings International, 7*, 16–35.

Akridge, J. M. (1998). High-albedo roof coatings – Impact on energy consumption (No. CONF-980123–). Atlanta, GA (United States): American Society of Heating, Refrigerating and Air-Conditioning Engineers, Inc.

Alahmad, M. A., Wheeler, P. G., Schwer, A., Eiden, J., & Brumbaugh, A. (2012). A comparative study of three feedback devices for residential real-time energy monitoring. *IEEE Transactions on Industrial Electronics, 59*, 2002–2013.

Allcott, H. (2011). Social norms and energy conservation. *Journal of Public Economics, 95*, 1082–1095.

Armstrong, A. W., Idriss, N. Z., & Kim, R. H. (2011). Effects of video-based, online education on behavioral and knowledge outcomes in sunscreen use: A randomized controlled trial. *Patient Education and Counseling, 83*, 273–277.

Ayres, I., Raseman, S., & Shih, A. (2013). Evidence from two large field experiments that peer comparison feedback can reduce residential energy usage. *Journal of Law, Economics, and Organization, 29*(5), 992–1022.

Azar, E., & Menassa, C. C. (2012). Agent-based modeling of occupants and their impact on energy use in commercial buildings. *Journal of Computing in Civil Engineering, 26*, 506–518.

Azar, E., & Menassa, C. C. (2014). A comprehensive framework to quantify energy savings potential from improved operations of commercial building stocks. *Energy Policy, 67*, 459–472.

Azar, E., & Menassa, C. C. (2015). Evaluating the impact of extreme energy use behavior on occupancy interventions in commercial buildings. *Energy and Buildings, 97*, 205–218.

Backer, T. E. (1991). Knowledge utilization the third wave. *Science Communication, 12*, 225–240.

Bernstein, H. M., & Russo, M. A. (2009). Green building retrofit and renovation: Rapidly expanding market opportunities through existing buildings. Smart Market Report, McGraw Hill Construction.

Bigné, E., Hernández, B., Ruiz, C., & Andreu, L. (2010). How motivation, opportunity and ability can drive online airline ticket purchases. *Journal of Air Transport Management, 16*, 346–349.

Blamey, A., Mutrie, N., & Tom, A. (1995). Health promotion by encouraged use of stairs. *British Medical Journal, 311*, 289–290.

Boyce, T. E., & Geller, E. S. (2001). Encouraging college students to support pro-environment behavior effects of direct versus indirect rewards. *Environment and Behavior, 33*, 107–125.

Braun, J. E. (1990). Reducing energy costs and peak electrical demand through optimal control of building thermal storage. *ASHRAE Transactions, 96*, 876–888.

Brug, J., Oenema, A., & Ferreira, I. (2005). Theory, evidence and intervention mapping to improve behavior nutrition and physical activity interventions. *International Journal of Behavioral Nutrition and Physical Activity, 2*(2). doi:10.1186/1479-5868-2-2.

Burn, S. M., & Oskamp, S. (1986). Increasing community recycling with persuasive communication and public commitment. *Journal of Applied Social Psychology, 16*, 29–41.

Buurma, H. (2001). Public policy marketing: Marketing exchange in the public sector. *European Journal of Marketing, 35*, 1287–1302.

Bzdel, L., Wither, C., & Graham, P. (2004). Knowledge utilization resource guide. Knowledge utilization and policy implementation.

Carrico, A. R., & Riemer, M. (2011). Motivating energy conservation in the workplace: An evaluation of the use of group-level feedback and peer education. *Journal of Environmental Psychology, 31*, 1–13.

Celsi, R. L., & Olson, J. C. (1988). The role of involvement in attention and comprehension processes. *Journal of Consumer Research, 15*, 210–224.

Chidiac, S. E., Catania, E. J. C., Morofsky, E., & Foo, S. (2010). A screening methodology for implementing cost effective energy retrofit measures in Canadian office buildings. *Energy and Buildings*, doi:10.1016/j.enbuild.2010.11.002.

Dann, S. (2010). Redefining social marketing with contemporary commercial marketing definitions. *Journal of Business Research*, 63(2), 147–153.

Darby, S. (2006). The effectiveness of feedback on energy consumption. A Review for DEFRA of the Literature on Metering, Billing and direct Displays 486, 2006.

Department of Energy. (2011). Join the commercial real estate energy alliance. Energy Efficiency and Renewable Energy Program. Retrieved February 16, 2016, from http://apps1.eere.energy.gov/buildings/publications/pdfs/alliances/creea_fs.pdf

Dietz, T., Gardner, G. T., Gilligan, J., Stern, P. C., & Vandenbergh, M. P. (2009). Household actions can provide a behavioral wedge to rapidly reduce US carbon emissions. *Proceedings of the National Academy of Sciences of the United States of America*, 106(44), 18452–18456.

Dolan, P., Hallsworth, M., Halpern, D., Dominic, K., & Vlaev, I. (2011). *Mindspace – influencing behavior through public policy: The practical guide*. London: Institute for Government – Cabinet Office. Retrieved November 28, 2011, from www.instituteforgovernment.org.uk

Dolan, P., & Metcalfe, R. (2013). *Neighbors, knowledge, and nuggets: Two natural field experiments on the role of incentives on energy conservation*. London: Centre for Economic Performance, LSE.

Entrop, A. G., Brouwers, H. J. H., & Reinders, A. H. M. E. (2010). Evaluation of energy performance indicators and financial aspects of energy saving techniques in residential real estate. *Energy and Buildings, Elsevier*, 42(2010), 618–629.

Environmental and Energy Study Institute. (2006). Energy-efficient buildings: Using whole building design to reduce energy consumption in homes and offices.

Environmental Information Administration. (2014). How much energy is consumed in residential and commercial buildings in the United States? Retrieved December 5, 2014, from http://www.eia.gov/tools/faqs/faq.cfm?id=86&t=1

Erickson, V. L., Achleitner, S., & Cerpa, A. E. (2013). *POEM: Power-efficient occupancy-based energy management system*. Proceedings of the 12th international conference on Information processing in sensor networks. ACM, pp. 203–216.

Erickson, V. L., & Cerpa, A. E. (2010). *Occupancy based demand response HVAC control strategy*. Proceedings of the 2nd ACM Workshop on Embedded Sensing Systems for Energy-Efficiency in Building. ACM, pp. 7–12.

Estabrooks, C. A., Thompson, D. S., Lovely, J. J. E., & Hofmeyer, A. (2006). A guide to knowledge translation theory. *Journal of Continuing Education in the Health Professions*, 26, 25–36.

Geller, E. S. (1981). Evaluating energy conservation programs: Is verbal report enough? *Journal of Consumer Research*, 8, 331–335.

Geller, E. S. (1995). Actively caring for the environment - An integration of behaviorism and humanism. *Environment and Behavior*, 27, 184–195.

Govindaraju, R., Hadining, A. F., & Chandra, D. R. (2013). Physicians' adoption of electronic medical records: Model development using ability–motivation-opportunity framework, information and communication technology. Springer, pp. 41–49.

Grier, S., & Bryant, C. A. (2005). Social marketing in public health. *Annual Review of Public Health*, 26, 319–339.

Hallahan, K. (2001). Enhancing motivation, ability, and opportunity to process public relations messages. *Public Relations Review*, 26, 463–480.

Hand, L.C. (2012). *Public policy design and assumptions about human behavior*. Western Political Science Association's Annual Conference, 1–39.

Handgraaf, M. J., Van Lidth de Jeude, M. A., & Appelt, K. C. (2013). Public praise vs. private pay: Effects of rewards on energy conservation in the workplace. *Ecological Economics*, 86, 86–92.

Harvey, L. D. (2009). Reducing energy use in the buildings sector: Measures, costs, and examples. *Energy Efficiency*, 2, 139–163.

Hastak, M., Mazis, M. B., & Morris, L. A. (2001). The role of consumer surveys in public policy decision making. *Journal of Public Policy and Marketing*, 20, 170–185.

Hayes, S. C., & Cone, J. D. (1977). Reducing residential electrical energy use: Payments, information, and feedback. *Journal of Applied Behavior Analysis*, 10, 425–435.

Heberlein, T. A., & Warriner, G. K. (1983). The influence of price and attitude on shifting residential electricity consumption from on-to off-peak periods. *Journal of Economic Psychology*, 4, 107–130.

Hedlund, J. (2000). Risky business: Safety regulations, risk compensation, and individual behavior. *Injury Prevention*, 6, 82–89.

Hoek, J., & Gendall, P. (2006). Advertising and obesity: A behavioral perspective. *Journal of Health Communication: International Perspectives*, 11, 409–423.

Houston, D. J., & Richardson, L. E. (2007). Risk compensation or risk reduction? Seatbelts, state laws, and traffic fatalities. *Social Science Quarterly*, 88, 913–936.

Hutton, R. B., & McNeill, D. L. (1981). The value of incentives in stimulating energy conservation. *Journal of Consumer Research*, 8, 291–298.

John, L. K., Loewenstein, G., Troxel, A. B., Norton, L., Fassbender, J. E., & Volpp, K. G. (2011). Financial incentives for extended weight loss: A randomized, controlled trial. *Journal of General Internal Medicine*, 26, 621–626.

Jollands, N., Waide, P., Ellis, M., Onoda, T., Laustsen, J., Tanaka, K., ... Meier, A. (2010). The 25 IEA energy efficiency policy recommendations to the G8 Gleneagles Plan of Action. *Energy Policy*, 38, 6409–6418.

Juan, Y. K., Gao, P., & Wang, J. (2010). A hybrid decision support system for sustainable office building renovation and energy performance improvement. *Energy and Buildings, Elsevier*, 42(2010), 290–297.

Katsev, R. D., & Johnson, T. R. (1984). Comparing the effects of monetary incentives and foot in the door strategies in promoting residential electricity conservation. *Applied Social Psychology*, 14, 12–27.

Kaufman, L., & Rousseeuw, P. J. (2009). *Finding groups in data: An introduction to cluster analysis*. Hoboken, NJ: John Wiley & Sons.

Kotler, P., Roberto, N., & Lee, N. (2002). *Social marketing: Improving the quality of life*. Thousand Oaks: Sage.

Kouveletsou, M., Sakkas, N., Garvin, S., Batic, M., Reccardo, D., & Sterling, R. (2012). Simulating energy use and energy pricing in buildings: The case of electricity. *Energy and Buildings*, 54, 96–104.

Levine, M., Urge-Vorsatz, D., Blok, K., Geng, L., Harvey, D., Lang, S., ... Novikova, A. (2007)). Residential and commercial buildings. Climate change 2007; Mitigation. Contribution of Working Group III to the Fourth Assessment Report of the IPCC. Cambridge University Press, Cambridge, United Kingdom and New York, NY, USA.

Lopes, M. A. R., Antunes, C. H., & Martins, N. (2012). Energy behaviours as promoters of energy efficiency: A 21st century review. *Renewable and Sustainable Energy Reviews*, 16, 4095–4104.

Machleit, K. A., Madden, T. J., & Allen, C. T. (1990). Measuring and modeling brand interest as an alternative ad effect with familiar brands. *Advances in Consumer Research*, 17, 223–230.

MacInnis, D. J., Moorman, C., & Jaworski, B. J. (1991). Enhancing and measuring consumers' motivation, opportunity, and ability to process brand information from ads. *Journal of Marketing, 55*, 32–53.

Mahdavi, A., & Pröglhöf, C. (2009). *Toward empirically-based models of people's presence and actions in buildings. Proceedings of building simulation*, pp. 537–544.

Marans, R. W., & Edelstein, J. Y. (2010). The human dimension of energy conservation and sustainability: A case study of the University of Michigan's energy conservation program. *International Journal of Sustainability in Higher Education, 11*, 6–18.

Mathews, E., Botha, C., Arndt, D., & Malan, A. (2001). HVAC control strategies to enhance comfort and minimise energy usage. *Energy and Buildings, 33*, 853–863.

McClelland, L., & Cook, S. W. (1980). Promoting energy conservation in master-metered apartments through group financial incentives. *Journal of Applied Social Psychology, 10*, 20–31.

McMakin, A. H., Malone, E. L., & Lundgren, R. E. (2002). Motivating residents to conserve energy without financial incentives. *Environment and Behavior, 34*, 848–863.

Midden, C. J., Meter, J. F., Weenig, M. H., & Zieverink, H. J. (1983). Using feedback, reinforcement and information to reduce energy consumption in households: A field-experiment. *Journal of Economic Psychology, 3*, 65–86.

Moorman, C. (1990). The effects of stimulus and consumer characteristics on the utilization of nutrition information. *Journal of Consumer Research, 17*, 362–374.

Nemry, F., Uihlein, A., Colodel, M. C., Wetzel, C., Braune, A., Wittstock, B., … Frech, Y. (2010). Options to reduce the environmental impacts of residential buildings in the European Union-Potential and Costs. *Energy and Buildings, Elsevier, 42*(7), 976–984.

Norcross, J. C., Koocher, G. P., & Garofalo, A. (2006). Discredited psychological treatments and tests: A Delphi poll. *Professional Psychology: Research and Practice, 37*, 515–522.

Parra-Lopez, E., Gutierrez-Tano, D., Diaz-Armas, R. J., & Bulchand-Gidumal, J. (2012)). Travellers 2.0: motivation, opportunity and ability to use social media. Social Media in Travel, Tourism and Hospitality, 171–185.

Peattie, K., & Peattie, S. (2009). Social marketing: A pathway to consumption reduction? *Journal of Business Research, 62*, 260–268.

Peschiera, G., Taylor, J. E., & Siegel, J. A. (2010). Response-relapse patterns of building occupant electricity consumption following exposure to personal, contextualized and occupant peer network utilization data. *Energy and Buildings, 42*(8), 1329–1336.

Polonsky, M. J., Binney, W., & Hall, J. (2004). Developing better public policy to motivate responsible environmental behavior–an examination of managers' attitudes and perceptions towards controlling introduced species. *Journal of Nonprofit and Public Sector Marketing, 12*, 93–107.

Prendergrast, J., Foley, B., Menne, V., & Isaac, A. K. (2008). *Creatures of habit?, The art of behaviour change*. London: The Social Market Foundation.

Richins, M. L., & Bloch, P. H. (1986). After the new wears off: The temporal context of product involvement. *Journal of Consumer Research, 13*, 280–285.

Rothschild, M. L. (1999). Carrots, sticks, and promises: A conceptual framework for the management of public health and social issue behaviors. *Journal of Marketing, 63*, 24–37.

Ruzelli, A. (2010). Proceedings of the 2nd ACM Workshop on Embedded Sensing Systems for Energy-Efficiency in Building. ACM, Zurich, Switzerland, p. 93.

Sahota, P., Rudolf, M. C., Dixey, R., Hill, A. J., Barth, J. H., & Cade, J. (2001). Randomised controlled trial of primary school based intervention to reduce risk factors for obesity. *British Medical Journal, 323*, 1029.

Schick, S., & Goodwin, S. (2011). Residential behavior based energy efficiency program profiles. Bonneville Power Administration.

Schneider, D., & Rode, P. (2010). Energy renaissance: Case study-empire state building, high performance buildings. American Society of Heating, Refrigerating and Air-Conditioning Engineers, Inc. (ASHRAE), 2010, pp. 20–32.

Schweiker, M., & Shukuya, M. (2010). Comparative effects of building envelope improvements and occupant behavioural changes on the exergy consumption for heating and cooling. *Energy Policy, 38*, 2976–2986.

Siero, F. W., Bakker, A. B., Dekker, G. B., & Van den Burg, M. T. (1996). Changing organizational energy consumption behaviour through comparative feedback. *Journal of Environmental Psychology, 16*, 235–246.

Spring, B., Walker, B., Brownson, R., Mullen, E., Newhouse, R., Satterfield, J., & Hitchcock, K. (2008). Definition and competencies for evidence-based behavioral practice (EBBP). Counsel for Training in Evidence-Based Behavioral Practice: Evanston, IL.

Steg, L., & Vlek, C. (2009). Encouraging pro-environmental behaviour: An integrative review and research agenda. *Journal of Environmental Psychology, 29*, 309–317.

United Nations Environment Programme. (2014). Sustainable buildings and climate initiative promoting policies and practices for sustainability. Retrieved December 31, 2014, from http://www.unep.org/sbci/AboutSBCI/Background.asp.

Van Houwelingen, J. H., & Van Raaij, W. F. (1989). The effect of goal-setting and daily electronic feedback on in-home energy use. *Journal of Consumer Research, 16*, 98–105.

Von Borgstede, C., Andersson, M., & Johnsson, F. (2013). Public attitudes to climate change and carbon mitigation—Implications for energy-associated behaviours. *Energy Policy, 57*, 182–193.

Wensing, M., Bosch, M., & Grol, R. (2010). Developing and selecting interventions for translating knowledge to action. *Canadian Medical Association Journal, 182*, E85–E88.

Whitsett, D. D., Justus, H. C., Steiner, E., & Duffy, K. (2013). Persistence of energy efficiency behaviors over time: Evidence from a community-based program.

Winett, R. A., Kagel, J. H., Battalio, R. C., & Winkler, R. C. (1978). Effects of monetary rebates, feedback, and information on residential electricity conservation. *Journal of Applied Psychology, 63*, 73–80.

Wong, N. H., Cheong, D., Yan, H., Soh, J., Ong, C., & Sia, A. (2003). The effects of rooftop garden on energy consumption of a commercial building in Singapore. *Energy and Buildings, 35*, 353–364.

Yudelson, J. (2010). Greening existing buildings, McGraw Hill – A Green Source Book.

Zaichkowsky, J. L. (1985). Measuring the involvement construct. *Journal of Consumer Research, 12*, 341–352.

Zografakis, N., Menegaki, A. N., & Tsagarakis, K. P. (2008). Effective education for energy efficiency. *Energy Policy, 36*, 3226–3232.

Improved governance for energy efficiency in housing

Henk Visscher, Frits Meijer, Daša Majcen and Laure Itard

Current practices show that the goals of energy saving and CO_2 reductions for creating an energy-neutral building stock can only be reached by strict and supportive governmental policies. In Europe the Energy Performance of Buildings Directive (EPBD) and the Energy Efficiency Directive (EED) are driving forces for member states to develop and strengthen energy performance regulations both for new buildings (via building approval procedures) and the existing building stock (via energy performance certificates or labels). The effectiveness of these current governance instruments and their impact on actual CO_2 reductions are found to be inadequate for ensuring actual (not hypothecated) energy performance is achieved. To realize the very ambitious energy-saving goals a radical rethink of regulatory systems and instruments is necessary. Building performance and the behaviour of the occupants is not well understood by policy-makers. Alternative forms of governance are needed that have more impact on the actual outcomes. Supportive governance to stimulate near-zero renovations in combination with performance guarantees is a promising approach. Furthermore, engagement with occupant practices and behaviours is needed. To ensure accurate outcomes-based governance, a better understanding of building performance and behaviours of occupants must be incorporated.

Introduction

Climate change mitigation is the most important driver for the ambitions to reduce the use of fossil fuels. There are also other reasons for implementing energy-efficiency policies in the European Union and its member states. These include the wish to diminish the dependency on fuel imports, the increasing costs and the fact that fuel resources are limited, as well as the impacts on public health. The European building sector is responsible for about 40% of the total primary energy consumption. To reduce this share, the European Commission has introduced the Energy Performance of Buildings Directive (EPBD) (2010/31/EC) and more recently the Energy Efficiency Directive (EED – 2012/27/EU). These frameworks require member states to develop energy performance regulations for new buildings, a system of energy performance certificates (EPCs) for all existing buildings and policy programmes that support actions to reach specific goals, *e.g.* building only 'nearly zero-energy buildings' (NZEB) by 2020 and realizing an almost carbon-neutral building stock by 2050.

Formulating ambitions and sharpening regulations are relatively easy to do. Technical solutions are currently available to realize the NZEB standard in building projects and an increasing number of NZEB projects are being built. However, substantial evidence exists that mainstream building projects do not realize the expected energy performance in practice – this is the building performance gap. What is perhaps even more important in this respect is that the focus predominantly should be on the existing building stock. About 75% of the buildings that will comprise the European housing stock in 2050 have already been built today. Therefore, it is important to have an insight into whether or not the EPCs provide reliable information (Hamilton et al., 2013).

Many researchers have found evidence of the performance gap (and the underpinning reasons for it). This paper briefly elaborates on this subject in next section. This is followed by reference to the results of research by Guerra Santin, Itard, and Visscher (2009) and Guerra Santin and Itard (2010) focusing on the situation in newly built houses. The fourth section presents the findings of Majcen, Itard, and Visscher (2013a, 2013b, 2015) who studied in detail the relation between the energy labels (EPCs) and the actual energy use in dwellings. The fifth section discusses these findings in order to answer the main question: What could be adequate policies and regulatory tools to control the actual energy use in houses? The conclusions are presented in the final section.

Performance gap

In Europe as much as 40% of all final energy use is used in the built environment. A total of 75% of this amount is for residential buildings (BPIE, 2011), therefore dwellings offer a large energy-saving potential. In the last few decades many European countries have introduced various energy-saving requirements in their national building codes. Before the 1970s there were few if any energy regulations for buildings, but after the first oil crises in the mid-1970s the demands on minimum U-values for walls were introduced into building regulations. In 1995 the Netherlands replaced the more prescriptive forms of regulations with energy performance requirements. In principle, energy performance regulations should create more latitude for finding and implementing innovative solutions to reduce the total amount of energy use of the building. Since the introduction of the EPBD in 2003 all European Union member states are required to establish and implement a form of energy performance regulations.

Building regulations are meant to prescribe a minimum accepted quality level for a building according to societal values and needs. However, the qualities (orientation, design, specification, built quality, maintenance, etc.) of a building can only partially influence energy use. The other significant influence is determined by the behaviour of the occupant. The design and materialization of a building can give better conditions for comfortable temperatures so these aspects are subject of the regulations (along with the lighting in the communal areas and use of lifts in multi-occupancy residential buildings). All other forms of energy use in dwellings (e.g. plug loads – refrigerators, washing machines, computers and cooking appliances) are not controlled by building regulations but covered by other legislation. In general, for older buildings the energy demand for space heating and cooling is dominant. In newer buildings with a very high level of insulation, the energy demand for appliances becomes dominant.

Regulations focus on the design and in the best cases there is even some control on the performance of a building at the end of the construction process. However, after the building is occupied there is currently little if any control over the energy use. The energy calculation methods used or referred to in regulations are based on models and parameters of the performance of construction types and materials, and on the expected or modelled heating behaviour of the inhabitants. It is clear that all these models and assumptions do not accurately portray the actual energy use. This can be called the performance gap. The performance gap occurs for several different reasons: the built artefact does not match the design (substitutions and poor built quality), the mechanical services are not commissioned correctly, the inhabitants do not understand how to operate the building, the inhabitants' behaviour and practices are not as expected, or the building does not fit the inhabitants' needs and capabilities.

Over the past 20 years numerous studies have compared the actual energy use with the expected or modelled energy use (Bordass, Cohen, Standeven, & Leaman, 2001; Bordass, Leaman, & Ruyssevelt, 2001; Branco, Lachal, Gallinelli, & Weber, 2004; Cayre, Allibe, Laurent, & Osso, 2011; Sorrell, Dimitropoulos, & Sommerville, 2009; Gram-Hanssen, 2010; Raslan & Davies, 2012; Hamilton et al., 2013). The general pattern that follows from these studies is that for dwellings with a good (theoretical) energy performance the actual energy use in general is higher than modelled. For the dwellings with a bad (theoretical) performance, the actual energy use is lower. There are various explanations for these findings. For the presumed good performance buildings it is a combination of underperformance of the building due to design and construction faults and changed behaviour of the occupants. This is partly the rebound effect (Berkhout, Muskens, & Velthuijsen, 2000; Galvin, 2015): if the conditions improve and the inhabitants think that the building is more energy efficient, they become less careful in their energy use behaviour (e.g. they use higher temperature settings, wear thinner clothing and operate the heating for longer periods). For the 'bad' performing buildings there is also evidence that the quality of the building could be underestimated. The U-values of solid walls in England were underestimated. Solid walls had been assumed to have a U-value of 2.1 W/m²K, however recent research has shown a value in the range of 1.6 W/m²K (Li et al., 2015). Rasooli, Itard, and Ferreira (2016) found similar results in a study in the Netherlands. In addition to this there is large impact by the behaviour of the occupants. The models assume an average heating of the whole building, however in poorly insulated buildings the occupants are frugal and use heat sparingly and they also tend to heat only the spaces that they actually use.

Energy performances of new dwellings in practice

In 1995 energy performance regulation for space heating and cooling of newly built constructions were introduced in the Netherlands. The regulation consists of a standard (norm) that prescribes the calculation method which is called the energy performance norm. The standard results in a non-dimensional figure called the energy performance coefficient. Every few years the level of this coefficient was decreased, representing a lower energy use demand for building-related energy use. In 2021 this coefficient will be on the level of nearly energy neutral according to the EPBD. Since the introduction of the energy performance regulation there has been little assessment of the regulation's effect on the actual energy use in the houses. Two studies found no statistical correlation between the energy performance coefficient level and the actual energy use per dwelling or per m². Analysis of the WoON (WoOn Energie, 2009) survey, which was undertaken on behalf of the Dutch government in 2006 and contained a representative sample of 5000 dwellings, also found no correlation between the different levels of the energy performance coefficient and the actual energy use per dwelling and per m². Guerra Santin, Itard and Visscher (2009) and Guerra Santin and Itard (2010) compared the actual and expected energy consumptions for 313 Dutch dwellings built after 1996. The method included an analysis of the original energy performance calculations that were submitted to the municipality as part of the building permit application, a detailed questionnaire and some day-to-day occupant diaries. These combined approaches generated very detailed and accurate data of the (intended) physical quality of the dwellings and installations, about the actual energy use (from the energy bills) and of the households and their behaviour.

The dwellings were categorized according to their energy performance coefficient. Due to the relatively small sample size, the differences between the actual heating energy of buildings with different energy performance coefficient values were insignificant. Nonetheless the average consumption was consistently lower in buildings with a lower energy performance coefficient, but not nearly as low as expected. In this sample, it was found that the increased level of the energy performance had very little effect on the actual energy use. Guerra Santin et al. found that building characteristics (including heating and ventilation installations) were responsible for 19–23% of the variation in energy used in the recently built building stock (Figure 1). Galvin (2016, p. 75) found that the average energy performance gap in low-energy dwellings averaged 64%, but was 45% for passive houses.

There are also indications that the gap is related to design and construction faults and that heating services operate in very different conditions than assumed beforehand. Nieman (2007) showed that in a sample of 154 dwellings, 25% did not meet the energy performance requirements in the design phase due to mistakes in the calculations. Nevertheless the building permit was issued. In 50% of the dwellings, the realization was not in accordance with the design. These results comply with other findings about inadequate performance of the building industry, but also by a low level of quality monitoring by the construction parties and the poor performance of the building control authorities in the Netherlands and other countries (Meijer, Visscher, & Sheridan, 2002; Meijer & Visscher, 2006, 2008; Van der Heijden, Meijer, & Vischer, 2007). Taking into account the above findings, it is doubtful whether further tightening of the energy performance regulations will lead to improvements in actual outcomes.

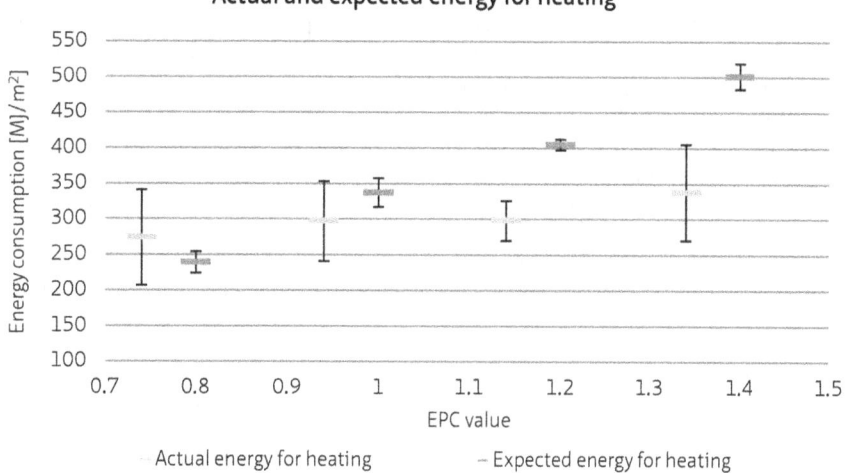

Figure 1 Mean and 95% confidence interval for the actual energy consumption (MJ/m²) and expected energy for heating (MJ/m²) per EPC value
Source: Guerra Santin et al. (2009)

Other and more efficient solutions exist to decrease the actual energy consumption of newly built dwellings. Important ingredients of the solution are: ensuring that appliances and installation are correctly installed, monitoring the calculated performances in practice; enlarging the know-how and skills of building professionals; and creating an effective and efficient building control and enforcement process. Monitoring the actual performance in the completed building becomes more important. The Dutch building control system is currently being reviewed. A new law on quality assurance for buildings will be introduced shortly. The main change will be that the responsibility for plan approval and site inspections will shift from municipal authorities to private parties. More significantly, the new emphasis will be on the assurance that the building 'as built' complies with the regulations. For many decades the main focus of building control in the Netherlands has always been on the design. So, this change can be considered as a step in the right direction. It is unclear yet how detailed such a compliance check on the completed building for the energy performance will be. For nearly energy-neutral buildings it would have to include a blower door test to check the air tightness and/or infrared scans to find thermal leakages.

Actual versus calculated energy use: existing dwellings

The largest energy saving potential is in the existing building stock. New dwellings add about 1% per year to the housing stock in Europe. The most important policy tool required by the EPBD in European member states is the issuing of EPCs. These give a hypothecated indication of the required energy to provide a certain average temperature in the building and depend on physical characteristics of the building. The certificate has no mandatory implications in the sense that owners could be forced to improve their buildings to certain levels. Nonetheless it is a crucial instrument for benchmarking and formulating policy goals. Building owners in all European Union member states have to obtain an EPC for a building at the moment it is sold or rented out. This is not yet current practice everywhere, mostly due to lack of enforcement. This especially applies to the private housing stock. In the Netherlands, however, the complete social housing stock is labelled with an EPC. The social sector in the Netherlands is still relatively large (35%) and well organized. For the social housing stock the EPCs are collected in a database called SHAERE (Sociale Huursector Audit en Evaluatie van Resultaten Energiebesparing, which translates as social rental sector audit and evaluation of energy saving results). With this database the progress of the renovation practices can be monitored.

The relation between the EPCs (with the calculated energy use) and the actual energy (based on measured data) use can be studied (Cohen & Bordass, 2015). A few years ago the sector formulated ambitious programmes, but these have been scaled down because of several reasons. The economic crisis of 2008–09 reduced the financial capacities of the housing associations and slowed the housing market. It also proved to be difficult to get approval of tenants for renovations that require an increase of the rents (70% of the tenants have to agree).

The Netherlands' social housing sector agreed in 2012 with the government and the National Tenants Union to a covenant about energy renovation goals. The most important goal is to reach an average label B in 2020 for the whole sector, which comprises 2.3 million dwellings (35% of the total stock). Research with the SHAERE database shows the progress in renovation. Figure 2 demonstrates the label steps over 2010–13. Note that most of the renovations have led to small improvements. If the current figures are extrapolated to 2020, it is evident that the goals of an average label B will not be reached (Figure 3). The label indexes relate the calculation of the energy index, which for label B is 1.25.

As noted previously, the actual domestic energy use is also influenced by the use and behaviour of the tenants (Gram-Hanssen, 2010). The dwellings with the worst EPC rating (category G) in practice use far less energy as expected, while the most energy-efficient dwellings (category A) use more. This is probably due to a combination of the rebound effect (an increase in comfort level of the dwellings), underperformance of the buildings and their equipment in the A- and B-labelled dwellings, as well as what Sunikka-Blank and Galvin (2012) called the prebound effect: less energy used then expected in the E–G labelled dwellings due to a lower comfort level and a better-then-estimated performance of the buildings. Figure 4 shows the actual and theoretical gas consumption per dwelling per EPC. These findings for the Dutch housing stock were first generated in a research project by Majcen et al. (2013a, 2013b).

The research by Majcen et al. (2013a, 2013b) was based on the Dutch energy labels issued in 2010 – a total of over 340,000 cases with 43 variables (regarding building location and technical characteristics, the properties of the label itself etc.). This dataset was derived from the publicly available database of the EPCs. These data were, on the basis of the addresses of the households, linked to actual energy use data. The energy data were provided by the CBS (Statistics Netherlands), which collected these data from the energy companies. The combined data file was then cleaned by deleting incomplete or obvious incorrect EPCs. This resulted in 193 856 usable cases.

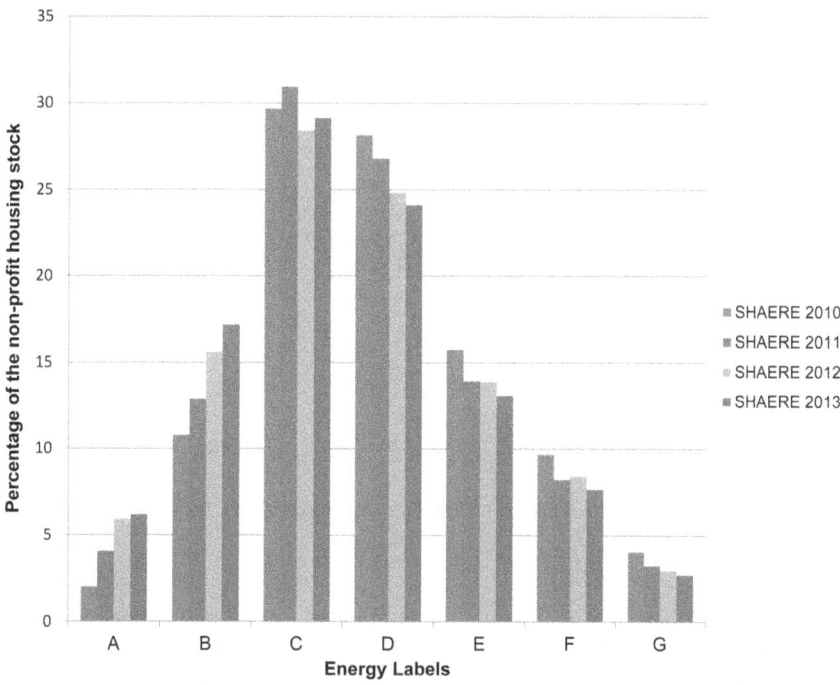

Figure 2   Distribution of the energy labels of the non-profit rented housing sector in the SHAERE database
Note: SHAERE = Sociale Huursector Audit en Evaluatie van Resultaten Energiebesparing (Social Rental Sector Audit and Evaluation of Energy Saving Results)
Source: Filippidou, Nieboer, and Visscher (2014)

This still large sample proved to be representative for all housing types and energy label classes.

To understand how the energy label relates to the discrepancies, the gas and electricity consumption in various label categories were examined and analysed. The actual and theoretical gas use per dwelling was

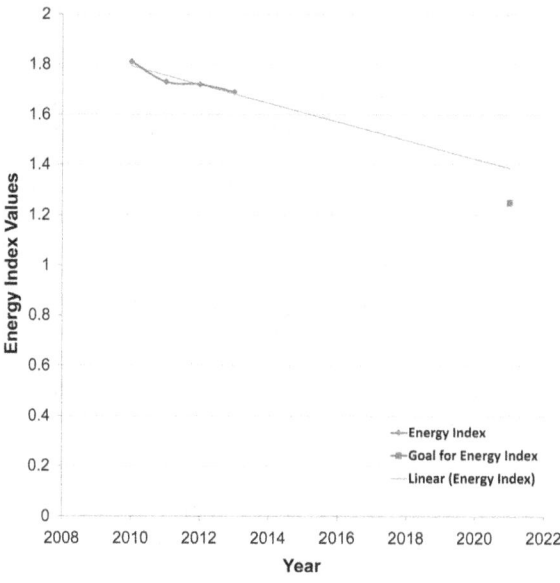

Figure 3   Development of the Energy Index in the Dutch non-profit housing sector since 2010
Source: Filippidou, Nieboer, and Visscher (2015)

compared and then analysed per m² of dwelling (Figure 4). Little difference exists between the actual and theoretical energy use calculated per dwellings and per m², except the difference in actual gas use between labels A and B. At the level of individual dwellings, the actual consumption was identical, but at the level of m² the dwellings in category A use less gas than dwellings in category B. This may relate directly to the fact that dwellings in label category A were found to be considerably larger than all other dwellings. From these figures it is clear that although better labels lead to higher actual gas consumption, there is a clear difference between the mean theoretical and mean actual gas consumption for each label. For the most energy-efficient categories (A, A+ and A++) and for category B, Figure 4 shows that the theoretical calculation underestimated the actual annual gas consumption. This is in contrast to the rest of the categories for which the theoretical calculation largely overestimated the actual annual gas consumption. So this validates the concept of the prebound effect.

This research indicates that the energy label has some predictive power for the actual gas consumption. However, according to the labels, dwellings in a better label category should use on average significantly less gas than dwellings with poorer labels, which is not the case. The actual heating energy consumption is on average lower than theoretical

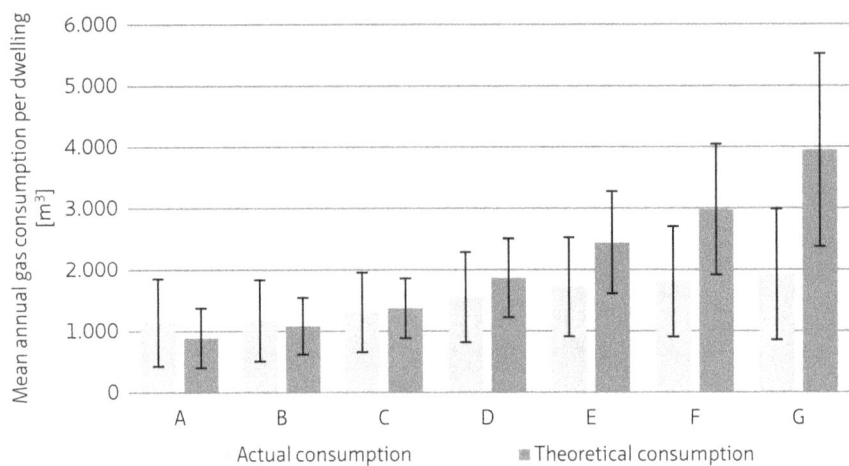

Figure 4 Actual and theoretical gas consumption in Dutch dwellings for each energy label – per m² dwelling area
Source: Majcen et al. (2013a)

consumption levels for most buildings (in this study for dwellings with labels C–G) as was observed previously by Guerra Santin, Itard, and Visscher (2009), Branco et al. (2004) Tigchelaar, Daniëls, and Maenkveld (2011), Cayre, Allibe, Laurent, and Osso (2011), and Hens, Parijs, and Deurinck (2010). Guerra Santin (2009) already suggested that at a lower EPC value the difference between the expected and actual consumption will be smaller. The study of Majcen et al. (2013a, 2013b) has proved this, and showed that even in very energy-efficient buildings actual gas consumption can exceed the predicted levels. These findings are also confirmed by Galvin (2016).

Discussion: alternative policies and regulations to control energy use

Dimensions of controlling domestic energy use

As mentioned above, controlling the demand for domestic energy has various dimensions:

- *Prescription*: traditional building regulations prescribe certain performances of the building envelope and the building services. Through building permit procedures a certain level of quality control is enforced.

- *Prediction*: regulations for new buildings and energy labels for existing buildings use simulations models to predict the energy use.

- *Measurement*: the actual energy use will be determined in the use phase by measuring.

For a long time there has been little research to couple together these aspects. This has changed in recent years since the energy-saving goals have become increasingly important. Insights from data of actual energy use and the actual performances of buildings (*e.g.* the real U-values) suggest that more specific forms of quality control are needed (Cohen & Bordass, 2015). These insights also allow for an improvement of the prediction models. It has also become clear that the behaviour of the inhabitants will significantly influence on the final energy use. At present, there is little done to control energy behaviours – pricing mechanisms for energy do not appear to be an effective deterrent and also have negative consequences on the poorest segments of society.

Traditional building regulations
Developments in building regulations and control

In the Netherlands as well as in many other countries, the building regulations are a recurring subject in the debate of governmental reviews as conflicting goals are evident. On the one hand, there is a desire to minimize regulations in order to reduce the administrative burden on citizens and businesses. On the other hand, there is a need to address new societal concerns through regulation and the establishment of standards to create a baseline for practice. Energy demand reduction and climate change mitigation are such concerns. The European Union and its member states have implemented regulations and enforcement schemes that should ensure very energy-efficient new buildings and have introduced instruments to improve the energy performance of the existing building stock. Although the general deregulation trend in Europe has led to less governmental intervention in the building sector, in the field of energy efficiency the number of regulations have increased and became more stringent. The desire for deregulation has led to a transfer of some responsibility to building owners and actual quality control has been transferred from the municipalities to private parties. At the same time, there has been a shift in the quality control from the design stage to the 'as built' situation. However, the increasing importance of assuring in-use energy performance has not been reflected in the regulations.

The research centre OTB (Research for the Built Environment at Delft University of Technology, the Netherlands) has been involved in studying alternative visions on building regulatory systems in international comparative projects (*e.g.* in Meijer, Visscher and Sheridan, 2002; Meijer and Visscher, 2006, 2008; and ECORYS, Delft University of Technology, 2015). What can be noticed in most countries are discussions (or sometimes even actual developments) where the balance slowly shifts:

- From command and control regulations towards more economic incentive-based policies.

- From public control and enforcement towards a more dominant role of private parties/building professionals (together with the creation of far more robust and reliable certification and accreditation schemes).

- From a strong focus on control of the design to the monitoring of the building process and testing of the quality of the final building and post occupancy monitoring.

For a successful transition towards energy neutral construction, stricter demands must be set for the knowledge and skills of the building professionals (designers, engineers, installers, constructors, etc.). Bordass and Leaman (2013) argue that professionals need to take increased responsibility for outcomes. Professionals will have to use new techniques and improve the quality and accuracy of the work. This means that they not only will have to improve their operating procedures but also have to provide performance guarantees. Maybe the competent persons scheme in England and Wales (*e.g.* in ECORYS, Delft University of Technology, 2015) could be an interesting example for this. Owners and users will require quality guarantees from the designers, installers and constructors. Certification and accreditation of parties, processes and products will become more important for building processes in general. For the realization of high energy performance standards, a reliable quality assurance system will be very important. In most countries that have some experiences with passive houses some form of performance guarantee and associated quality assurance scheme exists. In this respect Stevenson and Leaman (2010a, 2010b) give a good overview of methods that can be used for quality control and performance measurement.

Misfits of current regulatory approaches

It is evident that with the current approach found in general regulatory instruments can only partly influence the actual energy demand in houses. Regulations only address the energy use that is (partly) related to the physical condition of the building (as appliances and inhabitants are outside their scope). However, there needs to be a shift in focus to ascertain the 'as built' quality. For example, nearly zero-energy buildings would require airtightness tests and infrared scans (to highlight any thermal bridging). The adequate functioning and the capacity of ventilation systems also needs testing. A differentiation is needed in regulations to account for in-use performance. This is because any mistake during the construction process will lead to a reduction of the minimum required performance and efficiency, thereby negatively influencing the energy demand.

Analysing the actual energy use compared with the indications of the EPCs gives a clear insight in the under-prediction of the use in houses with good labels and large over-predictions in the house with bad labels. This also leads to wrong assumptions of payback times of the investments. Strict regulations for new houses and retrofits will improve the physical performance of the building, but have a limited influence on the actual energy use.

Alternative forms of governance

Given the limitations of current building regulations, what other forms of governance could be used to reduce the domestic energy use and CO_2 emissions?

Energy performance guarantees

An innovative approach for deep energy renovations to nearly zero in the Netherlands is called the net zero-energy renovation concept (Rovers, 2014). Houses from the 1960s and 1970s with a poor energy performance are retrofitted with a new highly insulated skin, air-source heat pump heating and photovoltaic (PV) panels. The renovation process is highly industrialized and the renovation time is limited to two weeks or less. Currently these deep retrofits are mostly done to social housing (houses from housing associations). A change in governance has been influential. A new law allows the housing associations to increase the rent by the cost of the average energy bill. After the retrofit, the tenants only pay a higher rent but no energy bill at all, provided their actual energy use within a prescribed limit of reasonable use. This only works if the theoretical estimations of the actual energy demand are correct.

Concurrent with this is the development of a new contractual obligation: an energy performance guarantee by the construction company. This is a kind of energy performance contract where the owner-occupant pays for the retrofit and gets a guarantee for a zero energy bill. In principle, the increase in rent should be offset by not having to pay an energy bill. The first evaluations are appearing now (Energiesprong, 2016), but they are only based on just a few cases. Some of the occupants are satisfied, but others are dissatisfied because the energy demand concept is below their expectation. It is based on lower indoor temperatures (20°C), short times for showering and an energy-sober lifestyle. If these occupants exceed the allowed level of energy use, then they have to pay extra for it. There will be much variation among users, but a reasonable baseline for normal energy demand may act as a positive influence on behaviours and practices. The near-zero concept of houses will help to reduce the variation, but still there will remain some variation and really zero cannot be guaranteed. This suggests that in addition to the physical aspects of creating a near-zero building, there are also social aspects relating to energy demand that need to be addressed. However, the overall impact on energy and CO_2 reductions from these buildings (and the underlying regime that created them) has been significant.

Energy performance guarantees are basically a voluntary development by market parties. Until now, the underpinning governance has been supportive. The initiative was developed by an agency (Energiesprong, which translates as 'energy jump') and financially supported by the government. Recently the support programme has stopped and the expectation now is that the market should further develop it.

Economic incentives

The policies for the existing stock are largely based on the EPCs. In the first place homeowners and occupants are informed about the energy performance of a dwelling. This can influence buyers and should stimulate owners to renovate their homes. Often the incentive schemes use the concept of payback times. This is based on the argument why not invest (e.g. €10 000) if this can be earned back in 10 years by a lower energy bill? The insights presented in this paper show that this hardly works for renovations on the skin of a building to reduce the heating demand. There is a slight reduction of energy use, but the comfort level increases (higher temperatures). To have a real impact on savings the retrofit should go to the level of near zero.

Another investment strategy is in on-site renewable energy generation. This is independent from the occupant behaviour. However, it is dependent on the energy price. The drastic reduction of the oil price in 2015 illustrates this. The feed-in tariffs for electricity are set by governments and are also unpredictable. In the Netherlands, homeowners can yearly feed in 3000 kW for the same price as the price they have to pay for electricity, which includes 75% taxes. This arrangement makes it very profitable to buy PV panels. However, the government is now considering a change to this regulation in 2020. This shows that taxes and incentives can be a very strong governance tool.

Murphy, Meijer and Visscher (2012) and Murphy (2014) investigated how owner-occupiers respond to various kinds of incentives by the national and local Dutch governments. Most of these incentives were connected to EPCs and advise on making houses more energy efficient. These forms of governance had only modest success. The willingness to invest in energy renovations is still limited, especially due to many uncertainties about the reliability of the contractors and the actual energy savings as found by Galvin (2014) in Germany.

Information

Other forms of information about actual energy use seem more promising. Quarterly energy use reports of the energy companies give better insights and are related to previous year's corresponding quarter and to neighbours' energy patterns. Smart meters are nowadays installed on a large scale. In a few years most homes in the Netherlands (and in several other countries) will have one. At the same time–energy management displays are increasingly used. Smart meters and these displays can be seen and used on smart phones. The insight will increase, but is this enough to stimulate energy-saving behaviour including renovation investments? Studies (e.g. Darby, 2008) about the potential of giving accurate feedback to inhabitants about their behaviour indicate that 5–10% savings might be achieved.

Cap and trade

Some lessons from the non-domestic sector may be appropriate. According to Nishida and Hua (2011) and Nishida, Hua and Okamoto (2016), the use of a cap-and-trade scheme in Tokyo is very promising, although its application to individual dwellings seems unlikely as it requires a competence that most homeowners do not have. However, a system that integrates building performance and energy demand into a simple package might be developed for the domestic sector. A system with packages that provide a fixed amount of monthly energy usage (e.g. similar to pricing systems for mobile phones), where a standard use would have a standard price, but an increased use would have a substantially higher price deserves further investigation. Such a scheme would require a definition of what a normal or average use would be for various classes of dwellings as well as the sizes and lifestyles of households.

Conclusions

To improve this situation for new residential buildings it has to be assured that constructions and installations are installed properly and in such way that they are not vulnerable for unpredictable or misuse by the occupants. This will set new demands on both the construction industry and the control/enforcement process. The public building regulations and enforcement systems will continue to have an important role.

The improvement of the existing building stock remains a major challenge. The potential energy savings are large, but the barriers to overcome are also high. Actual energy savings in renovated dwellings are below expectations due to rebound effects and lower than expected energy use in the old dwellings. Many owners believe that the benefits of the measures do not outweigh the costs.

For all kinds of governance policies and instruments, an accurate insight into the actual performances of buildings, actual energy use, behaviour and preferences of occupants will be essential. The unexplored territory for governance is how occupants' expectations, behaviours and social practices in using energy can be changed. Rational, economic incentives do not appear to be convincing or effective. Other levers, narratives and instruments are needed to monitor and encourage a frugal approach to energy demand and its management. To ensure the success of governance strategies and instruments, the support and engagement with occupants will have a vital role. This represents a new area for (national and municipal) governments and the construction supply side. For the latter, it will necessitate new forms of engagement with occupants to demonstrate optimized use and new social practices, as well as ensuring that energy performance guarantees deliver. The creation of positive feedback loops for inhabitants will be essential.

Disclosure statement

No potential conflict of interest was reported by the authors.

References

Berkhout, P., Muskens, J., & Velthuijsen, J. (2000). Defining the rebound effect. *Energy Policy, 28*(6–7), 425–432.

Bordass, B., Cohen, R., Standeven, M., & Leaman, A. (2001). Assessing building performance in use 2: Technical performance of Probe buildings. *Building Research and Information, 29*(2), 103–113. doi:10.1080/09613210010008027

Bordass, B., & Leaman, A. (2013). A new professionalism: Remedy or fantasy? *Building Research and Information, 41*(1), 1–7. doi:10.1080/09613218.2012.750572

Bordass, B., Leaman, A., & Ruyssevelt, P. (2001). Assessing building performance in use 5: Conclusions and implications. *Building Research and Information, 29*(2), 144–157. doi:10.1080/09613210010008054

Branco, G., Lachal, B., Gallinelli, P., & Weber, W. (2004). Predicted versus observed heat consumption of a low energy multifamily complex in Switzerland based on long-term experimental data. *Energy and Buildings, 36*(6), 543–555.

Buildings Performance Institute Europe (BPIE). (2011). *Europe's buildings under the microscope.* Brussels: BPIE. Retrieved from: http://bpie.eu/wp-content/uploads/2015/10/HR_EU_B_under_microscope_study.pdf

Cayre, E., Allibe, B., Laurent, M. H., & Osso, D. (2011). There are people in this house! How the results of purely technical analysis of residential energy consumption are misleading for energy policies, *Proceedings of the European Council for an Energy Efficient Economy (ECEEE) Summer School, 6–11 June 2011,* Belambra Presqu'île de Giens, France.

Cohen, R., & Bordass, B. (2015). Mandating transparency about building energy performance in use. *Building Research and Information, 43*(4), 534–552.

Darby, S. (2008). Energy feedback in buildings: Improving the infrastructure for demand reduction. *Building Research & Information, 36*(5), 499–508. doi:10.1080/09613210802028428

ECORYS, Delft University of Technology. (2015). Simplification and mutual recognition in the construction sector under the Services Directive, European Commission. Retrieved from http://bookshop.europa.eu/.

Energiesprong. (2016). Eerste ervaringen met prestatiegarantie voor nul op de meter woningen (First experiences with performance guarantees for zero on the meter dwellings). Retrieved from http://energielinq.nl/resources/monitoringprestatiegarantie.

Filippidou, F., Nieboer, N., & Visscher, H. (2014). The pace of energy improvement in the Dutch non-profit housing stock, in *Proceedings of World Sustainable Building 2014 Barcelona Conference* (volume 3, pp. 233–239). Barcelona: Green Building Council España. ISBN 978-84-697-1815-5. Retrieved from: http://www.wsb14barcelona.org/papers.html.

Filippidou, F., Nieboer, N., & Visscher, H. (2015). Energy efficiency measures implemented in the Dutch non-profit housing sector. In *Proceedings of the ECEEE 2015 Summer Study on Energy Efficiency.* Stockholm: European Council for an Energy Efficient Economy. Retrieved from http://proceedings.eceee.org/visabstrakt.php?event=5&doc=5-093-15.

Galvin, R. (2014). Why German homeowners are reluctant to retrofit. *Building Research & Information, 42*(4), 398–408.

Galvin, R. (2015). Integrating the rebound effect: Accurate predictors for upgrading domestic heating. *Building Research & Information, 43*(6), 710–722.

Galvin, R. (2016). *The rebound effect in home heating: A guide for policymakers and practitioners.* Abingdon: Earthscan from Routledge.

Gram-Hanssen, K. (2010). Residential heat comfort practices: Understanding users. *Building Research and Information, 38*(2), 175–186. doi:10.1080/09613210903541527

Guerra Santin, O., & Itard, L. (2010). Occupant behaviour in residential buildings in the Netherlands: Determinants and effects on energy consumption for heating. *Building Research & Information, 38*(3), 318–338.

Guerra Santin, O., Itard, L., & Visscher, H. J. (2009). The effect of occupancy and building characteristics on energy use for space and water heating in Dutch residential stock. *Energy and Buildings, 41*, 1223–1232.

Hamilton, I. G., Summerfield, A. J., Lowe, R., Ruyssevelt, P., Ewell, C. C., & Oreszczyn, T. (2013). Energy epidemiology: A new approach to end-use energy demand research. *Building Research and Information, 41*(4), 482–497.

Hens, H., Parijs, W., & Deurinck, M. (2010). Energy consumption for heating and rebound effects. *Energy and Buildings, 42*(1), 105–110.

Li, F. G. N., Smith, A. Z. P., Biddulph, P., Hamilton, I. G., Lowe, R., Mavrogianni, A., ... Oreszczyn, T. (2015). Solid-wall U-values: Heat flux measurements compared with standard assumptions. *Building Research & Information, 43*(2), 238–252. doi:10.1080/09613218.2014.967977

Majcen, D., Itard, I., & Visscher, H. J. (2013a). Theoretical vs. Actual energy consumption of labelled dwellings in the Netherlands: Discrepancies and policy implications. *Energy Policy, 54*, 125–136.

Majcen, D., Itard, L., & Visscher, H. J. (2013b). Actual and theoretical gas consumption in Dutch dwellings: What causes the differences? *Energy Policy, 61*, 460–471.

Majcen, D., Itard, L., & Visscher, H. J. (2015). Statistical model of the heating prediction gap in Dutch dwellings: Relative importance of building, household and behavioural characteristics. *Energy and Buildings, 105*, 43–59.

Meijer, F. M., & Visscher, H. J. (2006). Deregulation and privatisation of European building-control systems. *Environment and Planning B: Planning and Design, 33*(4), 491–501.

Meijer, F. M., & Visscher, H. J. (2008). Building regulations from an European perspective. In *Proceedings of COBRA 2008 – The construction and building research conference of the Royal Institution of Chartered Surveyors*, RICS, London.

Meijer, F. M., Visscher, H. J., & Sheridan, L. (2002). *Building regulations in Europe, part I: A comparison of systems of building control in eight European countries.* Housing and Urban Policy Studies 23. Delft: Delft University Press.

Murphy, L. C. (2014). The influence of energy audits on the energy efficiency investments of private owner-occupied households in the Netherlands. *Energy Policy, 65*, 398–407.

Murphy, L. C., Meijer, F. M., & Visscher, H. J. (2012). A qualitative evaluation of policy instruments used to improve energy performance of existing private dwellings in the Netherlands. *Energy Policy, 45*, 459–468.

Nieman. (2007). *Final report housing quality indoor environment in new built dwellings.* Arnhem: Vrom inspectie Regio Oost.

Nishida, Y., & Hua, Y. (2011). Motivating stakeholders to deliver change: Tokyo's Cap-and-Trade Program. *Building Research and Information, 39*(5), 518–533.

Nishida, Y., Hua, Y., & Okamoto, N. (2016). Alternative building emission-reduction measure: Tokyo Cap-and-Trade Programme (TCTP). *Building Research and Information, 44*(5–6). doi:10.1080/09613218.2016.1169475

Raslan, R., & Davies, M. (2012). Legislating building energy performance: Putting EU policy into practice. *Building Research and Information, 40*(3), 305–316. doi:10.1080/09613218.2012.681440

Rasooli, A., Itard, L., & Ferreira, C. I. (2016). A response factor-based method for the rapid In-situ determination of wall's thermal resistance in existing buildings. *Energy and Buildings, 119*, 51–61.

Rovers, R. (2014). New energy retrofit concept: 'Renovation trains' for mass housing. *Building Research and Information, 42*(6), 757–767.

Sorrell, S., Dimitropoulos, J., & Sommerville, M. (2009). Empirical estimates of the direct rebound effect: A review. *Energy Policy, 37*(4), 1356–1371. April 2009.

Stevenson, F., & Leaman, A. (2010a). Evaluating housing performance in relation to human behavior: New challenges. *Building Research and Information, 38*(5), 437–441.

Stevenson, F., & Leaman, A. (2010b). Developing occupancy feedback from a prototype to improve housing production. *Building Research and Information, 38*(5), 549–563.

Sunikka-Blank, M., & Galvin, R. (2012). Introducing the prebound effect: The gap between performance and actual energy consumption. *Building Research and Information, 40*(3), 260–273. doi:10.1080/09613218.2012.690952

Tigchelaar, C., Daniëls, B., & Maenkveld, M. (2011). *Obligations* in the existing housing stock: Who pays the bill?, *Proceedings of the European Council for an Energy Efficient Economy*, ECEEE Summer School, 6–11 June 2011, France.

Van der Heijden, J., Meijer, F., & Vischer, H. (2007). Problems in enforcing Dutch building regulations. *Structural Survey, 25*(3-4), 319–329.

WoON Energie. (2009). Woononderzoek Nederland, module energie, VROM.

RESEARCH PAPER

Energy efficiency and the policy mix

Jan Rosenow, Tina Fawcett, Nick Eyre and Vlasis Oikonomou

Energy efficiency policy is expected to play a key role for meeting the European Union's energy targets (particularly for reduced energy demand and reduced CO_2 emissions) using a range of policy instrument combinations. However, most analyses undertaken so far have focused on single-policy measures rather than developing a more generic framework for assessing to what extent a particular policy mix is effective and under which specific conditions. This paper both contributes to the theoretical literature on policy mixes and undertakes an empirical analysis of the current policy mixes in buildings efficiency policy in 14 European Union countries. Building on the existing literature, and using expert knowledge, an assessment of the interaction of 55 pairs of policies is presented. This identifies policy mixes likely to deliver more, less or the same energy savings in combination than singly. The theoretical assessment is compared with actual policy mixes present within the European Union, highlighting that combinations of multiple financial incentives may need further investigation. By bringing these forms of knowledge together, the paper suggests how buildings policy mixes could be made more effective, shows gaps in current knowledge and highlights key research needs.

Introduction

Buildings energy-efficiency policy is expected to play a key role for meeting the European Union's (EU) energy efficiency targets. EU member states have a common policy goal under the Energy Efficiency Directive (EED) (Rosenow et al., 2015) to deliver 20% efficiency improvement by 2020, most of which will be via efficiency improvements and reduced energy demand in buildings. Each member state faces different starting conditions and a different set of savings opportunities, influenced by history, geography, the nature of the current building stock, available fuels and energy-conversion technologies, and different cultural expectations and practices relating to thermal comfort. However, the range of technologies and techniques available to deliver efficiency savings is largely common across the EU. Member states and other policy actors face a choice of policies and policy mixes to try to deliver these savings.

A better understanding the effectiveness of different policy mixes is of practical as well as theoretical concern. In order to meet their efficiency targets, many member states are introducing additional policies into an often-already crowded policy space (ENSPOL, 2015a, 2015b) resulting in an increasing policy heterogeneity (Costantini, Crespi, & Palma, 2015). Recent analysis suggests that the National Energy Efficiency Action Plans (which describe how member states will meet their EED savings targets) put forward to date may not be adequate, and further policies may be required (Rosenow et al., 2015). As efficiency targets continue to become more stringent, the need for a well-functioning policy mix will also increase.

The importance of policy mixes in energy-efficiency policy has long been recognized, given the variety of instruments needed to overcome different 'barriers' or to support different technologies at various stages of development (Geller & Nadel, 1994; Uyterlinde & Jeeninga, 1999). Policy-making is becoming increasingly complex as power is redistributed from the national level to supra- and sub-national actors, but also outwards to quasi-state actors and non-state actors (Flanagan, Uyarra, & Laranja, 2011). This multilevel and multi-actor governance increasingly requires policy mixes to be designed to take into account decisions at other levels in order to achieve policy goals (Betsill & Bulkeley, 2006). However, most analyses performed so far have focused on single policy measures. There is currently a lack of research that describes the energy-efficiency policy mixes used in different countries or analyses complementarities and trade-offs between policy instruments.

In this paper the aim is both to contribute to the theoretical literature on policy mixes and to undertake an empirical analysis of the current policy mixes in buildings efficiency policy in 14 EU countries. By bringing these forms of knowledge together, it is possible to suggest how buildings policy mixes could be made more effective, show gaps in current knowledge and highlight key research needs. Building on the existing literature, a classification of current policy instruments is developed, and their interactions with each other in pairs described. The empirical analysis is based on a theory-based appraisal of policy instrument combinations as well as a survey carried out across 14 European member states. The empirical part of the paper illustrates which policy instrument combinations are most common rather than how effective they are. Such an undertaking would require data from ex-post evaluations using similar methodologies carried out in the countries under investigation. Whilst there are some ex-post evaluations for individual policy instruments (Wade & Eyre, 2015), the current authors are not aware of *ex-post* evaluations of policy mixes that provide insight into the effectiveness of different policy instrument combinations. Furthermore, the methodologies of existing evaluations are not consistent across Europe, which makes a comparative assessment of effectiveness very difficult, if not impossible. Hence, the intention of this paper is to develop a theory-based appraisal of hypothetical policy instrument combinations that indicates their potential effectiveness. This is contrasted with how policies are actually combined with the aim to understand how common are specific combinations and whether they are the ones expected to be most effective based on the theory.

The paper is structured as follows. First, the literature on the policy mix is reviewed, in particular with regard to energy efficiency. Second, a theory-based appraisal of policy instrument combinations is presented. Third, the results of the survey including which types of instruments are used and which instrument combinations are prevalent is presented. Finally, findings are discussed followed by conclusions.

Literature review
Public policy and the policy mix
The term 'policy mix' first emerged in the economic policy literature in the 1960s dealing with the relationship and interaction between fiscal and monetary policy (Mundell, 1962). It was mainly used within the economic policy literature until the late 1980s and early 1990s when the concept started being discussed by political scientists (Flanagan et al., 2011). Since then there has been an increasing use in other fields of public policy, particularly in environmental policy (*e.g.*, Fankhauser, Hepburn, & Park, 2010; Gunningham & Sinclair, 1999; Sorrell & Sijm, 2003).

Generally, there are two types of literature on the policy mix. First, there is a rich body of evidence on how policy, including the policy mix, emerges and changes over time (for a compendium of the most prominent theoretical approaches, see Sabatier, 2007). Second, there is an emerging literature on how to design an effective policy mix (Borras & Edquist, 2013; Organisation for Economic Co-operation and Development (OECD), 2010; Oikonomou & Jepma, 2008; Rogge & Reichardt, 2013). This paper focuses on the latter subject, and that literature is described below.

In an idealized world where policy-makers consider the optimal policy mix to address an issue area, three steps are involved in designing the policy mix (Borras & Edquist, 2013): (1) a primary selection of the specific instruments most suitable among the wide range of different possible instruments; (2) the concrete design and/or 'customization' of the instruments for the context in which they are supposed to operate; and (3) the design of an instrument mix, or set of different and complementary policy instruments, to address the problems identified. The underlying principle is the assumption for determining interactions originating from the theory of economic policy formulation, as described by Tinbergen (1952, 1954). The core assumption of the theory is that policy formulation should in principle target the maximization of the social welfare function, which can be replaced by prescribing fixed values of some variables and attribute them as targets. Tinbergen's theorem states that there should only be only one policy instrument per target in order to avoid redundancies in the policy framework.

In reality, the process of 'designing' policy mixes is much more complex and by definition inherently political (Bressers & O'Toole, 2005; Howlett & Rayner, 2013; Rogge & Reichardt, 2013); energy-efficiency policy is no exception (Varone & Aebischer, 2001). Interpretive flexibility of policy instrument types (Bijker, Hughes, & Pinch, 1989) leads to implementation in many different ways (*e.g.*, Slembeck, 1997). Policy instruments are often selected in an ad-hoc manner responding to issues that occur at the time of decision-making. Furthermore, policy instruments emerge depending on the political dynamics within a country with the potential of having unintended consequences which in turn impact on the development of the policy instrument itself (Béland, 2010). Policy mixes are usually not designed but emergent (Cunningham, Edler, Flanagan, & Laredo, 2013) and interactions between the different policy instruments in the mix are characterized by both complementarities and tensions resulting, for example, from conflicting goals in public policy.

Given the complex nature of policy mix formation, how can theoretical understandings best engage with this reality? Firstly, this paper follows Flanagan et al. (2011) in assessing the degree of 'optimality' of different policy mixes, following economic definitions of optimality, and it faces significant difficulties. A recent review within the context of climate policy mixes (including energy efficiency) supports this view (Görlach, 2013). Rather, evaluating which combinations are associated with synergies or trade-offs is more promising, and this paper builds on an existing body of literature in this tradition.

A number of studies have developed thinking on interactions between policy instruments. Recent work by the OECD (2010) emphasizes the coherence and appropriateness of the policy mix. An extensive literature review by Rogge and Reichardt (2013) concludes that coherence goes beyond consistency (the absence of contradictions) by focusing on synergies. Gunningham and Sinclair (1999) have developed typologies of different kinds of policy mixes: (1) mixes that are inherently complementary; (2) mixes that are inherently incompatible; (3) mixes that are complementary if sequenced; and (4) mixes whose complementarity or otherwise is essentially context specific. Howlett and del Río (2013) also develop policy-mix typologies proposing eight policy-mix types determined by whether or not the mix involves multiple governments, consists of multiple policies and addresses multiple goals.

The discussion above illustrates that there is an increasing theoretical understanding of the role of the policy mix that has yet to be applied to different policy domains including energy efficiency.

Energy efficiency and the policy mix
Theoretical contributions
Energy policy is probably the sector most studied regarding the policy mix and innovation (Cunningham et al., 2013), with a main focus on emissions trading schemes and renewable energy policies and, to a lesser extent, energy efficiency. However, even within this policy domain, papers analysing the policy mix rather than individual instruments are scarce. Lee and Yik (2004) and Oikonomou and van der Gaast (2008) deal explicitly with the policy mix and energy efficiency

in buildings. However, the analysis is largely based on theoretical expectations and undertaken within a cost–benefit framework, with little consideration of the interaction of policy instruments. Del Río (2010), Langniss and Klink (2006), Meran and Wittman (2008), and Perrels, Oranen, and Rajala (2006) investigate interactions between energy efficiency and renewable energy support schemes focusing on whether different support schemes and design elements lead to different interaction results. The analysis is carried out at an abstract level and considers potential theoretical policy combinations including their complementarities and trade-offs. Similar studies by Oikonomou, Flamos, and Grafakos (2010, 2014) assess renewable and energy efficiency policy as well as climate and energy instrument combinations against a number of criteria. However, the level of detail on combining different energy-efficiency policy instruments does not allow for a more sophisticated understanding of the complementarities and trade-offs either. Thus, understanding the interactions between building energy efficiency policies is an area requiring further work.

Another body of literature relevant to the policy mix and energy efficiency is market transformation research. Whilst not explicitly using the concept of the policy mix, Geller and Nadel (1994) analyse the role of different policies and programmes within the context of market transformation focusing on new energy-efficiency products. They conclude that four types of policies and programmes are typically used to achieve a higher penetration of energy efficiency products: (1) research and development (R&D) to develop new energy-efficiency measures, (2) market-pull or bulk purchase programmes to facilitate commercialization, (3) financial incentives to stimulate early adopters and (4) efficiency codes and standards to eliminate inefficient technologies and practices. Tholen and Thomas (2011) as well as Höfele and Thomas (2011) propose 10 criteria for judging the quality of energy-efficiency policy packages. Whilst they make important contributions around the different policy requirements depending on the maturity of energy-efficiency technologies and the different actors involved, they do not analyse the interactions between policy instruments. In addition, while market transformation theory provides useful insights, particularly around the timing of different policy types, the market for building refurbishment is different and more complex than that for products and appliances (Killip, 2013). So findings from this theoretical perspective must be applied with caution.

Empirical analysis
Few papers analyse case studies of the policy mix within the buildings energy efficiency policy domain. For example, that by Cunningham et al. (2013) contains a high-level discussion of the energy-efficiency policy mix approach of the French Agency for

Environment and Energy Management (ADEME). Shnapp (2015) looks at policy packages for the renovation of buildings, and presents best-practice examples worldwide, where best practice is defined in relation to a number of indicators. An 'ideal' policy package was developed in two stages: (1) desktop study and (2) a review by local experts. Vringer, van Middelkoop, and Hoogervorst (2015) carried out an impact assessment of Dutch policy to reduce the energy requirements of buildings and concluded that it was not possible to establish the effectiveness and efficiency of individual policy instruments due to a lack of evaluation studies. However, they were able to assess the total policy mix using high-level empirical data.

A number of researchers have made use of the MURE policy database for EU member states (see www. odyssee-mure.eu) and an International Energy Agency (IEA) database for the wider OECD (www. iea.org/policiesandmeasures/energyefficiency). Costantini et al. (2015) carried out an analysis of the frequency and timing of policy instruments targeting energy efficiency in OECD countries showing an increasing policy heterogeneity. Using the MURE database Eichhammer, Schlomann, and Rohde (2012) presented 'coherent combinations of policy instruments' to address the upfront cost barrier to energy-efficiency improvements. Boonekamp (2006) also used the MURE database, focusing on the Netherlands when investigating the interaction effects of various energy-efficiency policy measures. The paper systematically analyses whether or not combinations of policy instruments lead to higher or lower energy savings compared with using those instruments individually.

In summary, even though energy is probably the sector most studied regarding the policy mix, the evidence on energy efficiency and the policy mix is thin. In particular, there is a lack of research that (1) describes the policy mixes used in different countries, (2) derives typologies of typical policy mixes and (3) analyses complementarities and trade-offs between policy instruments.

Definition and characterization of policy instruments

A wide range of energy-efficiency policy instruments is used to deliver energy savings from buildings. However, the categorization of those policy instruments is not consistent and differs between studies and countries. In this study, a recent classification is used that is followed by all EU member states in order to comply with the EED, as listed in Article 7:

- energy efficiency obligations (EEOs)

- energy or CO_2 taxes

- grants

- loans

- on-bill finance

- tax rebates

- regulations

- voluntary agreements

- standards and norms (that aim at improving the energy efficiency of products and services)

- energy labelling schemes

This categorization is manageable in terms of potential instrument combinations whilst providing sufficient granularity. Furthermore, member states have used it in their reporting for Article 7 compliance, meaning there is relatively consistent empirical data that can be analysed. Note that member states can only notify energy-efficiency policies under Article 7 as long as they (1) deliver end-use energy savings and (2) are additional to existing energy efficiency minimum requirements resulting from EU law. In case measures such as building regulations are used, additionality to the existing requirements in the Energy Performance of Buildings Directive needs to be proven by the member state.

To inform the policy mix appraisal, various characteristics of each policy type are described in Table 1. For each policy type, its function, the underlying theory of change and behaviour type are described. These descriptions are a shorthand for a more complex and nuanced reality, and consider only the primary energy-related policy functions and theories of change. Most policy types solely affect purchase decisions, with two potentially also influencing habitual behaviours. EEOs are classified here as purchase subsidies as they typically involve a financial contribution from the obligated parties to the overall investment cost of energy-efficiency technologies/improvements. The remainder is paid by the beneficiary. Whilst there are exceptions to this, for example, if EEOs target low-income customers (Rosenow, Platt, & Flanagan, 2013), the majority of measures delivered by EEOs are only part-funded by the obligated parties (Rohde, Rosenow, Eyre, & Giraudet, 2014). From the perspective of the beneficiary, EEOs provide them with an economic incentive to install energy-efficiency measures.

By considering the 'policy function' and 'theories of change' columns it becomes clear that different policy types can deliver very similar interventions from the end user's point of view. For example, EEOs, grants and tax rebates all offer purchase subsidies to end

Table 1 Policy types and policy functions

Policy type	Policy function	Theory of change (for the end user)	Behaviour type	Policy class
energy or CO_2 taxes	To increase the price of energy or carbon-based energy in line with the polluter-pays principle	Response to economic incentives (dependent on elasticity of demand)	Purchase and habitual	Taxation
Energy efficiency obligations	To reduce the price of energy-efficient options (UK model)	Response to economic incentives	Purchase	Purchase subsidy
Grants	To reduce the price of energy-efficient options	Response to economic incentives	Purchase	Purchase subsidy
Tax rebates	To reduce the price of energy-efficient options to taxpayers	Response to economic incentives	Purchase	Purchase subsidy
Loans	To give people/organizations access to capital so they can buy energy-efficient options	Lack of access to capital/high cost of capital as a barrier to investment	Purchase	Access to capital
On-bill finance	To give people/organizations access to capital so they can buy energy-efficient options	Lack of access to capital/high cost of capital as a barrier to investment	Purchase	Access to capital
Regulations	To set legally enforceable minimum standards of energy efficiency for products, vehicles and buildings	Inefficient options no longer available	Purchase	Minimum standards
Voluntary agreements	To set minimum or fleet-average standards of energy efficiency for products, vehicles and buildings	Inefficient options no longer available	Purchase	Minimum standards
Standards and norms	To enable other efficiency policies to work	n.a.	Purchase	Underpinning measurement standards
Energy labelling schemes	To enable individuals and organizations to take account of energy in their purchase decision-making	Relevant information/advice provided at the right time can influence choices	Purchase	Information and feedback
Information, advice, billing feedback, smart metering	To enable individuals and organizations to take account of energy in their purchase decision-making and/or habitual behaviours/practices	Relevant information/advice provided at the right time can influence choices	Purchase and/or habitual (depends on instrument)	Information and feedback

users. While these policies may differ significantly in terms of total cost, public versus private cost, equity or who delivers the policy, because the theoretical analysis concerns the effectiveness of the policy mix (see the following section), the end-user perspective is key. Thus, the concept of a 'policy class' has been developed, based on both the policy function and theory of change columns, in order to help with the theoretical appraisal of policy mixes. Policy types that deliver the same type of intervention are assigned to the same policy class. The 11 policy types listed fit into six policy classes.

Theoretical appraisal of policy mixes

Methods

To present a systematic assessment of the interactions between the 11 policy types identified, it was necessary to simplify the analysis in two key ways:

- combinations of two policy types only are considered

- their interactions are analysed in terms of the net effect on energy savings

Even just looking at pair-wise combinations, there are 55 policy combinations to consider. If all possible combinations of three policy types are considered, this would result in 160 combinations – too large a number usefully to report on. Given that in the real world multiple policies are often combined, the current analysis is somewhat limited by focusing on bipartite policy mixes. However, the general principles being applied are also relevant for multi-instrument combinations.

The criterion against which policy combinations are judged is 'effectiveness', *i.e.*, how much energy is saved in combination compared with what policies would deliver individually. The literature review suggests this may be the only assessment criterion for which there is much evidence (with, for example, an analysis based on cost-effectiveness being very difficult because the figures are often provided without the underlying assumptions such as energy savings and cost of delivery of the programme). Following Boonekamp (2006), the effects on energy savings of combined measures can be:

- savings from a combination of instrument 1, instrument 2 and instrument n < savings from all instruments individually (overlap)

- savings from a combination of instrument 1, instrument 2 and instrument n > savings from both instruments individually (complementary)

- savings from a combination of instrument 1, instrument 2 and instrument n = savings from both instruments individually (neutral)

In this analysis the policy instruments are assumed to target the same sector and technologies at the same time (*i.e.*, direct interactions as per Boonekamp's classification). Boonekamp developed a logical framework for assessing instrument combinations in terms of net savings which he applied to a set of specific policies in the Netherlands. The current authors attempt to take this analysis to a more abstract level and provide general conclusions around the combinations of different types of policies rather than specific national policy instruments.

Using empirical evidence (alone) to measure the effectiveness of policy mixes is generally understood to be very problematic, given the lack of sufficiently good monitoring and evaluation of individual policies. Thus, these analyses have been made with reference to the literature, and using expert judgement from across the Energy Saving Policies and Energy Efficiency Obligation Schemes (ENSPOL) team, which includes experts from various EU member states who have experience with energy efficiency policy mixes.

This matrix is in line with similar efforts carried out, for instance, for the formulation of the Energy and Climate Policy Interactions Tool (Grafakos, Flamos, Zevgolis, & Oikonomou, 2010; Oikonomou, Flamos, & Grafakos, 2010; Oikonomou, Flamos, Zevgolis, & Grafakos, 2012).

The literature (including Gunningham & Sinclair, 1999; Kosonen & Nicodeme, 2009; Lee & Yik, 2004; Sorrell et al., 2003) suggests that combinations of policies fulfilling the same function (for the same technology and target group) are more likely to overlap than combinations that accomplish different

functions. This is the approach followed here; policy types within the same policy class are generally assumed to overlap, with minor exceptions.

Results

The matrix in Table 2 provides a high-level overview of the potential interaction effects. Table 2 shows that all policy types have positive interaction effects with some other policy types. It also shows that standards and norms, energy-labelling schemes and information measures have a reinforcing impact on all other policy types. Combinations of policy types providing financial incentives (in the purchase subsidy or access to capital policy classes) are more problematic and are expected to overlap with each other.

A number of findings about particular combinations are now highlighted:

- Generally, energy and CO_2 taxes are compatible with all other instruments as they increase the incentives for people and organizations to use financial incentives and implement regulations to reduce their energy consumption, and to adopt more efficient technologies.

- Combining purchase subsidies (EEOs, grants, tax rebates) with a provision of access to capital measures (loans, on-bill finance) for the same technologies is likely to deliver fewer savings compared with the sum of savings when those measures are used on their own. The same beneficiary can be overpaid for the same savings.

- Combining EEOs with additional financial incentives would not increase the savings beyond what would be delivered by the obligations on their own, as the policy design includes capped savings levels. Similarly, a combination with voluntary agreements targeting the same sector are unlikely to deliver additional savings beyond the targets. This assumes that the obligations levels are not influenced by the use of these other policies.

- Information and feedback policy types coupled with all other policy types, and other types in the same class, have reinforcing effects as they help facilitate effective implementation of all other instruments. This is because they influence decision-making in a different way from other instruments, using psychological or behavioural economics mechanisms rather than economic influences.

- Without standards and norms for measuring the efficiency of energy using equipment, buildings and building components, most policy instruments would not function. They are therefore not so

Table 2 Interaction effects of energy efficiency policy types

	Energy or CO$_2$ taxes	Grants	Loans	On-bill finance	Tax rebates	Regulations	Voluntary agreements	Standards and norms	Energy-labelling schemes	Information, advice, billing feedback, smart metering
Energy-efficiency obligations (EEOs)	+	–	–	–	–	0	–	+	+	+
Energy or CO$_2$ taxes		+	+	+	+	+	+	+	+	+
Grants			–	–	–	0	0	+	+	+
Loans				–	–	0	0	+	+	+
On-bill finance					–	0	0	+	+	+
Tax rebates						0	0	+	+	+
Regulations							–	+	+	+
Voluntary agreements								+	+	+
Standards and norms									+	+
Energy-labelling schemes										+

Notes: + = Complementary (savings from a combination of policies A and B greater than the sum of savings policies A and B);
0 = neutral (savings from a combination of policies A and B greater than the sum of savings policies A and B);
– = overlapping (savings from a combination of policies A and B less than the sum of savings policies A and B).

much complementary as foundational for all policy instruments.

The analysis in Table 2 considers only policy combinations undertaken at the same time, where the single-policy goal is to maximize energy savings. In reality, policy mixes are more complex than this: they must deliver multiple goals, which may include equity, providing an economic stimulus and social cohesion.

Further reflections on these results, and their interaction with the empirical analysis which follows, are included in the discussion section below.

Appraisal of existing policy mixes for energy efficiency

The empirical part of this paper is based on information provided by EU member states to the European Commission in order to comply with Article 7 of the Energy Efficiency Directive (EED) (2012/27/EU). The EED was designed to bring the EU back on track to achieve the 20% target and it is one of key steps identified by the Communication on the Energy Efficiency Plan 2011 and the Roadmap to 2025. Previous analysis by the European Commission has shown that existing energy-efficiency policy measures would not deliver the 20% target by 2020 and leave a significant gap of more than half the required reduction (Hoos, 2012).

Article 7 of the EED requires member states to establish either EEOs or alternative policy measures to achieve new energy savings each year over the 2014–20 period, amounting to 1.5% of the baseline annual energy sales to final customers. The European Commission expects that Article 7 will deliver an impact of around 10.5% by 2020 (European Commission, 2011). This equals more than half of the 20% target set by the EED. Therefore, it is the most important Article of the EED in terms of its estimated impact.

The member states had to notify to the European Commission by 5 December 2013 their detailed plans to reach the energy savings target under Article 7. These plans included, *inter alia*, the policy measures that member states plan to adopt and their implementation methodology. Further information on member states' plans was provided in their National Energy Efficiency Action Plans (NEEAPs), which had to be provided by 30 April 2014. It is this information that was used to investigate policy instrument combinations in the 14 countries analysed.

Method
Data on policy mixes in selected EU member states were obtained from national experts working within the

ENSPOL project. The countries analysed were Austria, Belgium, Bulgaria, Denmark, Estonia, France, Germany, Greece, Italy, the Netherlands, Poland, Spain, Sweden and the UK. For the 10 most important energy-efficiency policy instruments (in terms of expected energy savings provided in the Article 7 notifications on the European Commission's website), each country expert provided information on the following:

- *Type of policy measure*: using the categories listed above.

- *Subsector*: with regard to buildings, differentiating between residential (*i.e.*, domestic buildings) and service sector including public buildings (*i.e.*, non-domestic), new or existing buildings, appliances, heating cooling, and ventilation measures.

- *Technology focus*: this element focuses on whether or not the policy instrument supports specific technologies (*e.g.*, energy-efficient windows) or energy-efficiency improvements more broadly (*e.g.*, grants for whole-house retrofits only specifying the level of energy performance required).

- *New versus existing technology*: distinguishing between new or replacement/upgrade of existing technologies.

- *Cost of supported technology*: cost includes all cost involved (capital cost and ongoing cost, if applicable) regardless of how the cost may be shared across different actors. The cost categories are relative and refer to how a specific energy-efficiency technology/measure relates to other energy-efficiency technologies/measures.

- *Complexity of supported technology*: the complexity of implementing the technology/measure supported (not the policy measure supporting it). A boiler replacement, for example, would not be very complex, whereas a whole-house retrofit is deemed highly complex.

Data collection was completed using a matrix with drop-down menus to ensure a consistent approach. The data were then aggregated and analysed with the aim of identifying different policy mixes and instrument combinations (a full description of all policies per country used to meet Article 7 targets is available in ENSPOL publications; ENSPOL, 2015a, 2015b).

Policies adopted by countries to meet their Article 7 commitments cannot include those that are already required by other EU regulations. Therefore, standards and norms, energy labels and regulations are underrepresented in Article 7 notifications compared with their actual use in the full national policy mixes.

Results: instrument choice

All 14 countries had a mix of policies, with the exception of Denmark, which reported all its savings would come from EEOs. Figures 1 and 2 show the most commonly used policy types in the residential (*i.e.*, domestic) and non-residential (*i.e.*, non-domestic) buildings sectors. These are very similar, with grants being most popular across both sectors. The main difference is the absence of voluntary agreements in the non-residential sector. In the residential sector by far the most common instrument is grants (33%), followed by regulations (17%), loans (16%) and EEOs (11%).

Standards and norms are almost non-existent, which reflects the fact that they are mandatory at the EU level and not part of national policies to meet Article 7.

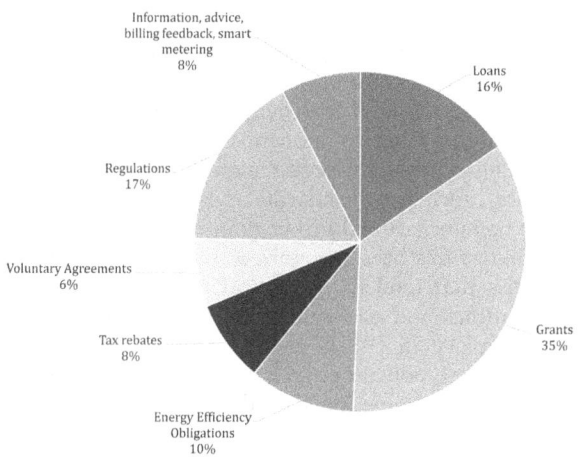

Figure 1. Most common policy instruments in the residential buildings sector

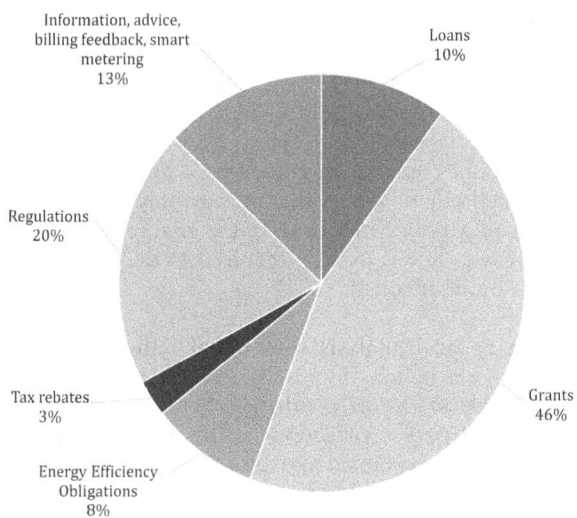

Figure 2. Most common policy instruments in the non-residential buildings sector

Results: policy type mixes

The next step of analysis is one level deeper. Interaction of policy instruments only happens in each subsector (*e.g.*, retrofits of existing residential buildings or changes to the heating system). An analysis was therefore carried out of the frequency of policy instrument combinations in each subsector within the residential and non-residential sector. For presentation purposes the results are presented across all subsectors in Table 3. In order to derive the number of policy instrument combinations, a model was used that allowed one to investigate the database of policy measures created and to identify which countries combine specific instruments to target the same subsector (*e.g.*, existing residential buildings). The figures in Table 3 represent the number of combinations within one country and in the same subsector. By restricting the combinations to the same subsector, it is possible to identify genuine combinations, *i.e.*, those mixes where policy interaction occurs.

The analysis of combinations shows the following:

- *Purchase subsidies*: there is a high frequency of purchase subsidies (provided through grants, loans, tax rebates and EEOs) combined with other policy instrument types and they are combined with all other instrument categories.

- *Voluntary agreements*: many purchase subsidies are combined with voluntary agreements.

- *Regulations*: regulations are combined with all other instrument types.

- Information measures: these are combined with many other instruments and specifically with regulations.

- *Standards and norms and energy-labelling schemes*: no combinations of those instruments with other policies were identified. This is because the EED does not allow member states to include policy measures which are non-additional as they are already a requirement under EU law.

- *Taxation*: measures are not combined with many other measures primarily because taxation is not used as much as other policy instrument types for compliance with Article 7 and because there are no taxes targeting specifically the energy consumption of buildings (most taxes are cross-cutting).

- *On-bill finance*: no combinations because on-bill finance is not used by many EU member states on a large enough scale to be a 'top 10' policy.

Table 3   Combinations of policy instruments across the building sector (only if in the same subsector)

	Grants	Loans	Tax rebates	Regulations	Voluntary agreements	Information, advice, billing feedback, smart metering
Energy-efficiency obligations (EEOs)	4	4	1	1		1
Grants		9	7	9	6	4
Loans			2	7	6	
Tax rebates				1	1	
Regulations					2	6

Discussion

This paper has contributed to thinking on policy mixes by developing and applying a method for considering policy interactions in theory, and by gathering and analysing empirical data on policy mixes for building energy efficiency across the EU. To assist the theoretical analysis, the concept of 'policy class' was developed. This was helpful to the authors when thinking through all 55 combinations systematically.

Given the simplifications made to undertake the theoretical analysis, particularly the need to look at one success criterion only, and to disregard many important contextual factors, it would be wrong to overclaim its usefulness. Factors not considered include context and calibration of individual policies, which are different in each country, and will affect energy savings as a single policy and in combination. Maximizing effectiveness of a policy mix misses out many other possible important policy goals – *e.g.*, cost-effectiveness or equity – and prioritizing these might lead to quite different policy mix decisions. Nevertheless, this systematic analysis offers a clear way of thinking about policy combinations, and identifies areas of potential under-performance. It provides a foundation for further, more detailed analysis.

Building on the existing literature, definitions for three possible interactions between policies – complementary, neutral and overlapping – were developed. These were used to analyse all possible two-way interactions between policy instruments. This analysis has showed which combinations are likely to increase policy effectiveness, decrease it or be neutral. All other things being equal, decision-makers should favour policies that are complementary, and try to avoid those that overlap. A number of policies appear always to interact in a complementary way with others: energy or CO_2 taxation, standards and norms, and those in the information and feedback class. There is a relatively small number of policy instruments combinations that can deliver less than the sum of their parts – and these are largely feature

policy types from the purchase subsidy and access to capital classes – policies largely under control of national governments, and which feature heavily in their Article 7 compliance plans.

The universally complementary policies, with the exception of taxation, are in many cases already in place at an EU level for energy-using products, buildings and building components. This includes energy-labelling schemes, a requirement to introduce smart meters, and test standards and procedures (which may be international rather than just EU level). Their usefulness has been recognized by policy-makers. The policies that tend to be neutral in their interactions, regulations and voluntary agreements also have a strong place in EU as well as national-level policy. Where these policies are missing for sectors or subsectors, their introduction should be considered.

While theoretical analysis suggests carbon or energy taxation would be complementary with all other policy types, countries take very different views on taxation of energy of different types and across different sectors (EUROSTAT, 2015). Some countries have high rates of taxation, while others are as low as EU legislation allows. Theoretically useful policies can be politically unacceptable (*e.g.*, the history of public opposition to household energy taxation in the UK (Dresner, Jackson, & Gilbert, 2006), or not fit with other policy goals. This illustrates one weakness of this method, which is that it can only consider the effect of a policy mix on a single goal (effectiveness), whereas policy is usually required to deliver multiple goals simultaneously.

The empirical analysis shows that nation-states rely strongly on policies in purchase subsidy and access to capital classes to deliver their Article 7 commitments. This is in part a reflection of the rules surrounding Article 7 (as explained above). Given that theory indicates that these combinations are at risk of overlapping, a more detailed examination of policy mix design could be helpful to nation-states in enabling them to deliver their targets. With the exception of

Denmark, all have chosen a policy mix, and it would be essential to understand the extent to which their policy mixes are designed to meet multiple goals.

One potential explanation for the use of EEOs with potentially overlapping instruments such as grants, finance mechanisms and voluntary agreements is that the scale of obligations may be dependent on the existence of other mechanisms. There is clearly overlap if other mechanisms simply help obligated parties deliver their obligations. However, in most cases the obligations will be the outcome of a policy negotiation with other key actors. If the scale of obligation can be increased by the use of other mechanisms in parallel, it may be more effective to have a combination. The extent to which this explains specific policy mixes in individual countries can only be determined by detailed case studies.

In addition, the scope of Article 7 is largely limited to policy instruments at the national level (although some member states have notified policy instruments at the regional level as well) and not all national policies are notified. At the regional and local governance level often a range of policy mixes are employed and combined with national policies (Van der Heijden, 2014). This means that other studies focusing on areas outside the scope of Article 7 are likely to find different policy mixes and effects.

Using the theoretical approach here, it is difficult to analyse explicitly combinations of more than two policy types for all combinations given the way in which the numbers escalate for combinations of three or more. However, there are ways to expand the analysis usefully. Firstly, using the policy class concept, the analysis could extend to a mix of four or five policy instruments, provided they originate from just two policy classes. Alternatively, expanding on the empirical analysis in Table 3, key combinations of more than two policies could be identified, and theoretical insights brought to bear on these mixes. The focus could be on mixes at risk of overlap to help identify risks of underachieving savings targets.

Conclusions

This paper has contributed to an emerging debate around the role of the policy mix within innovation studies and the energy-efficiency literature. Whilst the theoretical conceptualization of the policy mix has recently advanced, there are still few empirical studies accessing policy mixes within specific policy domains. This paper fills that gap for energy-efficiency policy in buildings. Developing a method based on the literature, combined with expert judgement, 55 different combinations of policy instrument types were assessed for their effectiveness.

Combinations that deliver more, less and the same as the sum of individual policies were identified. This analysis was based on the concept of effectiveness, which is of key importance in policy-making. However, theoretical analysis necessarily has to simplify reality, and is unable to incorporate the full complexity of multiple goals and contextual factors that affect national policy mixes.

In addition, empirical data were gathered and then used to describe and analyse the policy mixes in 14 EU member states. Reflections on how reality matches the policy combinations' theoretical analysis suggests were presented. The current buildings energy-efficiency policy mixes are dominated by combinations of purchasing subsidies providing a financial incentive to end-users to adopt more energy-efficient technologies. Theoretical analysis suggests that combining such instruments focusing on the same segment is likely to deliver fewer savings than using them individually.

Future research should analyse why certain policy instrument combinations are particularly prevalent, and what the political dynamics behind adopting those are, especially where the combinations adopted appear to have potential overlaps. Furthermore, research on the precise calibration of blends of purchasing subsidies in different countries is needed to understand better under which conditions such combinations are effective and why they may (or may not) be desirable, even though theoretically one would expect this is not necessarily to be the case. An important issue for energy-efficiency policy for buildings will also be the inclusion of the operation phase to ensure the potential energy savings from an interventions (*e.g.*, a technological change) are realized in practice and maintained over the long-term. Finally, the increasing Europeanization of energy-efficiency policy and its effect on the policy mix in EU member states constitutes a promising avenue for future research.

Acknowledgements

The authors are grateful to the Energy Saving Policies and Energy Efficiency Obligation Schemes (ENSPOL) project team for participating in the survey and for providing detailed information on the policy mix in European Union member states.

Disclosure statement

No potential conflict of interest was reported by the authors.

Funding

Parts of this paper are based on a report produced for the ENSPOL project Energy Saving Policies and Energy Efficiency Obligation Schemes funded by the European Commission (contract number IEE/13/824/SI2.675067).

References

Béland, D. (2010). Reconsidering policy feedback: How policies affect politics. *Administration & Society*, 42, 568–590. doi:10.1177/0095399710377444

Betsill, M. & Bulkeley, H. (2006). Cities and the multilevel governance of global climate change. *Global Governance*, 12, 141–159.

Bijker, W., Hughes, T., & Pinch, T. (Ed.). (1989). *The social construction of technological systems*. Cambridge, MA: MIT Press.

Boonekamp, P. G. M. (2006). Actual interaction effects between policy measures for energy efficiency – A qualitative matrix method and quantitative simulation results for households. *Energy*, 31(14), 2512–2537.

Borras, S., & Edquist, C. (2013). The choice of innovation policy instruments. *Technological Forecasting and Social Change*, 80(8), 1513–1522. doi:10.1016/j.techfore.2013.03.002

Bressers, H. A., & O'Toole, L. J. (2005). Instrument selection and implementation in a networked context. In P. Eliadis, M. Hill, & M. Howlett (Eds.), *Designing government: From instruments to Governance* (pp. 132–153). Montreal: McGill-Queens University Press.

Costantini, V., Crespi, F., & Palma, A. (2015). *Characterizing the policy mix and its impact on eco-innovation in energy-efficient technologies*. SEEDS Working Paper No. 11/2015.

Cunningham, P., Edler, J., Flanagan, K., & Laredo, P. (2013). *Innovation policy mix and instrument interaction: A review*. NESTA Working Paper No. 13/20.

Del Río, P. (2010). Analysing the interactions between renewable energy promotion and energy efficiency support schemes: The impact of different instruments and design elements. *Energy Policy*, 38, 4978–4989. doi:10.1016/j.enpol.2010.04.003

Dresner, S., Jackson, T., & Gilbert, N. (2006). History and social responses to environmental tax reform in the United Kingdom. *Energy Policy*, 34(8), 930–939. doi:10.1016/j.enpol.2004.08.046

Eichhammer, W., Schlomann, B., & Rohde, C. (2012). *Financing the energy efficient transformation of the building sector in the EU*. Paris: ADEME.

ENSPOL. (2015a). *Status quo of existing and planned energy efficiency obligation schemes*. Brussels: ENSPOL.

ENSPOL. (2015b). *Alternative schemes to EEOs*. Brussels: ENSPOL.

European Commission. (2011). Impact Assessment accompanying the document Directive of the European Parliament and of the Council on energy efficiency and amending and subsequently repealing Directives 2004/8/EC and 2006/32/EC {COM(2011) 370 final} {SEC(2011) 780 final}.

EUROSTAT. (2015). *Energy price statistics*. Retrieved August, 2015, from http://ec.europa.eu/eurostat/statistics-explained/index.php/Energy_price_statistics#Electricity_prices_for_household_consumers

Fankhauser, S., Hepburn, C., & Park, J. (2010). Combining multiple climate policy instruments: How not to do it. *Climate Change Economics*, 1(3), 209–225. doi:10.1142/S2010007810000169

Flanagan, K., Uyarra, E., & Laranja, M. (2011). Reconceptualising the 'policy mix' for innovation. *Research Policy*, 40(5), 702–713. doi:10.1016/j.respol.2011.02.005

Geller, H., & Nadel, S. (1994). Market transformation strategies to promote end-use efficiency. *Annual Review of Energy and the Environment*, 19, 301–346. doi:10.1146/annurev.eg.19.110194.001505

Görlach, B. (2013). *What constitutes an optimal climate policy mix?* Berlin: Ecologic Institute.

Grafakos, S., Flamos, A., Zevgolis, D., & Oikonomou, V. (2010). Multi-criteria analysis weighting methodology to incorporate stakeholders' preferences in energy and climate policy interactions. *International Journal of Energy Sector Management*, 4(3), 434–461. doi:10.1108/17506221011073851

Gunningham, N., & Sinclair, D. (1999). Regulatory pluralism: Designing policy mixes for environmental protection. *Law & Policy*, 21, 49–76. doi:10.1111/1467-9930.00065

Hoos, E. (2012). Energy Efficiency Directive: Making energy efficiency a priority. Unit C3 Energy Efficiency DG Energy, European Commission 5/10/2012 Stockholm GEODE Conference.

Howlett, M., & del Río, P. (2013, June). *Policy portfolios and their design: A meta-analysis*. Paper presented at the 1st International Conference on Public Policy Grenoble, France.

Howlett, M., & Rayner, J. (2013). Patching vs packaging in policy formulation: Assessing policy portfolio design. *Politics and Governance*, 1(2), 170–182. doi:10.17645/pag.v1i2.95

Höfele, V., & Thomas, S. (2011). Combining theoretical and empirical evidence: Policy packages to make energy savings in buildings happen. *Proceedings of the ECEEE Summer Study*, 1321–1327.

Killip, G. (2013). Transition management using a market transformation approach: Lessons for theory, research, and practice from the case of low-carbon housing refurbishment in the UK. *Environment and Planning C: Government and Policy*, 31(5), 876–892. doi:10.1068/c11336

Kosonen, K., & Nicodeme, G. (2009). *The role of fiscal instruments in environmental policy*. Taxation Papers. Working paper No. 19 2009. Luxembourg: Office for Official Publications of the European Communities.

Langniss, O., & Klink, J. (2006). White Certificate schemes and (national) green certificate schemes. EIE/04/123/S07.38640, prepared for the 'EUROWHITECERT' project for the European Commission Intelligent Energy Europe program.

Lee, W. L., & Yik, F. W. H. (2004). Regulatory and voluntary approaches for enhancing building energy efficiency. *Progress in Energy and Combustion Science*, 30, 477–499. doi:10.1016/j.pecs.2004.03.002

Meran, G., & Wittman, N. (2008). *Green, Brown, and now White certificates: Are three one too many? A micromodel of market interaction*. Berlin: Deutsches Institut fur Wirtschaftsforschung.

Mundell, R. (1962). The appropriate use of monetary and fiscal policy for internal and external stability. *IMF Staff Papers*, 9(1), 70–79.

Oikonomou, V., Flamos, A., & Grafakos, S. (2010). Is blending of energy and climate policy instruments always desirable? *Energy Policy*, 38(8), 4186–4195. doi:10.1016/j.enpol.2010.03.046

Oikonomou, V., Flamos, A., & Grafakos, S. (2014). Combination of energy policy instruments: Creation of added value or overlapping? *Energy Sources, Part B: Economics, Planning, and Policy*, 9(1), 46–56. doi:10.1080/15567241003716696

Oikonomou, V., Flamos, A., Zeugolis, D., & Grafakos, S. (2012). A qualitative assessment of EU policy interactions. *Energy*

Sources, Part B: Economics, Planning, and Policy, 7(2), 177–187. doi:10.1080/15567240902788996

Oikonomou, V., & van der Gaast, W. (2008). Integrating joint implementation projects for energy efficiency on the built environment with white certificates in the Netherlands. *Mitigation and Adaptation Strategies for Global Change,* 13(1), 61–85. doi:10.1007/s11027-007-9081-x

Oikonomou, V. & Jepma, C. (2008). A framework on interactions of climate and energy policy instruments. *Mitigation and Adaptation Strategies for Global Change,* 13(2), 131–156. doi:10.1007/s11027-007-9082-9

Organisation for Economic Co-operation and Development (OECD). (2010). The innovation policy mix. In OECD (Ed.), *OECD science, technology and industry outlook 2010* (pp. 251–279). Paris: OECD.

Perrels, A., Oranen, A., & Rajala, R. (2006). White Certificates and interactions with other policy instruments. EIE/04/123/S07.38640, prepared for the 'EUROWHITECERT' for the European Commission Intelligent Energy Europe program.

Rogge, K. S., & Reichardt, K. (2013). *Towards a more comprehensive policy mix concept for environmental technological change: A literature synthesis.* Working Paper. Karlsruhe: Fraunhofer ISI.

Rohde, C., Rosenow, J., Eyre, N., & Giraudet, L.-G. (2014). Energy saving obligations—cutting the Gordian Knot of leverage? *Energy Efficiency,* 8(1), 129–140. doi:10.1007/s12053-014-9279-1

Rosenow, J., Forster, D., Kampman, B., Leguijt, C., Pato, Z., Kaar, A.-L., & Eyre, N. (2015). Study evaluating the national policy measures and methodologies to implement Article 7 of the Energy Efficiency Directive. Study for the European Commission.

Rosenow, J., Platt, R., & Flanagan, B. (2013). Fuel poverty and energy efficiency obligations – A critical assessment of the supplier obligation in the UK. *Energy Policy,* 62, 1194–1203. doi:10.1016/j.enpol.2013.07.103

Sabatier, P. A. (2007). *Theories of the policy process.* Boulder, CO: Westview Press.

Shnapp, S. (2015, June). *Reducing energy demand in existing buildings: benchmarking tool for best practice renovation policies.* Proceedings of ECEEE 2015 Summer Study on Energy Efficiency, Toulon/Hyères, France.

Slembeck, T. (1997). The formation of economic policy: A cognitive–evolutionary approach to policy-making. *Constitutional Political Economy,* 8(3), 225–254. doi:10.1023/A:1009027913985

Sorrell, S., & Sijm, J. (2003). Carbon trading in the policy mix. *Oxford Review of Economic Policy,* 19(3), 420–437. doi:10.1093/oxrep/19.3.420

Sorrell, S., Smith, A., Betz, R., Walz, R., Boemare, C., Quirion, P., … Pilinis, C. (2003). *Interaction in EU climate policy.* Brighton: SPRU.

Tholen, L., & Thomas, S. (2011). *Combining theoretical and empirical evidence: Policy packages to make energy change in appliances happen.* EEDAL 2011 Session 2.

Tinbergen, J. (1952). *On the theory of economic policy.* Amsterdam: North Holland.

Tinbergen, J. (1954). *Centralization and decentralization of economic policy.* Amsterdam: North Holland Publishing.

Uyterlinde, M., & Jeeninga, H. (1999). Evaluation of energy efficiency policy instruments in households in five European countries. In Proceedings of the ECEEE Summer Study. Retrieved May 1, 2015, from http://www.eceee.org/library/conference_proceedings/eceee_Summer_Studies/1999/Panel_1/p1_20/paper

Van der Heijden, J. (2014). *Governance for sustainability and resilience: Responding to climate change and the relevance of the built environment.* Cheltenham: Edward Elgar.

Varone, F., & Aebischer, B. (2001). Energy efficiency: The challenges of policy design. *Energy Policy,* 29, 615–629. doi:10.1016/S0301-4215(00)00156-7

Vringer, K., van Middelkoop, M., & Hoogervorst, N. (2015, June). *An impact assessment of Dutch policy to reduce the energy requirements of buildings.* Proceedings of ECEEE 2015 Summer Study on Energy Efficiency, Toulon/Hyères, France.

Wade, J., & Eyre, N. (2015). *Energy efficiency evaluation: The evidence for real energy savings from energy efficiency programmes in the household sector.* A report by the UKERC Technology & Policy Assessment Function.

Alternative building emission-reduction measure: outcomes from the Tokyo Cap-and-Trade Program

Yuko Nishida, Ying Hua and Naomi Okamoto

The Tokyo Metropolitan Government has been actively promoting comprehensive policy measures to reduce carbon emissions from non-residential buildings. Among them, the Tokyo Cap-and-Trade Program (TCTP), implemented since 2010, is an important measure to accelerate the building sector's emission reduction to achieve Tokyo's greenhouse gas target, 25% reduction by 2020 from a year 2000 baseline level. Under TCTP, all large commercial and industrial facilities are required to achieve the 25% reduction in two compliance periods (FY2010–14 and FY2015–19). An emission-trading scheme was established that allows building owners to purchase carbon credits to compensate for shortfalls and sell excess reductions over the obligations. This paper assesses the effectiveness of TCTP based on extensive data obtained from 1300 facilities covered by the programme and surveys among facility owners from the first compliance phase. Data indicate that TCTP has been working effectively to reduce energy consumption in participating facilities to meet the ambitious emission reduction goals, to introduce new technologies, and to raise awareness and drive behavioural changes for energy demand reduction. TCTP is examined as an alternative policy instrument to building energy codes in a portfolio of sustainable building policies, highlighting its unique capacity for driving deeper and longer-term improvements for more ambitious mitigation targets.

Introduction

The Tokyo Cap-and-Trade Program (TCTP) was launched by the Tokyo Metropolitan Government (TMG) on 1 April 2010 to reduce energy consumption-related CO_2 emissions from existing buildings in the Tokyo metropolitan area. It is the world's first mandatory cap-and-trade programme focusing on buildings in commercial, industrial and public sectors. Approximately 1300 facilities within the Tokyo metropolitan area with annual energy consumption above 1500 kiloliters of crude oil equivalent are covered in this programme. The emission cap was set at 8% reduction for commercial facilities and 6% for industrial facilities and facilities using district

heating/cooling (DHC) systems for the first five-year compliance phase (FY2010–14), and 17% and 15% reductions respectively for the second phase (FY2015–19).

The baseline for reduction at each facility is the average of the annual emission levels of any three consecutive years within the period of FY2002–07 freely chosen by each facility owner. Facility owners are allowed to sell excess reductions above their annual obligation levels or purchase others' excess reductions depending on their needs. They can also purchase offset credits of four types: emission-reduction credits from small/medium facilities, credits from facilities outside the

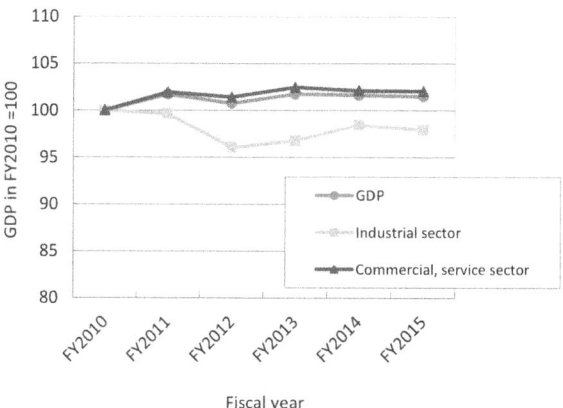

Figure 1 Gross domestic product (GDP) in Tokyo (FY2010–14)

Tokyo metropolitan area, renewable energy credits, and emission trading system (ETS) linkage with an adjacent prefecture (Saitama). Penalties for failing to meet the emission-reduction target include a fine, expenses and a surcharge for procurement of the emission-reduction gap, along with other administrative penalties. The launch of TCTP was built upon a mandatory Carbon Reduction Reporting Program in operation since 2002 that required large facility owners in the Tokyo metropolitan area to create their emission inventories and introduced them to basic emission-reduction measures. Nishida and Hua (2011) documented the design and planning process of TCTP, and summarized the effective approaches to motivate and engage stakeholders during the introduction phase of TCTP.

At the end of March 2015, TCTP completed its first five-year phase and entered the second compliance period. This paper examines the outcomes of the first compliance phase and assesses the effectiveness of a cap-and-trade mechanism for building energy consumption-related emission reduction, using data submitted by the 1300 facilities covered in the programme.

It is noteworthy that during TCTP's first compliance phase, the 11 March 2011 Tohoku Earthquake had a sizable impact on the Japanese economy and the energy supply. However, the gross domestic product (GDP) of Tokyo showed an overall trend of stabilization and recovery, reflected by the GDP trends in the commercial and service sectors (Figure 1). The total floor area of facilities covered by TCTP has also increased by 4% from their base years to FY2014.

Methods

Data from three sources are used for this analysis. The first is an annual report required by TMG from all facilities covered by TCTP. A Plan of global warming countermeasures (hereafter Plan) and a Checklist of measures (hereafter Checklist) are key submissions among the various required documents. Besides basic project information, such as facility address, name of owner, building use and floor area by use, the Plan contains Excel-format data of greenhouse gas (GHG) emissions and their chronological change, status of implemented measures for emission reduction, status of fulfilment of reduction obligation and the organizational settings for implementing measures. GHG emissions reported here were calculated from the energy consumption data from another required document. The data used in sections below entitled 'Energy efficiency improvement' and 'High-performance facilities' also come from the Plan. The information in the Plan document are in general disclosed to the public, except the status of obligation fulfilment.

Due to a strict third-party verification process, the accuracy of energy consumption and emissions data is high. All reported emissions data from mandated reduction facilities are then aggregated to calculate the total emissions. Year by year, the total number of facilities fluctuates slightly because of new entries and departures as well as some exemption cases. The exemption cases were due to significant changes in base emission level, which resulted from changes in building use, floor area, energy supply (regarding DHC plants), etc.

On the other hand, the Checklist is required to be submitted to highlight potential energy-efficiency measures and their effects. It includes a list of all equipment with their capacity and the year in which they were bought, as well as the status of their contributions to the 62 common measures identified by TMG for building energy-efficiency improvement. By completing this checklist, some of the common measures that have not yet been adopted in certain facilities are highlighted and the potential energy/emission reductions shown. The individual Checklists submitted to TMG are not disclosed; however, the aggregated results and some analysis results are reported in the feedback report. The analysis in the section below 'Mechanism to achieve reduction targets' is based on the Checklist data. Although the Checklists are fairly well completed, data accuracy may vary because verification is not required for this document.

Secondly, TMG monitors emission trading and discloses data on credits issued and transferred every month. As no exchange channel has been established for TCTP, TMG is providing price evaluation twice a year with survey results showing the market trend for those who intend to trade. The survey is outsourced to Argus Media Ltd, which evaluates prices based on interviews with the covered facilities related to trading cases and intermediary agencies. The section below entitled 'Emission trading' is based on these data and survey results.

Thirdly, TMG administered surveys targeting all facilities covered by TCTP after March 2011 to study how they coped with the energy crisis after the earthquake, in particular with the peak-power cut in summer. The survey was conducted every year from 2011 to 2014 by e-mail and the response rate was 40%, 38%, 32% and 63% for each respective year. In 2014 the survey was conducted not only to study the response to the energy crisis but also to obtain an overall view of TCTP (hereafter 'the 2014 survey'). The section below entitled 'Impact from the 2011 crisis' used the results from these surveys.

Statistics from the first five-year phase (FY2010-14) implementation of TCTP

CO₂ emission reduction
CO_2 emission reduction is the main goal of TCTP. Data showed that the five-year average reduction level is over 20% from the base-year level, and a total reduction of 14 281 000 tCO₂e has been achieved. Figure 2 shows the annual total amount and the percentage of reduction from the baseline over the five years. Compared with the programme's cap level, i.e., the total obligation of reductions for the covered facilities as a whole, this result is far beyond expectation. The target CO_2 emission reduction of 8% for the first five-year compliance phase has been achieved.

These reductions are realized in most of the covered (i.e., participating) facilities. In total, 91% of facilities have fulfilled the obligations. Figure 3 shows that since the second year of programme implementation, more than 90% of facilities have surpassed their obligations.

CO_2 emission intensity (per m² of built area) has also improved steadily. Among the building types of the covered facilities, office buildings are the most dominant. This category's emission intensity saw a 27%

drop in FY2013 compared with the base year (Figure 4). Other types of buildings have also achieved emission-intensity reductions between 10% and 30%. Although most categories showed the similar trend for steady reductions, the only exception is the information and communication centres, which increased their emission intensity in FY2010, but in FY2013 it showed a drastic reduction of 10%. This type of facility had some struggles in reducing energy consumption against the increasing demand, with seven new data centres added to the existing stock of 21 facilities.

Energy efficiency improvement
Regarding the energy efficiency of buildings, the same type of benchmarking dataset is available. Figure 5 shows the energy intensity in commercial and service sector buildings decreased for three years, and a very similar picture is indicated. A 27% improvement in energy intensity has been realized in office buildings.

In the TCTP scheme, the emission factor of fuels (including electricity) is fixed over the compliance period, so reductions in energy consumption directly account for CO_2 emission reductions. The difference between CO_2 emission reductions and energy consumption reductions comes from the introduction of renewable energy on site or from the use of the ETS. The offset amount utilizing the ETS is very limited, thus currently the only significant factor is renewable energy installations in buildings. Such differences in two major types of buildings are shown in Figure 6. The trend appears to be very similar.

High-performance facilities
Under TCTP, it is essential to set a cap, set a minimum reduction target for all facilities and its subsequent

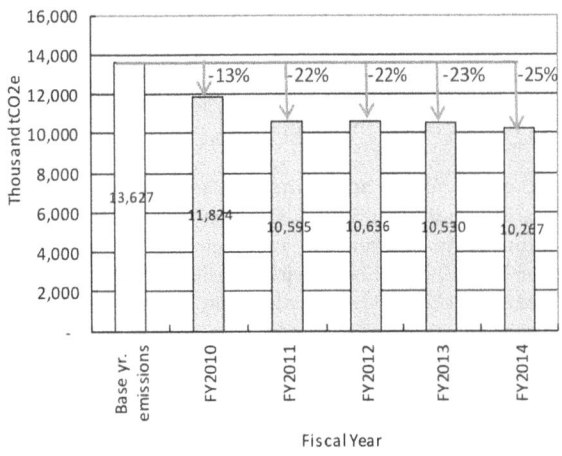

Figure 2 Total CO₂ emission reduction from covered facilities

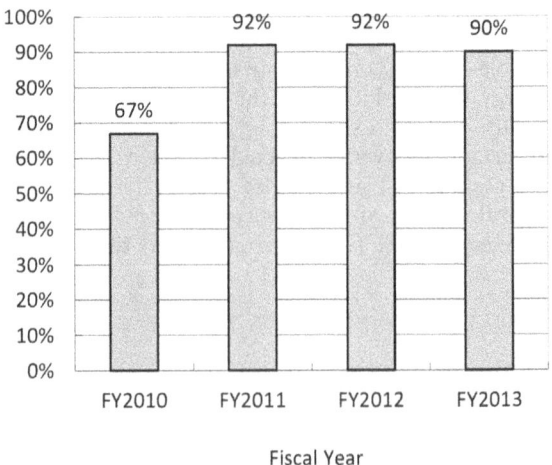

Figure 3 Percentage of facilities surpassing their obligations

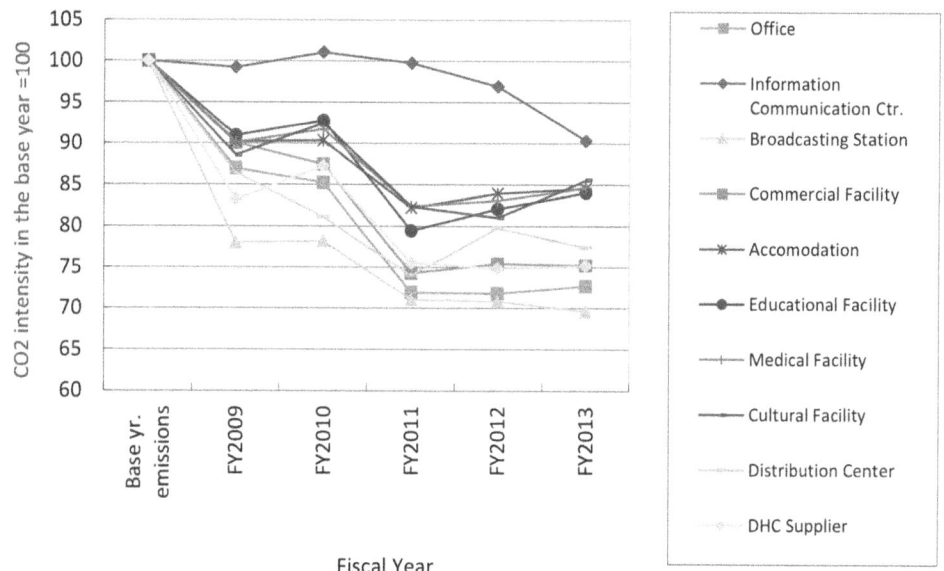

Figure 4 CO₂ emission intensity reduction by facility type

enforcement, and encouraging facilities that are already energy efficient to improve further also plays an important role.

Theoretically, the latter part of the work is supposed to be carried out through the ETS under TCTP. However, the ETS has not been heavily utilized. Instead, another

scheme unique to TCTP has been working well. The Top Level Facility Certification system (hereafter TOP system) was introduced through the stakeholder negotiation process to appreciate very high performance facilities, and enables leading developers to gain a reputation and advantages in the property (real estate) market. The certification system is very demanding

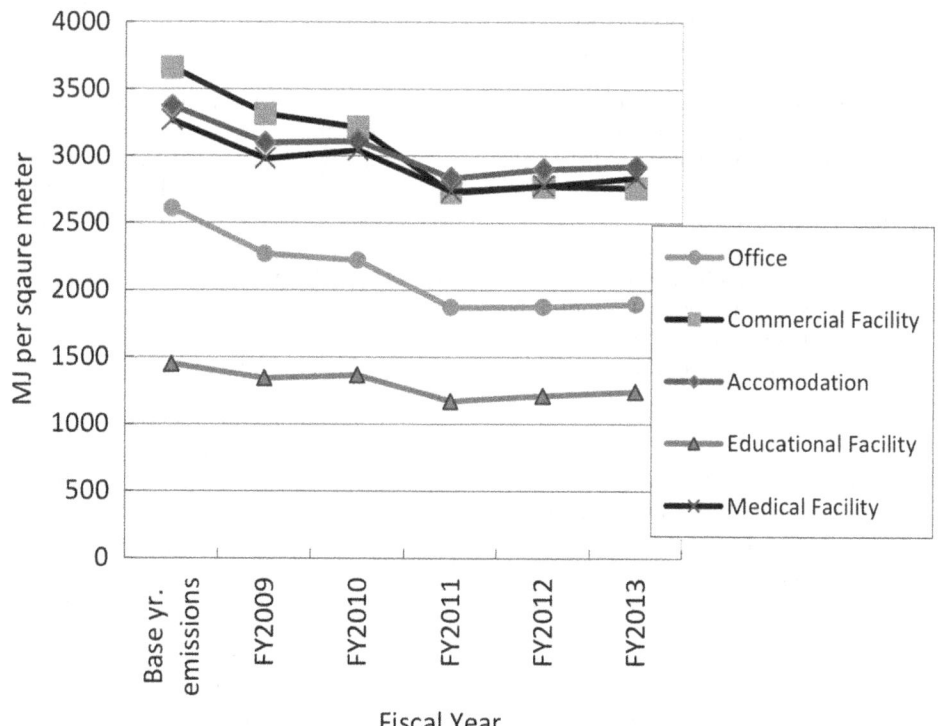

Figure 5 Energy intensity

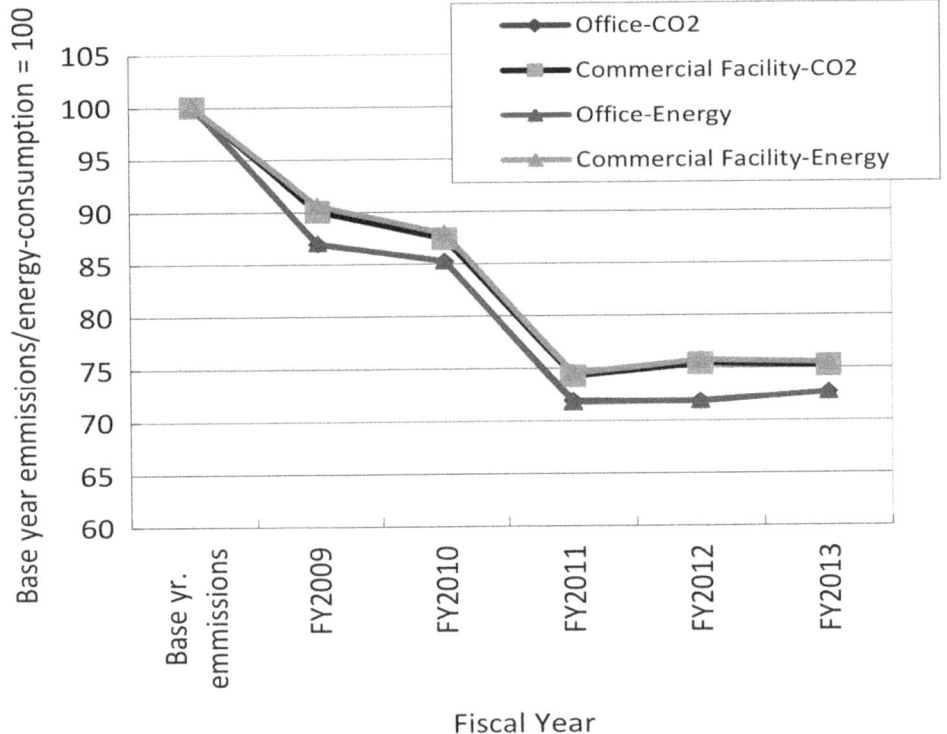

Figure 6   Comparison of energy consumption and CO_2 emission trends

with more than 100 items to be checked and verified to prove the highest level of equipment and performance. Certified facilities are allowed reduced compliance factors of 50% and 75% (for top-level and near-top-level facilities, respectively) of the reduction obligation.

By the end of FY2014, 87 facilities were certified as top-level or near-top-level facilities, and their performance was published by TMG to show recognition as well as to disseminate the advanced technologies employed and ambitious actions taken by them.

Overall, data suggest that TCTP has been working effectively as a policy tool to reduce energy consumption in existing buildings in the Tokyo metropolitan area in order to meet the ambitious emission reduction goal.

Mechanism to achieve reduction targets

This section analyzes how the reduction target was achieved. In particular, two questions are addressed: What kinds of measures have been chosen by facilities? and How have they changed over time with the impact of the energy crisis in 2011?

TMG policy context

TCTP has its roots in the mandatory reporting programme since FY2002 in Tokyo. Therefore, emission reductions have been driven by mandatory reduction

goals with supportive measures rather than by the trading scheme. The TMG implemented the mandatory reporting programme twice, for three years from 2002 and for five years from 2005, targeting the same large emitters as TCTP. The first programme mainly required facility owners to report their emission amount and the implementation status of emission-reduction measures. Disclosure of the report was only done by request. In the latter programme, reporting contents were improved and expanded, and various promotion tools for implementation were added, such as feedback with benchmarking data, strongly recommending 12 quick-payback measures, as well as rating and disclosing each facility's performance.

During this reporting term, TMG consistently valued the establishment of organizational setups in each facility to promote energy efficiency and information/data exchange between TMG and the facility through the responsible person/contact point. In the final stage, TMG required the facility to designate a general manager for climate change countermeasures, who is responsible for the promotion and implementation of measures, as well as a technical adviser (allowed to be outsourced). These managers and advisers are required to participate and complete seminars provided by TMG. For tenant buildings, TMG required each major tenant to designate a representative to promote measures. Additionally, TMG values the monitoring, assessment and rating of performance, and feedback with benchmarking data, which will be discussed below.

In the process of TCTP introductory discussions, it turned out that such an organizational approach, which values the organization and network in building management with incentives through ratings, was also favoured by businesses. On the other hand, the ETS (market mechanism) was somehow perceived with some scepticism, partly because the news from the first phase of the European Union Emissions Trading System (EU ETS) reported its difficulties, including those of allocating allowance, and caution for speculation.

Another aspect is that by designating a responsible person and maintaining the connection with TMG, facility energy managers can improve their positions in their organization and bring more visibility to their profession as an engineer. At the very least it seemed to create a better environment for them to promote measures for energy efficiency and/or conservation, setting off a trend that facility management should always serve building users (even against better energy control).

Based on such experiences in the previous programme, TCTP was designed not only to rely on market mechanisms but also to utilize continuously an organizational approach with performance disclosure and the top-level certification system. This historical context affected the design of TCTP, and also helps explain the achievement of keeping emissions below the cap.

Measures taken and being planned

Measures adopted by the covered facilities are well documented in the reports submitted by facilities. TCTP requires detailed information on what kinds of measures have been introduced since the base year, as well as those planned for the future. The effectiveness of each measure for emission reduction is calculated. Figure 7 shows that the estimated reductions by reported measures (taken/planned) have been growing steadily.

Table 1 lists the major measures reported with estimated CO_2 emission reductions in commercial facilities (excluding industrial facilities). These measures are the most frequently reported items categorized by major energy usage, although the 'other' category is very large. Covered facilities introducing measures in the category of heat source/air-conditioning (AC) during the first compliance period (FY2010–14) account for 65% of the total, and those that introduced lighting-related measures account for 75%. The measures in the two categories, heat source/AC and lighting, are relatively large in number, which is understandable because these are the common major end uses of energy in large-scale commercial buildings.

The data in Table 1 come from the narrative reporting section from the Checklist. Since the ways of describing measures taken or planned varied, and actions spread over many categories or with some of them, such as particular detailed operational measures, difficult to categorize, a significant share of measures were put into the 'Other' category in the annual data summaries.

Investment in equipment

Among the measures adopted, installation of high-efficiency equipment, including heat source, pumps, AC systems, fans and lighting fixtures, account for a large portion. As these measures are associated with certain costs, it is safe to say that a steady investment was created to install high-energy-efficiency equipment.

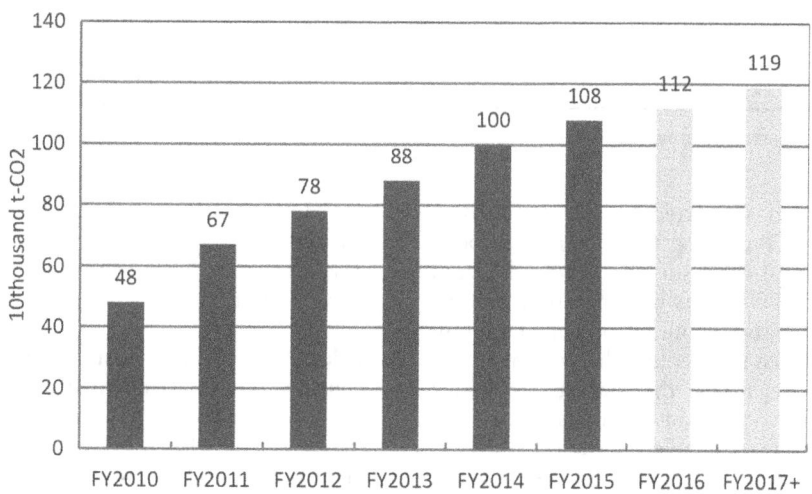

Figure 7 Estimated reduction by measures taken and planned by covered facilities

Table 1 Measures taken and planned by covered facilities

Measures (taken and planned after the base year)		Number of measures	Estimated reductions (tonnes CO_2e)
Heat source, air-conditioning (AC)			
I[a]	High-efficiency heat source	317	127 583
I	High-efficiency pumps and energy-saving control	308	28 182
I	High-efficiency AC system	293	27 101
I	High-efficiency packaged AC system	58	2172
I	Variable air volume AC	26	5024
I/O	Cooling with outdoor air	233	19 943
I	Fresh-air intake control based on CO_2 concentration	99	15 808
I	Total heat exchanger	41	3286
I	High-efficiency fans	230	12 149
O	Control indoor temperature setting in summer and 'Cool Biz' (campaign not to wear ties and jackets to avoid low-temperature settings)	86	10 032
O	Warming up control	28	476
O	Optimization of AC starting time	120	12 245
Lighting			
I	High-efficiency lighting and energy-saving control	1208	84 018
I	(LED lighting)	(954)	(66 376)
I	(High frequency - HF - lighting)	(90)	(7694)
I	(Sensors)	(80)	(3005)
O	Relaxed lux standards	256	18 953
O	Partial light-out and shorter lighting hours	24	768
Other			
I	Elevator energy-saving control	93	1954
I/O	Building energy management system	39	5726
O	(Energy consumption visualization)	(10)	(1153)
I	Demand control system	5	532
	Other	5953	740 204
Total	10 211		1 194 382

[a]*Note*: I = installation, O = operation and maintenance

This tendency is also supported by the results of a TMG survey of TCTP covered facilities in October–November 2014. Of the surveys received, 82% answered that after TCTP was launched they became 'very positive' or 'positive' about introducing high-efficiency equipment when they replace this equipment. In the same survey, regarding the standard payback period for investments, 65% of them answered more than four years was acceptable. A period of seven years or longer was reported by 32% of respondents, which was surprising considering that a three-year payback was assumed acceptable. TCTP's direct impacts are limited here, but 16% answered that the payback period has become longer after TCTP implementation.

Taking a closer look at such investment in equipment, Figure 8 shows the trend in installation of turbo chillers before and after TCTP. Improvement of energy efficiency can be clearly seen after FY2008 when TMG ordinance for TCTP was established. Comparing the term before TCTP (FY2002–08) with that after

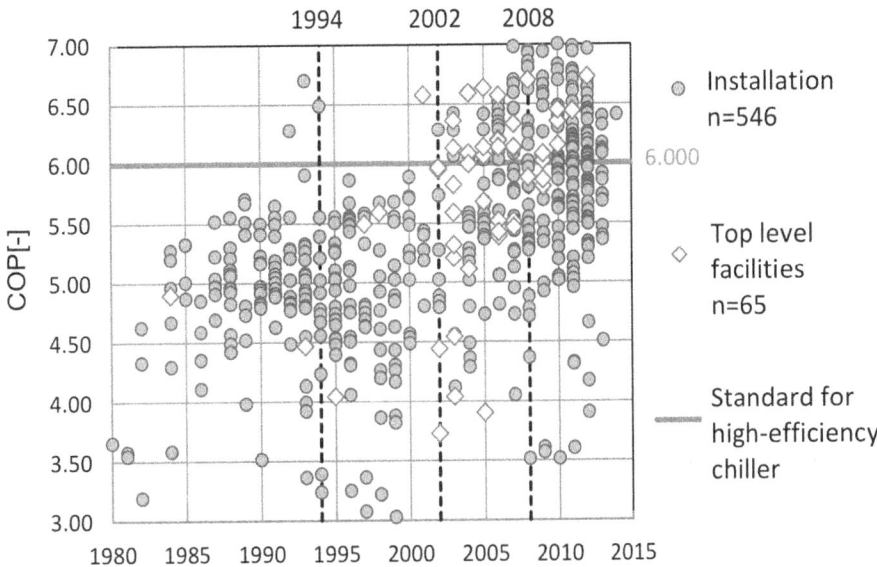

Figure 8   Turbo chiller installation

TCTP (FY2008–14), the installation rate of high-efficiency turbo chillers increased from 53% to 70% among top-level facilities, and from 31% to 48% among other facilities under TCTP.

Technology shifts

The upgrade tendency among covered facilities not only showed energy efficiency but also featured a shift of technology. Figure 9 shows the installations of lighting equipment and a huge jump in 2010 and 2011 can be clearly seen in LED installations. The change from 2009 to 2010 is significant, and that from 2010 to 2011 is also rather drastic. After that, conventional lighting as well as other types of high-efficiency lighting (except LED) have almost disappeared from the scene. Similarly, but less drastically, among heat sources for cooling, particularly absorption

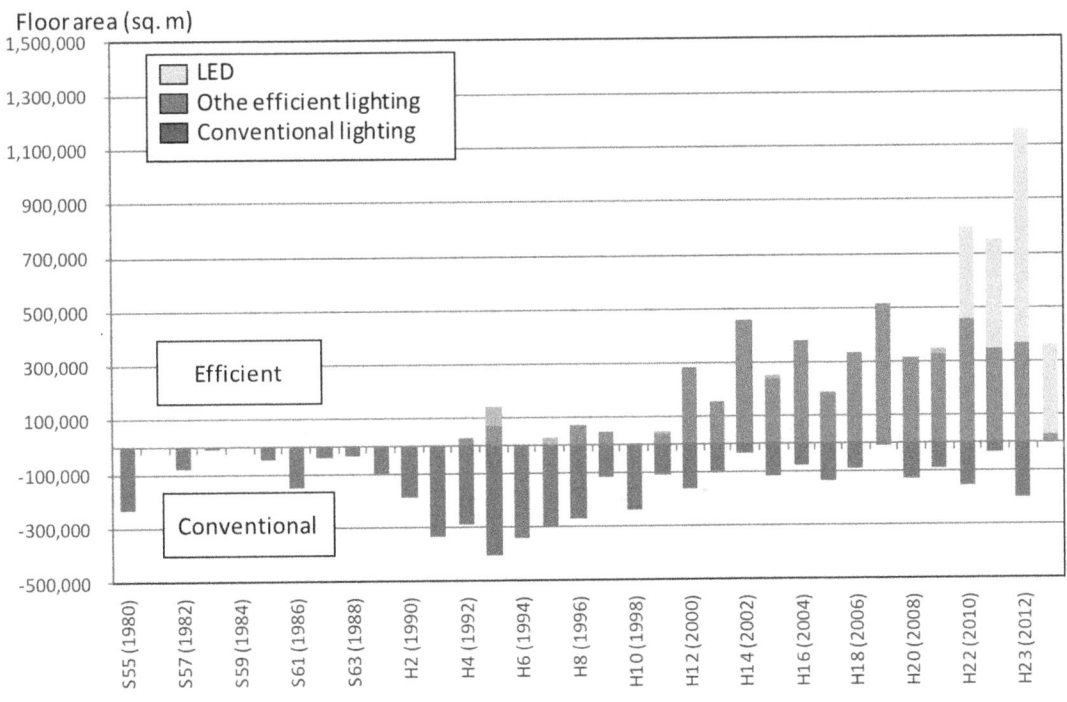

Figure 9   Installation of lighting systems

chiller installations have decreased greatly and instead turbo chillers have taken their place.

As mentioned above, renewable energy installation is limited, which is understandable because most of the covered facilities are located in high-density urban areas, which makes it difficult to install large-scale renewable energy systems. In addition, renewable energy installation on site is much more expensive than other energy-efficiency-improvement technologies. The influence of TCTP on renewable energy installation is limited thus far. The influence through ETS has not yet been determined. In other words, data showed that energy-efficiency improvements in the covered facilities have been cost-effective and a lower hanging fruit compared with the installation of renewables in urban settings.

Operational measures

Along with the measures with investment, operational measures without financial cost or with minimum cost deserve great attention. Some operational measures, including optimization of AC starting time, relaxing light-level standards and cooling with outdoor air, can be found. The accumulated reductions estimated here are not as high as those from measures with investments. However, operational measures are a collection of small actions with promising cost-effectiveness.

On the other hand, TMG identified 62 common measures as a checklist, which forms part of the yearly reporting documents, in order to raise awareness of potential measures for building owners and managers to consider. In the 'Energy-saving feedback report' that TMG send to facility owners, each measure is indicated with data, such as what percentage of covered facilities adopted this measure in what portion of their facility, estimated efficacy, introduction rate among certified top-level/near-top-level facilities, and so on. Among these 62 measures, 27 (44%) are categorized as operational measures. According to this checklist, most operational measures are positively introduced in a 'large portion' of the facilities.

Impact from the 2011 energy crisis

In order to describe how energy savings have proceeded and changed, it is crucial to look at the influence of the energy crisis after the Great East Japan earthquake and Fukushima nuclear reactor incident in March 2011. The scheduled rolling power cut, which was implemented in only a part of the planned area for two weeks, was not implemented in Tokyo's central business district (CBD) area. Most buildings and facilities started to work on energy savings immediately after the incident. As the power peak in Tokyo occurs in summer, the national government issued energy controls, which mandated a peak cut of 15% in large power users (whose power contract is over 500 kW) for two months. A variety of programmes and campaigns were carried out. A significant

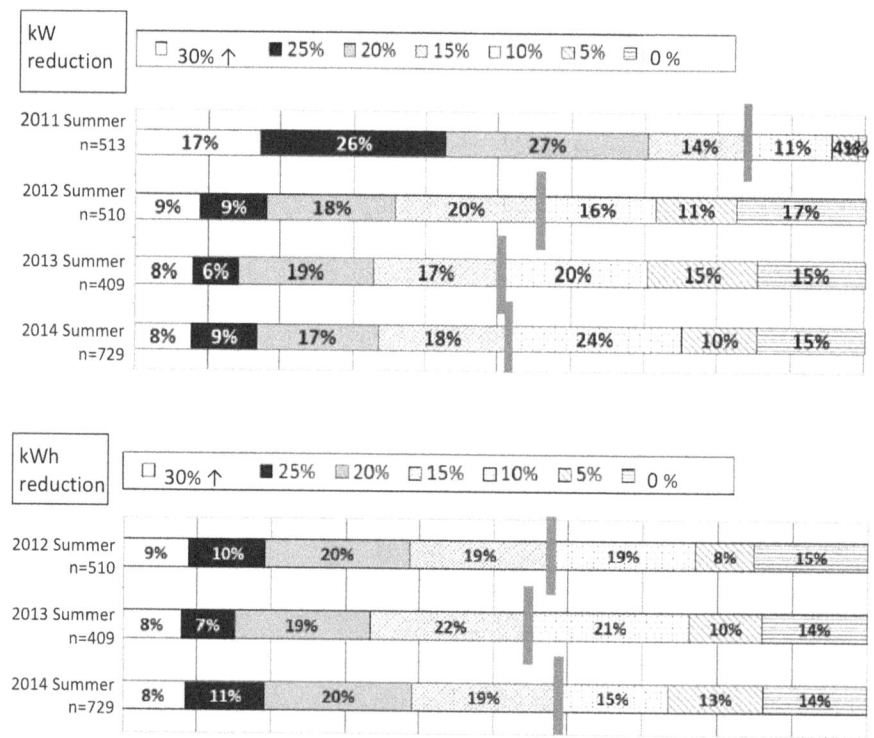

Figure 10 Peak power and electricity consumption reductions

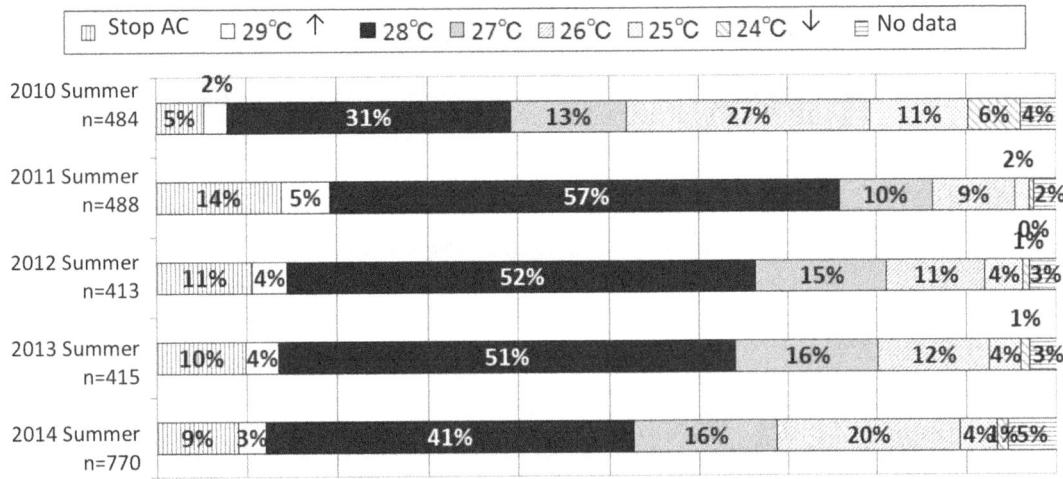

Figure 11 Indoor temperature control

response was seen from buildings, and the immediate results in peak power (in kW) and electricity consumption (in kWh) reductions are presented in TMG's survey results in Figure 10.

Survey results also suggested that in summer 2011, due to the sudden necessity for a cut in peak power, many facilities took multiple operational measures to a deep level, which resulted in significant peak shaving. However, some measures, particularly 'frugal' energy-saving measures, were difficult to continue due to the burden on users; therefore, they have been loosened or discontinued over recent years. On the other hand, electricity consumption reductions have remained stable. Systematic control installations were accelerated during that term. LED installation was a good example of a measure showing a significant increase in equipment introduction, as Figure 10 shows. At the same time, indoor

temperature control is a clear example of a loosened measure in recent years (Figure 11). Another interesting example is light-level control, which has been continuously implemented and is now common (Figure 12).

The impact of the earthquake and energy crisis also caused some concern when assessing TCTP outcomes. Energy saving actions in the wake of the earthquake naturally reduced power consumption and resulted in large emission reductions. It is also difficult to divide the effects among causes. However, some findings indicate that TCTP effectively contributed to emission reductions over the years, and more importantly, TCTP contributed positively to dealing with the energy crisis.

Firstly, before the earthquake, the majority of emission reductions from the first year are attributable to TCTP, particularly in the major building types of office buildings

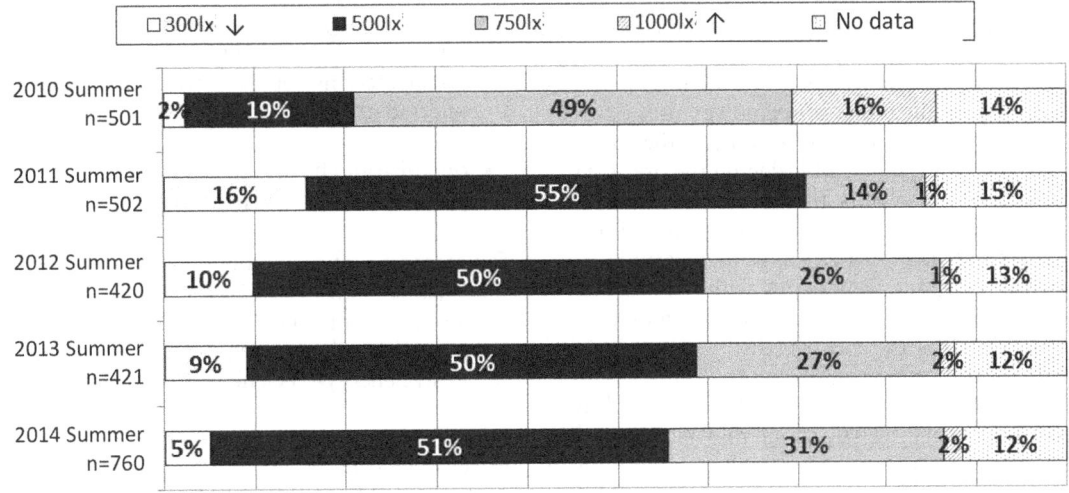

Figure 12 Indoor light level control

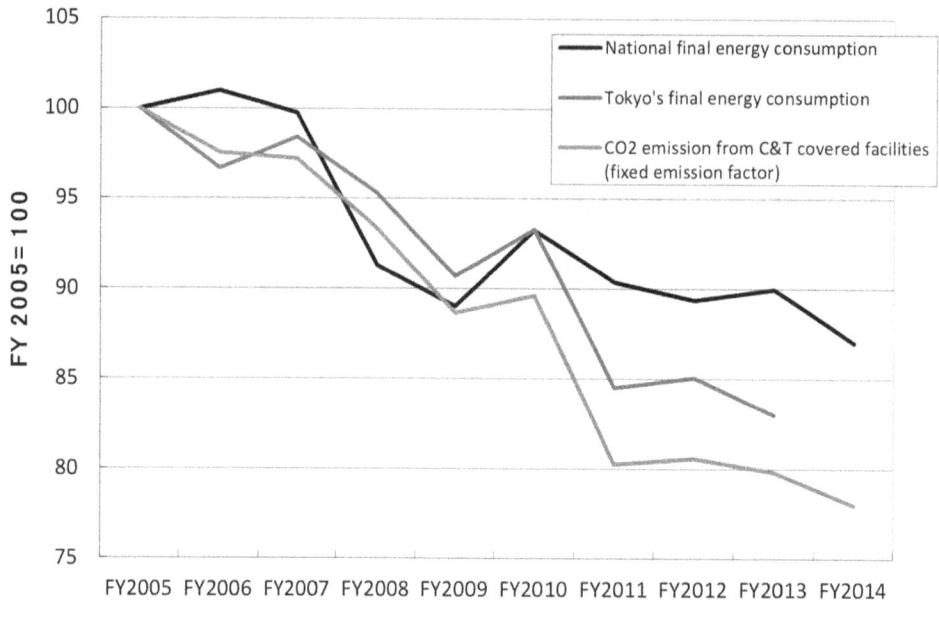

Figure 13 CO$_2$ emissions comparison with national average

and commercial buildings. Other evidence comes from a comparison with the national average. Under the assumption that CO$_2$ emissions show a similar trend to energy consumption and can thus be seen as a proxy of total energy consumption, Figure 13 shows a comparison of the CO$_2$ emission levels in TCTP covered facilities, alongside the final energy consumption trends in Tokyo and at the national level. This indicates that TCTP facilities have surpassed the national average energy consumption (or CO$_2$ emission) reduction trend.

Secondly, in the third and fourth years, after the energy crisis had been overcome, high-level emission reductions were maintained or further improved. Immediately after the earthquake, energy savings were mainly carried out by operational improvements to deal with the emergency situation, such as reducing light levels, using a higher temperature setting for AC and strictly managing unnecessary power use. After the immediate urgency had passed, some of the emergency responses causing much burden on the user side (*i.e.*, stopping some elevators and escalators or setting ACs at 28°C) were expected to be retired. Continuous energy reductions require the focus of measures to move towards substantial measures with investment. Figures 11 and 12 show the changing trends of measures taken before and after the energy crisis.

Finally, and most importantly, TCTP contributed to coping with the energy crisis, which may be counted as one of the most important co-benefits. According to a survey of TCTP facilities (answered by 789 facilities out of 1267) in November 2014, 81% of

those who participated agreed that they used or highly utilized their organizational network established under the TCTP requirement in order to promote energy savings and cope with the energy crisis in their buildings. Prior to the earthquake, most of the TCTP facilities had already created and used such an organizational network.

TCTP requires that facility owners:

- establish an organization and/or network of related personnel in their buildings to promote energy efficiency

- designate leaders for promoting energy efficiency, one of whom is a representative of the facility owner (usually a facility manager) and another is an engineering representative; the names of those selected are submitted to TMG

- organize and have regular meetings with tenants to promote energy-efficiency actions

The reason to set such requirements in the TCTP ordinance came from the lessons learned from the experiences of the prior reporting programme that a human network with a clear target is crucial for pursuing comprehensive energy-efficiency improvements.

Through a series of interviews of best practices of energy-saving efforts after the energy crisis, various levels of communication with differing entities, such as key personnel in the company, in-house engineers, engineering/consulting companies, tenants, users, energy consultants and local governments, was pointed out as one of the key factors for success.

Emission trading

The trading scheme in TCTP is working on a small scale. As of 31 March 2016, accumulated trade for five years was 112 cases, totalling 163 870 t-CO_2e, of which 30 652 t-CO_2e have been surrendered as reduction obligations. On the other hand, issued credits for the same five years totalled 2 678 217 t-CO_2e. This is made up of 2 370 625 tonnes from excess reductions of covered facilities, 270 115 tonnes from renewable energy credits, 37 477 tonnes from small and medium-sized facilities' credits, etc. There is an apparent surplus in credits. This is an inevitable result of the fact that 91% of covered facilities have already reduced more than the obligations.

TMG provided the demand and supply prospects for the first compliance period (which ended in FY2014 and is now in the verification and transfer consolidation period), which estimated the demand for credits at about 200 000–300 000 tonnes, while excess reduction credits, including those not yet issued, were estimated at 9.5 million tonnes. The owners of 1.3 million tonnes of issued credits are willing to sell them on the market, and those owning 520 000 tonnes are very much willing to sell according to the latest TMG survey on TCTP price valuation (as of October 2015) by Argus Media Ltd. Such excess supply affects the carbon price, which was initially JPY9600/tonne (yen) but has now fallen to JPY3500. The same survey explained the status of those who own the credits for sale:

> Sellers want to sell as much as they can, however, they are not in a rush to sell. For example, they: have not yet established a section for selling, feel unrealistic to display the sale as one product of their business line-up, are ready to some extent, or not selling at all.
>
> (Argus Media Ltd, 2015)

Some intermediary agencies point out that when the price become lower than JPY2000–3000/tonne, some will decide not to sell their excess credit. This suggests that trading profit is relatively small compared with their main business, so they are not pressing to sell. At least many will bank their excess reductions to the second compliance period.

Whether the emission-trading scheme of TCTP is going to play a bigger role in the second compliance period (FY2015–19) depends on how much more demand is created. However, even in the next period, emission trading will remain small for the following reasons if TMG do not introduce any new policies:

- The emission trade scheme in TCTP was designed as a supplemental measure utilized only when the covered facility cannot meet the reduction obligations:

  - Compliance periods are set as five years, which does not require trading to meet the requirement on a yearly basis.

  - Tradable emissions are limited only within each facility's excess reductions (lack of distribution of total emission allowance as tradable credits in the market).

- Scepticism exists in the carbon market. In particular, during the introductory stakeholder discussions, business groups criticized the ETS for encouraging speculation and disturbing long-term investments and technical developments:

  - Such arguments back the supplemental design of trading in TCTP.

  - The ordinance provides TMG with the authority to launch measures in cases of extreme carbon prices.

- Many facilities have already reached the reduction requirement level for the second compliance period, and their fulfilment to meet the obligation level is not so difficult. Therefore, excess supply will probably continue during the second period, and carbon price increases seem unlikely.

Overall, the trading scheme in TCTP has played an important role in enabling the mandatory emission-reduction programme with an ambitious target for the covered sector; to provide participants flexibility when taking measures to fulfil their obligation; and to demonstrate putting a price on carbon emissions and preparing a full-scale carbon market for the future. At the current stage, ETS remains a supplemental measure rather than a major one in TCTP. This is obviously different from some other cap-and-trade programmes in which trading plays a much stronger role to lead allowances to their highest valued use, *i.e.*, covering those emissions that are most costly to reduce. However, when Tokyo moves forward to the next stage in carbon emissions mitigation, the ETS scheme will be crucial to fulfil the target.

On the other hand, among the expected positive impacts from ETS, the following aspects are not well covered by ETS:

- to give incentives and encourage those who have already achieved high performance to be more energy efficient

- to enable facility owners/managers to choose the most cost-effective measures, including offsets in the carbon market

However, in TCTP, instead of ETS, other sub-programmes and arrangements are covering these functions. For example, Top Level Facility Certification is working well to incentivize high-performance facilities. Setting longer compliance period allows facility owners the highest flexibility to choose measures, including operational measures with minimum cost.

Behavioural change

At the same time, an awareness of the need for energy saving has increased and behaviours have changed among persons involved, including owners, facility managers, operators and tenants. According to a TMG survey of TCTP covered facilities in October–November 2014 ($n = 794$), since TCTP launched:

- 72% answered the level of interest from senior management has increased

- 75% answered the motivation of those who engaged in energy savings has risen

- 73% answered that it has become easier to gain cooperation from employees in their facilities

- 65% answered that it has become easier to gain cooperation from tenants

Although not as significant, some tangible gains were noticed on the following points:

- 55% answered the number of hours spent on meetings on energy savings has increased (10% reported having frequent meetings on the issue even before TCTP)

- 46% answered the number of people involved has increased

The 2011 energy crisis seems attributable to a part of such drastic awareness and behavioural change; however, the initial intention from the TMG side to involve the corporate decision-makers in the process has been successfully realized. The broader participation from senior management to employees in facilities as well as tenants has spread. For this, the requirement by TCTP to establish explicit organizational arrangements in facilities, a tenant reporting system and other programme settings seems to play a significant role.

Critical governance issues
Monitoring and assessment

As mentioned in the trading section above, TCTP relies on the ETS as an important supplemental measure in the programme. The TMG put an emphasis on the mandatory total emission reduction from each facility. In addition, there is a legacy from the previous Carbon Reduction Reporting Program. These factors are reflected in monitoring, the assessment system and supporting programmes as follows:

- *Yearly reporting and its contents*
  Although TCTP sets five-year compliance periods, facility owners are required to submit yearly reporting not only of data on carbon emissions (including other gasses), energy consumption and their trading status, but also of their organizational arrangement, including the responsible manager and engineer, their carbon/energy reduction plan (measures) with their estimated reductions, and the checklist to find out if they are carrying out the 62 common measures highlighted. The contents of these reports are the data source for the feedback report and for monitoring the entire programme.

- *Feedback reports with benchmarking data*
  Energy-saving diagnosis reports are sent back to the facility owners, presenting benchmark data and status based on a comparison with other facilities in the same usage category.

- *Listing common measures and presenting potential measures for the facility*
  In the feedback report, a chart of potential measures is included with a large amount of information on each measure.

- *Checking each facility's operation management*
  To avoid weak points of the grandfathering method, if covered facilities would like to choose their own historical emissions as the base emission, they have to satisfy standards and rules on energy management/operation based on TMG guidelines. A report with supporting documents/data must be verified by a third-party verifier and submitted to TMG. This process helps to raise the basic management quality in facilities.

- *Identifying best practices and disseminating this information*
  Through the reports, various best practices have been identified. Such great examples are relevant for facility owners/managers/operators and users. Various methods are used to disseminate these examples and experiences, including seminars, symposiums and online publication of case studies, etc.

Target setting versus prescriptive regulation

TCTP is categorized as a target-setting type of regulation that allows facility owners/managers to decide their best approach to achieve the target. Conventional regulations are prescriptive, represented by building energy codes. Compared with such regulations, TCTP has several unique features and may be a prototype cap-and-trade scheme, though it is currently the only programme that covers buildings and includes buildings' entire energy consumption. TCTP can be an alternative governance approach because it can cope well with the major challenges of conventional regulations:

- *TCTP targets existing buildings*
  Approximately 1300 existing buildings and facilities, accounting for 20% of total carbon emissions in Tokyo, are covered by TCTP. TCTP directly approaches building owners/managers and drives them to immediate action for energy savings. For existing buildings, energy codes can only be applied or checked when buildings or major equipment are being renovated. Building energy codes are rather suitable to regulate new constructions.

- *TCTP directly mandates a certain amount of emission reductions instead of regulating design specifications*
  All covered buildings must reduce energy consumption and CO_2 emissions by a certain percentage. With some penalties, the effectiveness of total emission reductions is highly assured. On the other hand, energy codes mandate the introduction of design/equipment with certain specifications. The gap between design and real performance is highly recognized nowadays. Various factors, including operation/maintenance, user awareness, and even the difference of designed and real performance of equipment, affect the results.

- *Facility operation and controls are crucial for energy savings*
  Under TCTP, such operational measures are very much appreciated due to their cost-effectiveness. As the survey indicated, TCTP enhances the value of building facility engineers/managers, which in turn promotes better energy management. In contrast, it is challenging for energy codes to regulate operation or encourage building managers/engineers.

- *TCTP can set ambitious reduction targets by utilizing the emission-trading scheme*
  The TCTP cap was set as 8% for the first period and as 17% for the second period in order to satisfy Tokyo's emission-reduction target of '25% GHG reductions by 2020 from 2000 levels'. The cap setting is the most difficult part of the cap-and-trade scheme. In order to realize an ambitious target, it is crucial to link tangible targets of the community, city and nation as a whole.

- *Cost-effectiveness*
  Target setting-type regulation allows building owners/managers to decide their measures and their timing according to their business plan and also their facility's update schedule, which enables them to take cost-effective solutions. A longer compliance period under TCTP supports building owners to take measures best suited to their business/facility plan. ETS also contributes to the cost-effectiveness, although in TCTP ETS functionality is limited.

Administrative costs for TCTP are mostly attributable to associated staffing costs. Looking at the TMG staff assignment, the number of personnel assigned to the basic cap-and-trade scheme, including programme management and ETS, is rather small. Rather, staff members are mainly allocated to TCTP-specific work, in particular supporting facility owners/managers and tenants.

Baseline setting

The selection of emission baseline is a unique feature of TCTP to encourage participants' early actions. However, it has also been a cause of some disagreement on whether or not the emission reductions were attributable to TCTP or to the previous reporting programme. Under TCTP rules, each facility under TCTP can set its own base-year emission by the average emissions of three consecutive years from 2002 to 2007.

This was one of the crucial points of negotiation with stakeholders during programme development because some of the facilities had already voluntarily introduced measures and invested in their buildings' energy efficiency during the previous programme by the time negotiations took place on TCTP's introduction (mainly in 2007 and 2008). The TMG also wanted to avoid any interruption in energy-efficiency improvements even during the discussion or preparation phases. It was strongly anticipated that if the base year were set based on emissions after FY2008, before the launch of TCTP in 2010, facilities might idle their efforts and emit more to be on the safe side. Therefore, the base year must be before FY2008, and had to include the previous programme term to show an appreciation for their response to the voluntary programme. That resulted in an announcement in 2007 that 'TCTP will take account of emission reductions even during the previous programme and will also account for reductions before the launch in 2010'.

According to TMG officers, this policy has been more effective than originally expected, and emission reduction efforts were continuously executed after the announcement through to the launch of TCTP. Some experts reported that because of the TCTP introductory discussions and the decisions made in 2008, a positive mood and further actions to improve building energy efficiency were induced. It is difficult to confirm whether such influences were the result of stakeholder engagement. However, even by omitting the first-year results, TCTP clearly demonstrated significant reductions in the latter years, indicating a significant influence.

Tenant engagement

Another crucial issue is tenant engagement. It was a big decision to include energy consumption from tenant-occupied space as a responsibility of the facility owner. How to establish good governance in a building from the perspective of energy management is the key for tenant buildings. More importantly, tenant rules in TCTP were significantly enhanced during the stakeholder discussion process. For example, tenants' responsibility for energy reduction and their cooperation with facility owners are clearly stipulated in the ordinance. Large tenants are required to report their carbon emissions/energy consumption and submit to TMG via facility owners. They also have to set up an organizational arrangement for their own energy savings and participate in activities initiated from the owner side.

In 2014, an evaluation programme on tenant activities was launched for further tenant action promotion. In this programme, approximately 800 large tenants are evaluated by their emission-reduction performance and their measures/activities taken based on their report. Feedback reports with benchmarking data and checklist charts showing potential measures were sent from TMG directly to tenants. Rated results are disclosed on the TMG website to encourage tenants to participate. This programme aims to encourage further those who engage in energy efficiency actions from the user side.

Overall, TCTP governance is sustained by detailed monitoring, assessment and support processes.

Importance of medium or long-term continuity/consistency

TCTP was designed as a medium to long-term approach from the very beginning, which was clearly communicated to facility owners/managers and related stakeholders in all dialogues during the policy planning and implementation processes. This common understanding of continuity provides businesses with the necessary confidence and flexibility to invest in making changes to meet the targets. Several features of the programme, including the five-year compliance phases, announcement of the declining cap in the long run at the launch of the programme, and efforts to keep carbon credits stable, are crucial for facilitating businesses to plan ahead, align facility energy goals with business plans, and take actions relatively early to invest in equipment and facilities, as well as in carbon credits. It is also necessary to assure facility owners of the consistency of the policy by showing long-term political commitment by the government.

Conclusions

No single policy is sufficient to achieve the potential energy savings (IPCC, 2014). Building codes and appliance standards with strict energy-efficiency requirements when well enforced, tightened over time, and made appropriate to the local climate and other context have been among the most effective in terms of the environment and cost-effectiveness (IPCC, 2014). However, at the same time, considering the risk of the lock-in effect, there is an urgent need to take ambitious and immediate measures. In Tokyo, the Energy Saving law plays a limited role, while the synergistic effects between TCTP for existing buildings and green building programmes (rating and performance requirements) for new constructions are driving the emissions reduction in the building sector. Overall, the data and survey results from the first five-year compliance phase suggest that TCTP has been working effectively as a policy instrument to reduce energy consumption in existing buildings in order to meet TMG's ambitious emission reduction goals, to introduce new technologies and to drive behavioural change.

The design and implementation of TCTP reflects a clear approach of using environmental policies to increase the perceived market pay-off, an attention to maximizing flexibility in compliance and a strong intention to diffuse energy-efficient technologies in buildings, which are all important features of effective approaches to nurture market success of environmentally friendly technologies and mitigate carbon emissions. As TCTP experienced, a declining cap and its announcement at the launch of the programme backed by long-term political commitment to environmental goals have been the crucial success factors.

Improving energy efficiency of the vast number of existing buildings requires a long time for planning and implementation. Major renovations are an excellent opportunity to introduce deep retrofits. However, it takes a very long period of time for all existing building owners to take actions. Considering the effective results of the TCTP, at least for large commercial buildings, a Tokyo-type cap and trade scheme would be one of the promising policy measures to be packaged into a portfolio of building regulations and programmes.

During the years of programme planning and implementation, TMG has developed a systematic series of strategies and techniques and an extensive network to address stakeholder concerns, some of which directly affect the outcome of TCTP, such as uncertainty perceived by facility owners in the costs of policy abatement (Aldy & Stavins, 2012). Now that TCTP has entered its second phase, it is crucial to continue fostering communication among stakeholders around the buildings and convince facility owners/managers to invest in their facilities over the long-term for maximizing the economic and environmental values.

Investments in energy efficiency, including reductions in building energy consumption, continue to face multiple internal barriers in companies, despite the apparent benefits of these investments, and a societal drive for greater energy efficiency sometimes can be an opportunity for innovation (DeCanio, 1993). Policy tools like TCTP can be seen as such a drive to support corporate decision-making and priority-setting, with informational and organizational services that are not in the scope of traditional building energy regulatory policy tools.

In urban settings, buildings tend to be high density and their floor-area ratio very high. In this context, it is extremely difficult to install on-site renewables to generate sufficient energy to meet a building's energy demand. In the long-run, to achieve more aggressive emission reduction goals, such as net-zero-energy buildings, some kind of trading and/or offsetting

scheme to link off-site renewables with more ETS utilized seem to be necessary, adding more significance to the TCTP-type scheme and importance of evaluation studies like this one.

Challenges still exist. Now the cap for the first phase under TCTP is met and a large surplus of credits is created, it is time to announce the future direction and level of ambition. This is clearly indicated in survey results in which 54% of facilities answered that they wanted to know more details concerning the third compliance period, for which a higher target needs to be set. Another challenge for TCTP is reflected by the answers to the same question in the 2014 survey. The streamlining of TCTP was demanded by 56% of respondents. The issue of increased complexity partially comes from the aforementioned supporting characteristics of TCTP, which is a strength of TCTP but which at the same time increases the complexity and places a burden on the facility side, and calls for continuous policy innovation to solve.

References

Aldy, J. E., & Stavins, R. N. (2012). The promise and problems of pricing carbon: Theory and experience. *Journal of Environment and Development, 21*(2), 152–180.

Argus Media Ltd. (18 November, 2015). Trading price valuation result of trading price in mandatory total emission reductions and emission trading program (TCTP). Retrieved from http://www.kankyo.metro.tokyo.jp/climate/large_scale/attachement/151118shiryou5.pdf

DeCanio, S. J. (1993). Barriers within firms to energy-efficient investments. *Energy Policy, 21*(9), 906–914.

IPCC. (2014). *Climate change 2014: Mitigation of climate change. Working Group III contribution to the Fifth Assessment Report of the Intergovernmental Panel on Climate Change.* New York: Cambridge University Press.

Nishida, Y., & Hua, Y. (2011). Motivating stakeholders to deliver change: Tokyo's Cap-and-Trade Program. *Building Information & Research, 39*(5), 518–533.

Governance strategies to achieve zero-energy buildings in China

Jingjing Zhang, Nan Zhou, Adam Hinge, Wei Feng and Shicong Zhang

In response to climate change, governments are developing policies to move toward ultra-low-energy or 'zero-energy' buildings (ZEBs). Policies, codes, and governance structures vary among regions, and there is no universally accepted definition of a ZEB. These variables make it difficult, for countries such as China that wish to set similar goals, to determine an optimum approach. This paper reviews ZEBs policies, programmes, and governance approaches in two jurisdictions that are leading ZEBs development: Denmark and the state of California in the United States. Different modes of governance (hierarchy: principal–agent relations, market: self organizing and network: independent actors) are examined specifically in relation to policy instruments (prescriptive, performance or outcome-based). The analysis highlights differences in institutional conditions and examines available data on energy performance resulting from a building policy framework. The purpose is to identify ZEBs governance and implementation deficits in China and analyse alternative governance approaches that could be employed in China, which is currently developing ZEBs targets and policies. Conclusions suggest that the ZEBs governance structure in China could benefit from widened participation by all societal actors involved in achieving ZEBs targets. China's ZEBs policies would benefit from employing a more balanced hybrid governance approach.

Introduction

In the developed world, buildings consume more energy than any other sector. In response to climate change and growing resource shortages, major world regions are developing policies to move toward zero-energy buildings (ZEBs). Building energy performance policies and codes, definitions of ZEBs, and governance structures for implementing policies vary among regions. Currently, the European Union (EU) Energy Performance of Buildings Directive (EPBD) requires 'nearly-zero-energy buildings' (nZEBs) target for all EU member countries' new construction by 2020, and the United States is pursuing a 'net-zero-energy buildings' (NZEBs) goal. Despite the apparent similarity in these terms, there are significant differences in the definitions, policies and support mechanisms associated with them. This variation makes it difficult not only to understand and evaluate global progress toward ZEBs but also for countries such as China that are considering setting similar goals to assess the optimum approach to adopt. For simplicity, in this paper the umbrella term 'zero-energy buildings' (ZEBs) is used to refer to all the various types of low-energy buildings encompassed in the various initiatives studied. However, when referring to policies specifically tied to EU EPBD initiatives, or US net-zero activities, the abbreviations 'nZEB' and 'NZEB', respectively, are used.

Defining a clear target is the first step in the ZEBs governance process. ZEBs definitions have been widely studied from a technical perspective, but there has been little examination of the governance of ZEBs. Moreover, in the building industry, the concept of governance is commonly understood as referring to only regulations and administrative instruments (*e.g.*, mandatory or voluntary codes). This paper aims to fill a knowledge gap by providing guidance toward ZEBs definitions and governance strategies in China and elsewhere, based on an understanding of ZEBs targets and policy instruments adopted in leading world regions, as well as overall governance approaches that influence their policy design and outcomes. That is, a jurisdiction's governance mode affects how a ZEB target is interpreted and implemented. This issue has several dimensions. One is the overall governance approach (*e.g.*, whether policies are formulated and imposed by government or in deliberations among societal participants/stakeholders). Another is the numerous elements that must be integrated to achieve a ZEB (*e.g.*, building energy efficiency, plug loads, renewable energy sources, etc.), which requires consensus among various stakeholders. Related to this complex integration is that traditional building codes cannot address all the elements necessary to achieving a ZEB. Finally, a strategic governance process will be always partly technical and objective, and partly political and subjective (Simmons, 2015).

This paper reviews ZEBs targets in leading world regions and ZEBs policies and governance modes in two jurisdictions where ZEBs targets have been successfully implemented: Denmark in the EU and California in the US. These two jurisdictions exemplify different governance modes. The selection of these two cases is primarily based on their successful implementation of ZEBs policies. Denmark is one of the first two countries in the EU that established a binding national target for ZEBs. In 2014, one-third of all ZEBs in the US were located in California. The experiences in these jurisdictions are relevant to China where similar ZEB goals have just recently been proposed. An often-cited challenge is that China's green governance and implementation do not match the government's stated green objectives (The World Bank and Development Research Center of the State Council of China, 2014).

The aim of the current paper is to provide guidance for China and other regions interested in developing ZEBs goals. Given the different institutional conditions in the selected jurisdictions, our guidance is not necessarily focused on directly transferring policies from region to region. Based on the assumption that the choice of specific policy instruments is likely dictated by the prevailing governance processes, the aim is to develop guidance for other regions based on identification of the governance and implementation strengths and deficits in these jurisdictions, and an understanding of how they designed and implemented ZEBs targets appropriate to their local institutional conditions.

The remainder of this paper discusses major features of ZEBs definitions in leading world regions, gives an overview of governance modes and identifies governance strategies in Denmark and California, and reviews building energy-efficiency and renewable-energy (mainly solar photovoltaic – PV) elements of the ZEBs policies in these two regions. The analysis emphasizes the overall governance strategy and existing policy framework to support compliance with and enforcement of ZEBs targets. Finally, the paper describes current ZEBs activities and governance in China, and concludes key features of China's prospective ZEBs targets, and key governance strategies for China to achieve the ZEBs goal.

Building energy code approaches

Globally, there are two main types of building energy performance codes: 'prescriptive' and 'performance' based. The prescriptive-based code approach specifies how a building must be constructed, including specifying energy performance values for building elements such as window U-values. In contrast, the performance-based approach specifies total energy consumption for new construction, based on modelled energy use. In other words, performance codes would state a minimum and/or maximum target (generally excluding the method by which this is achieved), whereas the prescriptive method defines a specific target and/or method to comply. In the US, some jurisdictions are moving toward a third type of code, an 'outcome-based' approach that requires a measured, post-occupancy energy performance level (IPEEC, 2015). The outcome-based approach is similar to a performance-based code in that it specifies requirements for building design, construction and commissioning. The main difference is that an outcome-based approach also includes requirements for the building's actual operation. An advantage of an outcome-based code is that by assessing actual measured energy demand rather than predicted energy consumption, it addresses the impacts of plug loads and occupant behaviour, which are currently not covered by most building regulations. In China, new energy-consumption standards (energy quotas) are being developed in the official public review copy of draft 'Standard for Energy Consumption of Buildings', which states a maximum energy intensity by building type and climate/location. The quota is similar to an outcome-based code in defining a maximum energy consumption intensity by building type over a specified period (usually 12 consecutive months) (Xin, Lu, Zhu, & Wu, 2012). However, the quota applies only to actual energy use for existing buildings (that is, it is outcome-based). For new buildings, the quota is based on predicted energy performance (performance-based).

The type of code – prescriptive, performance or outcome-based – dictates the approach by which the ZEBs goal will be achieved. For instance, the prescriptive approach would require minimum levels for different building components such as U-values, airtightness and the share of renewable energy use, while the performance approach would specify that the building's *predicted* annual final/primary energy demand or modelled carbon emissions be offset by renewable energy produced on-/off-site or by credits purchased from the grid. Under an outcome-based approach, the ZEBs target would be achieved in a manner similar to that of the performance approach, *i.e.*, by measuring and balancing energy use and carbon emissions, except that the measurement and calculation of the *actual* energy/carbon balance would occur during the post-occupancy period. Implementing outcome-based ZEBs codes would require some market actors to alter their roles and strategies. For example, because occupant behaviour affects compliance with outcome-based codes, designers would likely engage more with energy management and building occupiers than is currently typical, and officials who currently typically only evaluate designs and plans would need to extend their involvement into the period after a building is operating (Denniston, Frankel, & Hewitt, 2011). Capacity building and education, benchmarking, measurement, verification, and incentives would be critical elements of an outcome-based approach (Evans, Halverson, Delgado, & Yu, 2014).

There are currently few international experiences with outcome-based codes. Most countries require that building designs be reviewed for compliance with building energy codes, but only a very few jurisdictions also inspect buildings to ensure code compliance, for example, Swedish building code enforcement includes an option of verification of actual energy consumption once the building is operational (Global Buildings Performance Network, 2014). In most jurisdictions, actual energy use in buildings is usually considered in programmes such as benchmarking, labelling, energy audits and commissioning. These existing programmes provide feedback mechanisms to support code compliance and documentation of actual energy savings. This paper's focus is not whether or not China should adopt an outcome-based code but instead it considers the positive and negative aspects of different ZEBs

definitions, code compliances approaches and governance strategies that China might adopt.

Zero-energy buildings' definitions and targets

This section reviews the ZEBs definitions and targets adopted in leading world regions. The purpose of this review is to identify key considerations for the Chinese ZEBs definition, not to recommend a single type of ZEBs definition.

Definitions of zero-energy buildings

ZEBs definitions for policy purposes encompass numerous location-specific factors, including the technologies to be adopted in ZEBs, the economic viability of ZEBs, the feasibility of implementing ZEBs technologies, impacts on the regional electricity grid and benefits to different stakeholders. These factors, in turn, influence the deployment of ZEBs technologies.

There is currently no universally accepted definition of a ZEB. One aim of the joint International Energy Agency (IEA) SHC TASK40/ECBCS Annex 52, was to illuminate different international ZEBs definitions. Existing definitions correspond to regional political and other considerations associated with promoting ZEBs (Sartori, Napolitano, & Voss, 2012). A key element underlying a ZEB definition is the balance between weighted energy demand and supply (Figure 1). Currently, four definitions of this balance are commonly found around the world: net-zero site energy, net-zero source energy, net-zero energy costs and net-zero energy emissions (Table 1). In practice, approximately 75 methodologies are used around the world to determine this balance. Some jurisdictions consider more than one metric (Ecofys, Politecnico di Milano, & University of Wuppertal, 2013). This wide variety of metrics indicates the regional diversity in the interests and motivations of ZEBs market actors.

Figure 1 underscores the importance of energy efficiency in achieving the ZEBs goal. The ZEBs design concept generally requires more than passive design strategies. Table 2 summarizes the differences among ZEBs definitions in selected world regions.

Table 2 shows that different regions interpret factors differently, particularly the weighting factors of energy supply and demand, the ZEBs boundary (whether on- or off-site renewable energy is included within the ZEBs 'footprint' and whether a zero-energy goal must be achieved by a single building or can be achieved over a group of buildings), what energy use is accounted for in the ZEBs, minimum ZEBs requirements, and building types. Newly proposed building codes must take into account existing

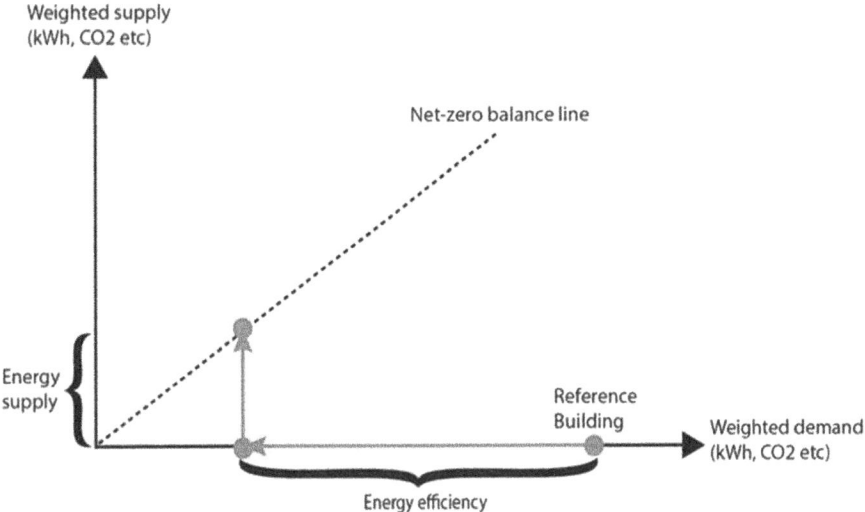

Figure 1 Zero-energy buildings' (ZEBs) balance concept.
Source: Sartori et al. (2012).

building codes and voluntary efficiency standards. For example, in Europe, the Passive House standard and other voluntary codes are widely disseminated and form a baseline for developing the ZEBs concept.

Some general findings about the range of definitions can be summarized as follows:

- The primary energy metric (source energy) is more commonly used in the EU, while the site energy metric (final energy) has been used extensively in the US, although the US Department of Energy (DOE) recently defined the primary energy metric for ZEBs (DOE, 2015).

- Plug loads are generally not included in European definitions but are required in US definitions.

- To help meet the EU's long-term climate target, some EU definitions include both minimum

energy-efficiency and renewable-energy requirements; renewable energy is less prominent in US definitions.

- Embodied energy is considered only in Norway's and Switzerland's definitions. When buildings become much more efficient, the share of embodied energy from materials and manufacturing can be as high as 62%, according to a case study in California (Faludi, Lepech, & Loisos, 2012).

- EU building code requirements have gradually shifted from prescriptive to performance-based. In the US, codes generally offer both prescriptive and performance-based options; the US building industry has also begun to experiment with outcome-based codes.

- In East Asia, *e.g.*, South Korea, where high-rise buildings are more common than in Western countries, the system boundaries for high-rise buildings and community-level ZEBs are extended to include off-site renewable options.

EU and US zero-energy buildings' targets

Europe aims to reduce building-sector greenhouse gas (GHG) emissions by 88–91% by 2050 compared with 1990 levels (Boermans & Grözinger, 2011). The revised EPBD requires that all new EU buildings occupied and owned by public authorities are nZEBs after 31 December 2018, and all new buildings are nZEBs by 31 December 2020. The EPBD, the Energy-Efficiency Directive (EED) 2012/27/EU, the directive on renewable energy sources (209/28/EC), and directives on ecodesign and energy labelling (2009/125/EC and 2010/30/EU) provide the legal framework for the EU's nZEBs target. As required by EED Article 4 (2012), member states must also establish long-term

Table 1 Common zero-energy building (ZEB) definitions.

Concept	Definition[a]
Net-zero site energy	$m - r - r0 \leq 0$
Net-zero source energy	$m + g - r - r0 \leq 0$
Net-zero energy costs	$\$m - \$r \leq 0$
Net-zero energy emissions	$CO_{2m} - CO_{2r} - CO_{2g} \leq 0$

Note: [a] m = Final energy use at the meter; r = renewable energy produced on-site; $r0$ = renewable energy supply off-site; g = energy transmission lost; $\$m$ = bought energy cost; $\$r$ = on-site renewable production sales; CO_{2m} = CO_2 emissions from final energy use; CO_{2r} = offset CO_2 emissions from on-site renewable energy production; CO_{2g} = CO_2 emissions offset through carbon-trading schemes.
Source: Ji and Guo (2013).

Table 2 Zero-energy building (ZEB) definitions in leading world regions

Region	Definition/label	Metric			System boundary		End uses and life-cycle stages included					Minimum requirements		Single building types	
		Primary (source) energy	Final (site) energy	Carbon emissions	On-site	Off-site	HVAC[a]	DHW[a]	Lighting	Plug load/appliance	Embodied energy	EE[a]	RE[a] share	New	Existing
EU	EPBD[a]	✓			✓	✓	✓	✓	✓			✓	✓	✓	✓
DE[a]	EffizienzhausPlus	✓			✓	✓	✓	✓	✓	✓		✓	✓	✓	✓
DK[a]	BR10	✓			✓	✓	✓	✓	✓			✓		✓	
CH[a]	Minergie-A	✓			✓		✓	✓	✓	✓	✓	✓	✓	✓	
NO[a]	Zero-emission building			✓	✓	✓	✓	✓	✓	✓	✓		✓	✓	
UK[a]	Zero-carbon standard			✓	✓	✓	✓	✓	✓			✓	✓	✓	
US[b]	Zero-net-energy building	✓	✓		✓	✓	✓	✓	✓	✓		✓		✓	
SK[a]	Nearly zero-energy building	✓			✓	✓[c]	✓	✓	✓			✓		✓	
JP[a]	Net-zero-energy building	✓			✓		✓	✓	✓	✓		✓		✓	

Notes: [a]HVAC = heating, ventilation and air-conditioning; DHW = domestic hot water; EE = energy efficiency; RE = renewable energy; EBPD = Energy Performance of Buildings Directive; DE = Germany; DK = Denmark; CH = Switzerland; NO = Norway; UK = United Kingdom; SK = South Korea; JP = Japan.
[b]In the US, California has proposed to use the time-dependent valuation (TDV) metric in a definition of ZEBs, which is a conversion of final energy metric; at the national level, the Department of Energy (DOE) recently provided the primary energy metric (source energy) as an option for a definition of ZEBs.
[c]In South Korea, the system boundaries for high-rise buildings and community-level ZEBs are extended to include off-site renewables options.

The definitions/labels listed are provided mostly by public authorities. Others from the research community or voluntary sources are not included here. The ZEB target is usually required in new buildings. Building code regulations usually refer to single buildings; however, the ZEB concept can also be realized in cluster of buildings. End uses and life-cycle stages included (accounting system); electro mobility is generally not included, except in DE. Balance period: this is typically based on the operational year; the life-cycle perspective is only considered by definitions in CH and NO. Minimum requirements also include indoor climate and comfort elements that are not listed here. The minimum renewable integration requirements in UK definition are met through carbon compliance minimum levels.

Sources: Zhang, Xu, Jiang, Feng, and Sun (2014); Ecofys et al. (2013), Dokka, Sartori, Thyholt, Lien, & Lindberg (2013), Atanasiu, Kunkel, & Kouloumpi (2013), Zero Carbon Hug (2014), UK Green Building Council (2014), APEC Energy Working Group (2014).

strategies for mobilizing investment to renovate national building stocks, with emphasis on 'deep' renovations to achieve the nZEBs goal. In contrast to 2013 when only five member states had nZEBs definitions in place (Groezinger et al., 2014), by 2014 a majority of EU member states had adopted or were in the process of approving definitions. Denmark was one of the few countries that first established a legally binding (coercion regulation) national nZEBs plan and definitions for new buildings by 2020. The EU directive requires that the nZEBs goal be achieved but does not dictate the means by which this must be accomplished.

In the US, based on the Energy Independence and Security Act of 2007 and Executive Order 13514, the DOE's goal is to create the technology and knowledge base for cost-effective zero-energy commercial buildings by 2025. The DOE formalized a NZEB definition in September 2015 (DOE, 2015). To date, there are 39 NZEB verified or building clusters that, as a group, achieve zero-energy targets. These projects have been documented to meet, over the course of a year, all net energy demand through on-site renewable energy sources (New Buildings Institute, 2015). Several US state and local jurisdictions have set NZEBs goals. For example, the California Energy Commission plans to update Title 24, the energy-efficiency portion of the California building code, to include the NZEBs goal (CPUC, 2008). About one-third of US NZEBs are located in California, which is recognized as a leader in NZEBs market transformation (Cortese, Higgins, Lyles, & Hamilton, 2014). Overall, NZEBs targets in the US have been initially advocated by a non-governmental organization, NGO Architecture 2030, using a bottom-up approach. The US NZEBs goal is non-binding but includes detailed recommendations (targeting regulation).

Method and theoretical framework for analysing governance modes

This study's conclusions relate to two main themes: ZEBs governing targets and ZEBs governance strategies that might be adopted in China and elsewhere. The research material collected for this study is largely based on a literature review of ZEBs definitions (Table 2) and relevant policy instruments (see the supplemental data online). The conclusion about governing targets for ZEBs in China is based on a review of ZEBs technical definitions in leading world regions as well as the current status of building codes and energy use in China. For governance strategies, it is not sufficient to draw conclusions based on transfer of successful policy instruments from one jurisdiction to another, because the institutional conditions in each region are fundamentally different from one other. Ultimately, China needs to develop its own

governance approaches that are based on its own institutional environments – the simple adaptation of approaches from other regions is unlikely to be directly transferable. Nevertheless, there are broad governance strategies that can be drawn from governance theory and case studies for China to consider. More specific step-by-step governing approaches or a proposed roadmap are beyond the scope of this paper; such an analysis would require a more in-depth investigation of China's case and its evolving policy-making and implementation processes. The conclusions about alternative governance strategies are based on an understanding of the theoretical background of governance concepts and governance modes, and their adoption in the ZEBs policy design and implementation in the selected jurisdictions. It is important to emphasize the concept of governance first and identify governance deficits. By understanding how different governance modes influence policy design and implementation, it is possible to identify the linkage between empirical policy instruments and overall governance strategies. Our central claim is that the policy instruments adopted to achieve ZEBs targets are embedded in the dominant governance mode in each region, rather than the other way around. In this paper, the governance strategies specifically refer to building energy performance, without taking into account other types of requirements for buildings, such as health or life safety building codes. In the remainder of this subsection, the theoretical background of governance concept and modes of governance are introduced.

The core meaning of governance is the means employed to steer and coordinate a region's economy and society (Pierre & Peters, 2000). Much debate centres around definitions of governance and modes of governance processes. Governance comprises actors from all segments of society working together to pursue collective goals. General categories of governance modes are state-centric, society-centric, and hybrid forms that are a mix of state intervention and societal autonomy. In reality, hybrid governance is the most common type and it often operates in the shadow of hierarchical and administrative forms (Hildingsson, 2015). Governing modes can be discerned from the dimensions of polity (a region's institutional properties), political (actor constellations), and policy (policy instruments) (Treib, Bähr, & Falkner, 2007).

The paper examines modes of governance in relation to policy instruments. In political science literature, much discussion concerns what constitutes legitimate input (*i.e.*, accountability, participation, transparency) to governance, referred to in aggregate as the quality of the decision-making process. Whether a governing mode results in legitimate output (*e.g.*, effectiveness of institutions, environmental performance) is not well researched (Bäckstrand, Khan, Kronsell, &

Lövbrand, 2010), for example, with regard to problem-solving capacity or effectiveness. The current analysis does not focus on which modes of governance are preferable but on describing policy alternatives under different governance modes. The modes of governance considered in this paper are:

- hierarchy (principal–agent relations): a state centric mode of governance that normally operates through administrive orders and sets of rules and is exercised by states, governments and bureaucracies in relation to societal actors (Bäckstrand et al., 2010)

- market (self-organizing): a society-centric mode with explicit focus on non-state actors (Hildingsson, 2015)

- network (interdependent actors): based on resource dependencies between private and public actors and/or individuals (Bäckstrand et al., 2010)

Each governance mode is made up of different combinations of three elements: administrative, economic and deliberative rationality (Bäckstrand et al., 2010). It is essential to note that we are not looking at governance modes and their combinations of types of rationalities as if those constituted a checklist that could be used to design a governance strategy. Instead, governance modes and their combinations of rationalities are an overarching conceptual lens to guide the analysis of empirical policy documents.

As an alternative to analysing output legitimacy, this analysis focuses on effective policy outcome and distinguishes policy output and policy outcome. Policy output is the direct result of policies, *e.g.*, more energy-efficient design, integration of renewable energy. Policy output is a means to achieve policy outcome. Examples of policy outcome are achieving ZEBs targets or reducing GHG emissions (Lucia, 2012). Policy outputs are the main resources by which we can understand the policy background in each region. These can be found in the supplemental data online for this paper as well as the 'ZEBs policy background' subsections for each jurisdiction discussed below. Policy outcomes are discussed in the subsection 'Implications for building regulations to achieve ZEB outcomes' below. The governing process influences how a policy is implemented or enforced. For example, a policy can be implemented by coercion (detailed standards with little leeway in implementation), voluntary compliance (non-binding instruments that only define broad goals), targets (non-binding instruments that offer detailed recommendations), or framework regulations (binding law, with leeway in implementation) (Treib et al., 2007).

Governance of zero-energy buildings in Denmark and California

This subsection identifies the ZEBs policies and governance approaches in Denmark and California, two regions leading ZEBs development. California has been leading the US ZEBs market in terms of total built projects, and Denmark, which has among the most progressive and ambitious energy and climate policies in the world, was one of the first two EU member states to establish a national ZEBs plan. For this analysis, the research team disaggregated Denmark's and California's governance frameworks into eight policy outputs: targets and building codes, certification, economic instruments, compliance and enforcement, information tools, demonstration projects, education and training, and research and development (R&D) (Ecofys et al., 2013). This policy output broadly covers the relevant elements of ZEBs, including building energy efficiency, solar PV and plug load. Details of these components can be found in the supplemental data online. The subsections below summarize the general ZEBs-relevant governance strategy in Denmark and California.

Denmark
ZEBs policy background
Denmark's ZEBs goal has been seen as an opportunity to increase the renewable energy share in the national energy mix (Marszal et al., 2010), and, in the long-term, to achieve a fossil fuel-free country by 2050. Therefore, ZEBs policies in Denmark tend to apply to multiple sectors. The ZEBs target is accepted by a broad majority in Denmark's parliament and thus has long-term credibility (Low Carbon Transition Unit, 2013). Denmark's ZEBs goal covers new and existing buildings as well as the electricity and heat production systems. The Danish government has signalled to the construction industry that the ZEBs goal will be mandatory in the building code by 2020. Since 2006, the Danish building code has provided three different performance levels that builders can choose. This approach makes clear that the industry needs to prepare for the future ZEBs market. The national policy has inspired several municipalities to adopt ambitious building performance targets. This policy process is considered 'dynamic' in that clear energy-savings targets have been set within a long-term time frame, building energy codes are regularly revised to move buildings toward those goals, and the overall approach encourages developers to go beyond the minimum code requirements (Global Buildings Performance Network, 2014).

The ZEBs policy package in Denmark has a regulatory focus (PRC Bouwcentrum International and Delft University of Technology, 2011), consistent with Denmark's history since the oil crisis of the 1970s of levying energy and carbon taxes and establishing building regulations to mandate energy efficiency. In general, since the 1970s, Denmark's financial mechanisms to encourage energy-efficient buildings have been heavy energy and carbon taxes. Today, Denmark has a comprehensive set of taxes, charges and other fiscal instruments on energy, transportation, pollution and resources. Most of these taxes and charges flow into the general governmental budget, with exception of a 'public service obligation' that is charged as a tariff collected by a state-owned non-profit enterprise (Energi-net.dk), which is independent of the government budget. Renewable-energy deployment in the building sector is supported by a feed-in tariff and net-metering rules (Ministry of Climate Energy and Building of Denmark, 2013). In 2012, solar energy grew exponentially in Denmark; the country achieved its 2020 solar goal in just one year (Solarplaza, 2013). Solar PV is only one of the renewable alternatives for achieving ZEBs in Denmark. Regarding plug loads, Danish appliance energy-efficiency standards generally depend on EU codes (The Danish Government, 2011).

Governance mode in relation to ZEBs policy

Overall, Denmark's ZEBs policy package is balanced. The governance mode for ZEBs can be described as a mix of the hierarchical, market and network forms defined above. Standards, taxes, subsidies and labelling schemes were applied in steering Danish society to adopt ZEBs goals, which are incorporated as part of broader energy policy. The ZEBs goal has been implemented in connection with local planning that also takes into account district heating and renewable energy development (Quitzau, Hoffmann, & Elle, 2012). In addition, the institutional frameworks and lines of responsibility between central and local governments are well defined for municipal energy planning (Sperling, Hvelplund, & Mathiesen, 2011). This governance strategy creates a legitimate way for interdependent actors to participate in establishing ZEBs policies, and communication and information-sharing are imperatives. This approach has helped to elicit high-level political support and unified action (World Green Building Council, 2013). Denmark's ability to act as a society is viewed as a key factor in the country's successful energy transition since 1970s (Lund, 2010).

Key building regulations for energy code compliance

The essential elements for complying with Denmark's ZEBs regulations are: capacity-building and education, benchmarking and disclosure, measurement, verification, and incentives. Denmark was the first country to introduce benchmarking legislation in 1997. The legislation requires energy labelling (energy performance certificates) for new construction and for existing homes at the time of sale (Kjaerbye, 2009). Building owners and property (real estate) agencies face

financial penalties for violating energy performance requirements (European Commission, 2013). Formal legislation passed in 2013 (Law No. 642) authorizes rolling out smart meters to 100% of customers by 2020 (European Commission, 2014). Denmark's energy performance certificate scheme and smart metering programme create preconditions for measuring and verifying building performance. Compliance measures include mandatory and voluntary inspections and sample testing of boilers, appliances, airtightness etc. (European Commission, 2013; Low Carbon Transition Unit, 2013). In particular, periodic furnace/boiler performance checks are mandatory (Evans et al., 2014). For new buildings meeting 2015 or 2020 (ZEBs) requirements, the air-tightness sample size must be 100%. Many municipalities have already applied these requirements to all new buildings (Low Carbon Transition Unit, 2013). Denmark's ZEBs target is performance based. Under the current building regulations, post-occupancy performance requirements are linked to the energy performance certificate scheme. Overall, Denmark has robust programmes to ensure building code implementation and compliance. On issue of building energy code, the compliance procedures are also adjusted over time (Yu, Evans, Delgado, & Northwest, 2014).

California
ZEBs policy background

As a frontrunner in US energy-efficiency and renewable-energy policy and technology deployment, California already has a substantial policy package supporting ZEBs, explicitly outlined in California's ZEBs action plans (CPUC, 2012, 2013). In addition, the state's mild climate and mature solar market help make ZEBs technically and financially feasible. The state's targets are new residential ZEBs by 2020 and new commercial ZEBs by 2030. To date, the California Energy Commission and California Public Utility Commission (CPUC) have, with stakeholders, developed ZEBs action plans for research and technology; codes and standards; heating, ventilation and air-conditioning (HVAC); lighting; and other areas (CPUC, 2012, 2013).

California's overarching goal is bottom-up, market-driven development and deployment of the ZEBs concept. The market-based approach is evident in the economic instruments devised to support ZEBs, e.g., solar PV incentives include tax credits, low-interest loans and rebates. These economic incentives, along with the state renewable portfolio standard and net metering and other interconnection standards, help eliminate the up-front costs of ZEBs technologies. As the solar cost makes up the majority of the incremental cost of ZEBs, the variety of regulatory and financial programmes supporting solar technology helps to motivate builders and building owners to move toward ZEBs. In some cases, commercial ZEBs paid no extra up-front costs (PG&E, 2012) when integrated design trade-offs were made appropriately. However, from the regulatory perspective, there are still barriers to mandating a ZEB goal. The California Energy Commission can only use the building energy-efficiency standard, Title 24, as a regulatory vehicle for this mandate if the measures needed to achieve the ZEBs goal are deemed cost-effective (Heschong Mahone Group Inc., 2012). Therefore, the ZEBs goal is currently not legally binding. In addition, revisions to the state's appliance standard, Title 20, to achieve ZEBs face legal challenges in part because California's appliance standard is subject to federal pre-emption (Chase, McHugh, & Eilert, 2012).

Governing mode in relation to ZEBs policy

Overall, the market-based governance mode dominates in California. Tax credits and rebates are employed rather than subsidies and tax charges, and voluntary certificates have been established. These are in line with the broader US political–economic context of free market (neo-liberal) policies in the 1980s and 1990s, which weakened states' abilities to govern the public good (Kallis, Kiparsky, & Norgaard, 2009). California's administrative structure exhibits characteristics of hierarchical, market and network governance in the form, respectively, of renewable portfolio standards and voluntary targets, expert and NGO networks, and decisions deliberatively made through advisory boards, multi-stakeholder panels and campaigns.

Key building regulations for energy code compliance

A mandatory benchmarking and disclosure policy for commercial buildings adopted in 2007 and widespread smart meters in the three investor-owned utilities' (IOUs) service territories provide a baseline for measuring and verifying actual building performance. California has a strong inspection programme for both prescriptive and performance building codes. To obtain a construction permit, builders must submit plans to the local building department for code compliance review. A code official also conducts a field inspection prior to issuing a certificate of occupancy (Misuriello, Penney, Eldridge, & Foster, 2010). Other efforts to ensure code compliance include simplifying the regulatory process and providing training (California Energy Commission, 2014). However, none of these measures is linked to a post-occupancy energy performance requirement for buildings. In the US, some voluntary initiatives already encompass an outcome-based compliance path, including the Living Building Challenge certificate and the new IgCC outcome-based code. The US Environmental Protection Agency (EPA) Energy Star voluntary programme (for commercial buildings) and the American Society

of Heating, Refrigerating, and Air-Conditioning Engineers (ASHRAE) Building Energy Quotient programme (for commercial buildings) also include post-occupancy requirements (Khanna, Romankiewicz, Zhou, Feng, & Ye, 2014). Currently, California is participating in the national discussion of outcome-based building codes (Denniston et al., 2011). A major challenge for building energy code enforcement and compliance in California is the lack of prioritization of energy codes relative to health and safety codes (Misuriello et al., 2010). This problem is commonly faced by other jurisdictions as well (Misuriello et al., 2010).

Zero-energy-building governance in China
ZEBs policy background

In 2013, China added nearly 3.9 billion m^2 of new buildings (National Bureau of Statistics of China, 2014). This represents about half the world's annual new construction. Therefore, implementing a clear ZEB target along with the associated governance strategies will have a very significant impact on future building energy consumption in China. Since the ZEBs concept has just recently been introduced in China, as of the end of 2015, fewer than 10 demonstration projects have been reported (China Academy of Building Research, 2014; Ye, Zuo, & Hu, 2013). Most of these projects are commercial buildings and not certified as ZEBs. These cases were mainly supported by leading building developers and building research institutes that intended to test their design and building components. Other types of ultra-low energy buildings, such as those meeting the Passive House standard, have been developed and tested more than ZEBs in China. For instance, there have been 40 buildings proposed by 28 public/private actors to meet the Passive House standard by the end of 2015 (Zhang, 2015). Considering the enormous size of the new building stock growth in China, these practices represent just the beginning of ZEBs development.

Currently, national building codes require that both commercial and residential buildings be 65% more efficient than the baseline from the early 1980s; some province and city codes are even stricter, requiring a 75% energy-efficiency improvement for residential buildings. These codes are currently prescriptive. Performance-based codes have only recently been proposed by China's Ministry of Housing, Urban and Rural Development (MOHURD). These proposed codes define energy quotas for different types of buildings in four climate zones. Some pilot cities, *e.g.*, Beijing, Shanghai and Shenzhen, have adopted commercial-building energy quotas. In Shenzhen, the proposed annual on-site electricity intensity quota is 90 kWh/m^2 for governmental buildings and 120 kWh/m^2 for office buildings (Housing and

Construction Bureau of Shenzhen Municipality, 2015). Existing buildings must meet energy quotas based on actual energy measurements; for new buildings, the requirement is based on predicted energy performance. When a new building is completed and begins operating, the energy quota is regarded as the rated value of energy use, which can be used as a benchmark to manage building energy performance (MOHURD, 2014).

China is facing a choice of direction for the next phase of building-sector energy requirements. Discussion about a building sector roadmap by 2030 has been initiated within MOHURD and among building industry leaders. Although a final ZEBs target has not yet been released, a conference presentation by roadmap working group participants described elements that are being considered for inclusion in the roadmap. A building sector road map would be the first signal from Chinese officials to the market of the goal of realizing ZEBs by 2030 (Wu et al., 2015). In contrast to the EU and US ZEBs goals, the Chinese ZEBs goal is part of the building sector roadmap that is planned to encompass not only energy-efficiency requirements but also water and material efficiency, indoor environment enhancement, and waste reduction goals. Currently, the draft plan does not explicitly state a ZEB goal but refers to the Chinese Passive House standard (Wu et al., 2015) and proposes a non-binding targeting regulation.

China's existing policies related to ZEBs are summarized in the supplemental data online. Because ZEBs have just appeared on the Chinese market, policy instruments such as certification, information tools, demonstration projects, education and training, and R&D are lacking. In general, China's building energy-efficiency policies impose regulations that grow incrementally more stringent over time, with stricter building codes and energy consumption caps for different types of buildings. In addition to these requirements, a variety of financial incentives encourage high-performance building pilot projects, *e.g.*, green buildings, highly energy-efficient buildings and energy-efficiency retrofits of existing commercial buildings. For PV, the most relevant policies are MOHURD subsidies initiated in 2006 for building-integrated renewables, the Golden Sun demonstration subsidy programme launched by the Ministry of Finance and the National Development and Reform Commission (NDRC) in 2009, and the feed-in-tariff for renewables (NDRC, 2013). Traditionally, China's renewable-energy policy has paid more attention to wind than solar, and the PV policy has mostly been formulated from the supply side (Zhi, Sun, Li, Xu, & Su, 2014). Solar PV projects did not achieve a large scale in China until 2009. In response to international trade barriers to Chinese PV products, policies are shifting slowly to expand the domestic market. Regarding

plug loads, the newly proposed energy quota can be seen as an attempt to regulate actual electricity use once a building is operating.

Governance mode in relation to ZEBs policy

Overall, the hierarchical governance mode dominates in China, exhibited in a variety of building codes and updates, governmental subsidies, periodic inspections and a 'target responsibility' system to ensure compliance. The decision-making process for ZEBs policies and codes is restricted to experts on advisory boards. There are some market governance-mode elements, such as voluntary standards, industrial alliances and incorporation of the building sector into a carbon cap-and-trade scheme. Campaign-style approaches also exist, but they are limited to efforts to address emergent local environmental crises (Liu, Lo, Zhan, & Wang, 2014). From a governance perspective, this approach of formulating and implementing ZEBs policies means a collective and complex goal is being pursued without participation by all societal actors who will directly influence the policy outcome.

Key building regulations for energy code compliance

China is increasingly investing in real-time, on-line building energy monitoring, and MOHURD has committed to incorporating benchmarking into the 13th Five-Year Plan (Szum, 2014). In addition, the voluntary Chinese green building codes provide design and operational rating options. Together with China's energy quotas, these elements create the preconditions for performance- or outcome-based code compliance. In China, building code compliance is verified by local governments and third parties. The central government relies on annual inspections and the target responsibility system to ensure compliance. In contrast to the compliance checks in the US, the annual inspection check in China also involves enforcement, and corrections are required, if needed, to meet the corresponding energy codes (Yu et al., 2014). Local governments are usually subjected to publicity pressure about violations of the target. In some pilot cities, *e.g.*, Beijing and Shenzhen, commercial building owners face legal and financial penalties if their buildings exceed the energy quota (Dong, 2014). This approach indicates that the code compliance path is beginning to achieve legitimacy.

Governance target and policy implications for China

Zero-energy buildings' terminology in the Chinese context

Table 2 shows a broad range of US and European ZEBs definitions. This range of definitions tailored to local conditions suggests that China could develop its own

ZEBs definitions to suit the country's diverse climate zones and societal goals (*e.g.*, emissions reductions goals). Currently, increasing numbers of ZEBs projects in China have adopted the 'site energy' definition, with on-site renewables only. However, although there have been a few attempts in the research community to define a concept tailored to the Chinese context, a clear, official ZEBs definition is still lacking (Zhang et al., 2013). Several aspects of ZEBs in China might differ from features of other international examples. These aspects, discussed below, could be considered in the process of officially defining China's ZEBs concept.

First, achieving ZEBs in China's urban areas will be challenging if the ZEB system boundary is limited to on-site renewables. This is because of the predominance of urban high-rise buildings that have a limited roof area for PV panels. To address this challenge, the ZEBs definition could take into account energy use at the neighbourhood or community level rather than at the individual-building level; it could allow renewable-energy credits to offset energy use in urban high rises, or it could allow for other diverse alternative energy sources. A preliminary technical study by Chinese researchers using the site-energy ZEBs definition shows that, in theory, it is possible to build an 8–10-storey commercial ZEB and a 9–11-storey residential ZEB in China's three climate zones (*e.g.*, Beijing, Shanghai and Guangzhou) (Huang, 2014).

Second, many regions in Europe developed ZEBs definitions based on existing codes, *e.g.*, Minergie-A in Switzerland, or as a progression from the Passive House standard. In China, MOHURD has already gained significant experience implementing green building codes and has recently started to experiment with the Passive House concept and the energy quota described above. These experiences could be considered in developing the Chinese ZEBs concept.

Finally, most European and US ZEBs definitions adopt one operational year as the balance period for determining net energy use. This suggests that the Chinese code needs to move from being prescriptive to being performance- and outcome-based. China's proposed energy quota policy is a step toward addressing the actual measured energy performance of buildings. Monitoring and new mechanisms to help ensure compliance need to be developed, *e.g.*, fines or denial of occupancy permits (IPEEC, 2015).

Policy implications for China

This analysis of governance strategies for implementing ZEBs targets focuses on the policy dimension. In many cases, more than one policy instrument can be used, however the choice of specific instruments is affected by the prevailing governance process in each jurisdiction (Young et al., 2015).

Strategic governance of zero-energy buildings in China

Large variation exists within any governance system, so generalizations are by definition simplifications. Nonetheless, it is possible to identify dominant features in the governance processes. For instance, defining a clear, stable and long-term collective target is the prerequisite for governance, and leadership is important in all governance systems, whether from government, market actors or civil society. The policy options that have been adopted in each region are consistent with the governance modes in the region, which are various balanced hybrid forms made up of a mixture of hierarchy, market and network. The governance mode in China is distinctly hierarchical, although some market elements have been introduced. Deliberative forms of governance involving non-governmental actors are less evident in China's decentralized authoritarian system. Energy efficiency, renewable energy and other environmental issues are understood to be the government's responsibility in China. While a strong state that takes progressive steps toward climate change mitigation policies is an important element of successful energy policy, the absence of participation from societal actors who directly influence ZEBs policy outcome is a barrier. By contrast, in California where ZEBs policies have succeeded, responsibility is shared among government (at all levels), the market and society. Similarly, a key lesson from Denmark is that the ability to act as a unified society is a key factor in the Danish energy transition that has taken place since the 1970s. And state institutions implemented these high-level agreements to ease the governance process. Thus, from a governance perspective, the key question for China's developing ZEB policies appears to be not how government can better control the issue (Zhou, Ai, & Lian, 2012; Zhou & Lian, 2012), but how government can adapt to and encompass participation by the diverse, complex set of stakeholders whose contributions are essential to the success of ZEBs policies.

In addition to lack of participation from all societal actors, China's policy implementation outcomes are also often affected by differences between national and local priorities. In China's fragmented vertical and horizontal governance structure ('*tiaotiao kuaikuai*'), mandatory standards and hard targets have been the most important measures (Kostka & Hobbs, 2012). These are interconnected to a cadre evaluation system for government workers, in which performance evaluation and political promotion are based primarily on short-term economic growth. Projects that favour the local economy and are initiated by developers are usually preferred. At the same time, urban planning guidelines, including building regulations, may be ignored (Lu, 2012). In this context, building energy-efficiency incentive implementation is solely through a top-down target responsibility system. The target responsibility system has been found problematic (*e.g.*, unscientific targets that are inflated and difficult to verify, with compliance responsibility placed on less powerful bureaus compared with the bureaus in charge of economic growth, among other issues) (Lo, 2014). Researchers have suggested that, under the current governance mode, policy bundling, interest bundling and policy framing could facilitate effective implementation by local governments (Kostka & Hobbs, 2012). Kostka and Hobbs (2012) noted that bundling processes are carried out informally; local officials and enterprises cooperate on energy efficiency in exchange for compensatory benefits for each enterprise on other issues. Policy framing can be accomplished using reframed language to incentivize participants. To address coordination issues, Ran (2015) suggested that environmental policies can be effective when overseen by local party secretaries or mayors. These are only a few possible remedies to improve policy effectiveness under the current governance mode. More studies are needed to analyse how China's very long governance history has resulted in the country's current system of bureaucracy and how that, in turn, affects the options for new policies and transitions in governance mode. Applying a general environmental target to implement ZEBs would likely face similar, if not greater, barriers at the local level as are faced by the current target responsibility system.

Implications for building regulations to achieve ZEBs outcomes

Various code enforcement and compliance evaluation tools exist in Denmark, California and China. However, curbing building energy use in China is particularly challenging. Building energy use intensity in some EU and North American jurisdictions has decreased during the past 10 years. By contrast, it has been challenging to reduce building energy use in China, even with implementation of energy-efficiency policies, because of increased saturation of energy-using appliances and improved interior comfort conditions as incomes rise. In some EU and North American jurisdictions that achieved near-total saturation of the most energy-efficient appliances between 2000 and 2012, electricity use per household decreased and residential total final energy use per floor area dropped during the same period (IEA and IPEEC, 2015). In Denmark, for example, residential energy use per capita decreased 7.5% (climate adjusted) from 2000 to 2012 (ODYSSEE, 2015; Statistics Denmark, 2015). The same metric in California decreased 3.15% from 2000 to 2012 (US Census Bureau, 2015; US EIA, 2015).

However, during the same period, the per capita energy use in the Chinese residential sector (both urban and rural) increased substantially by 137.5% (National Bureau of Statistics of China, 2015), even as district heating energy intensity has been continuously decreasing because of significant building energy conservation work. And, further complicating the picture, the current average whole-building energy intensity in China is still significantly lower than in the Western world and thus has potential to continue growing as China's living standards increase (IEA and IPEEC, 2015). These conditions indicate that successfully implementing a ZEB target in China will be challenging, with efforts to make buildings more energy efficient potentially offset by growing energy demand from increasing living standards.

In the two regions studied in this paper, code compliance has been facilitated through regular inspection and testing of building components. Benchmarking and disclosure programmes along with widespread smart-metering programmes have provided feedback on code compliance. At the same time, voluntary codes and labels reflect actual post-occupancy energy consumption.

In China, the voluntary green building evaluation and labelling programme consists of a design and operation rating label; reporting of actual energy consumption once the building is operating is not required (Khanna et al., 2014). Currently, China is investing in real-time, on-line building energy monitoring. MOHURD is also committed to incorporating benchmarking into the 13th Five-Year Plan. And a transition from prescriptive to performance (energy quota) codes has started in some pilot cities where associated incentives and penalties ensure code compliance. This infrastructure provides the basis for adopting a future outcome-based code if appropriate. Given the challenges in transitioning from prescriptive codes to performance- and outcome-based codes, it is likely that all three types of codes would co-exist for a period of time.

Although governance modes in Denmark, California and China differ, building regulations and code enforcement and compliance in these three regions share many common features, such as capacity-building and education, benchmarking, verification and code compliance checks. In China, the question is how these similar strategies can be expanded and strengthened, how policies can be implemented effectively at the local level, and how to address implementation gaps between central and local governments.

Summary and conclusions

A clear, stable, long-term, collective target is the basis for governance. China's building energy code has begun to transition from prescriptive to performance-based, and in some cases outcome-based (the new

Table 3 Summary of key zero-energy buildings (ZEBs) policy governance considerations

Defining a clear, long-term governing target for ZEBs	• ZEBs definition boundary could consider going beyond the individual-building level • Develop ZEBs definitions based on existing codes and industry practices • The current Chinese code needs to move from being prescriptive to being performance- and outcome-based. Monitoring and new mechanisms to help ensure compliance need to be developed
Key building regulations for energy code compliance	Building regulations, code enforcement and compliance measures have been established in China. These include capacity-building, education, benchmarking, verification and code compliance checks. The challenges lie in how these strategies can be expanded and strengthened for monitoring real ZEBs energy performance, and how the relevant policies can be implemented effectively at the local level
Key governing strategies	China could benefit by shifting to a more balanced, hybrid governance system that includes hierarchical, market and network forms. Other than administrative and market instruments, encompassing more deliberative elements in decision-making and implementation processes could help to reduce the current governance deficits. The complexity of ZEBs targets requires broader participation from a range of societal actors who will directly influence the policy outcome, including influential local officials, developers, consumers, building owners and occupants

energy quota policy). China's development of a ZEB definition could benefit from the example of China's existing voluntary codes as well as innovative definitions used in leading world regions. Building regulations and code enforcement and compliance in the EU, US and China share many common features, although there is a need to expand and strengthen these mechanisms in China (as well as other countries). One key element to consider is that traditional building regulations cannot address all key ZEBs issues; governance to implement a ZEB target needs to take a multi-sector perspective (including both the building and energy sectors), which also requires political consensus among various stakeholders. Most important, based on an understanding of governance concepts and an empirical analysis of ZEBs governing strategies in the selected regions and China, it appears clear that China's focus needs to shift to a governance process, in which all relevant societal actors participate in ZEBs policy development and implementation. That is, China could benefit by shifting to a balanced, hybrid governance system that includes hierarchical, market and network elements such as those illustrated in the cases of California and Denmark. Governance of the ZEBs target in China is currently limited by the structures and issues associated with the target responsibility system and the use of economic levers as the main policy instruments under the country's predominantly hierarchical governance mode. The complexity of ZEBs targets requires much more broad participation from societal actors who will directly influence the policy outcome, including influential local officers, developers, consumers, building owners and occupants (Table 3).

Finally, it is important to note that this paper only takes a snapshot of current systems of ZEBs governance, how these systems evolve were not taken into account by the study. An adaptive capacity is needed

that allows for experimentation, failure, learning and recognition that any governance system operates at a less-than-optimal level. Due to the broad international scope (in terms of both geography and complexity) of ZEBs policies and activities, many examples and strategies are not included in this paper. Future research should study, in depth, the institutional and structural challenges to deploying the ZEBs concept in China.

Disclosure statement

No potential conflict of interest was reported by the authors.

Funding

This work was supported by the US Department of Energy [contract number DE-AC02-05CH11231]; by faculty funding from the Department of Environmental and Energy Systems Studies at Lund University; by a research exchange scholarship from the Swedish Royal Physiographic Society; by the Energy Foundation; and by the China Academy of Building Research Program (P. R. China).

Supplemental data

Supplemental data for this article can be accessed at 10.1080/09613218.2016.1157345

References

APEC Energy Working Group. (2014). *APEC Project report: Nearly (Net) zero energy building*. Beijing, China: APEC.

Atanasiu, B., Kunkel, S., & Kouloumpi, L. (2013). *NZEB criteria for typical single-family home renovations in various countries*. Brussels, Belgium: BPIE.

Bäckstrand, K., Khan, J., Kronsell, A., & Lövbrand, E. (2010). Environmental politics and deliberative democracy: Examining promise of new modes of governance (p. 29). In K.

Bäckstrand, J. Khan, A. Kronsell, & E. Lövbrand (Eds.), Massachusetts, US: Edward Elgar Publishing.

Boermans, T., & Grözinger, J. (2011). *Economic effects of investing in EE in buildings – the BEAM² model background paper for EC workshop*. Köln, Germany: Ecofy.

California Energy Commission. (2014). Title 24 training. Retrieved August 2, 2014, from http://www.energy.ca.gov/title24/training/

California Public Utilities Commission. (2008). *California long-term energy efficiency strategic plan*.

California Public Utilities Commission. (2013). Lighting Action Plan 2013–2015.

California Public Utility Commission. (2012). HVAC Action Plan 2010–2012.

Chase, A., McHugh, J., & Eilert, P. (2012). Federal appliance standards should be the floor, not the ceiling?: Strategies for innovative state codes & standards. In *2012 ACEEE summer study on energy efficiency in buildings* (pp. 13–36). Pacific Grove, CA: ACEEE.

China Academy of Building Research. (2014). Introduction of CABR nearly zero energy building. In *2014 US–China CERC-BEE Annual Workshop*. Berkeley, CA. Retrieved from https://cercbee.lbl.gov/sites/all/files/attachments/Day1-Panel4-EEStandards-Wei.FINAL_NEW_pdf

Cortese, A., Higgins, C., Lyles, M., & Hamilton, B. (2014). *Getting status update 2014?: A look at the projects; policies and programs driving zero net energy performance in commercial buildings*. Portland, Oregon: New Buildings Institute.

Danish Government. (2011). *Our future energy*. Copenhagen, Denmark: Danish Ministry of Climate, Energy and Buildings.

Denniston, S., Frankel, M., & Hewitt, D. (2011). Outcome-based energy codes on the way to net zero (white paper). Retrieved from http://newbuildings.org/sites/default/files/Hewitt-Outcome_to_Zero.pdf

Dokka, T. H., Sartori, I., Thyholt, M., Lien, K., & Lindberg, K. B. (2013). A Norwegian zero emission building definition. In *Passivhus Norden 2013*.

Dong, S. (2014). Building energy efficiency policy gradually transit to legalization (In Chinese). *China Building Metal Structure*, 11, 18–20. Retrieved from http://mall.cnki.net/magazine/Article/ZJJS201411003.htm

Ecofys, Politecnico di Milano, & University of Wuppertal. (2013). *Towards nearly zero-energy buildings: Definition of common principles under the EPBD*. Köln, Germany: Ecofys.

European Commission. (2013). *Implementing the Energy Performance of Building Directive (EPBD) – featuring country reports 2012*. Brussels, Belgium.

European Commission. (2014). Cost–benefit analyses & state of play of smart metering deployment in the EU-27. Brussels: European Commission.

Evans, M., Halverson, M., Delgado, A., & Yu, S. (2014). Building energy code compliance in developing countries?: The potential role of outcomes-based codes in India. In *2014 ACEEE summer study on energy efficiency in buildings* (pp. 61–74). Monterey, CA. Retrieved from http://aceee.org/files/proceedings/2014/data/papers/8-283.pdf

Faludi, J., Lepech, M. D., & Loisos, G. (2012). Using life cycle assessment methods to guide architectural decision-making for sustainable prefabricated modular buildings. *Journal of Green Building*, 7(3), 151–170. doi:10.3992/jgb.7.3.151

Global Buildings Performance Network. (2014). Designing and implementing best practice building codes: Insights from policy makers. Retrieved from http://www.gbpn.org/sites/default/files/05_Design and implementation of best practice building codes_1.pdf

Groezinger, J., Boermans, T., John, A., Seehusen, J., Wehringer, F., & Scherberich, M. (2014). *Overview of member states information on NZEBs working version of the progress report – final report*. Cologne, Germany: Ecofys.

Heschong Mahone Group Inc. (2012). *The road to ZNE: Mapping pathways to ZNE buildings in California*. Gold River, California: PG&E.

Hildingsson, R. (2015). *Governing decarbonisation the state and the new politics of climate change*. Lund, Sweden: Lund University.

Housing and Construction Bureau of Shenzhen Municipality. (2015). *Standard for energy quota of office building in Shenzhen (proposal in Chinese)*. Shenzhen, China: Housing and Construction Bureau of Shenzhen Municipality.

Huang, K. (2014). *Net-zero energy building feasibility study in China (in Chinese)*. Tianjin, China: Tianjin University.

IEA and IPEEC. (2015). *Building energy performance metrics*. Paris, France: IEA Publications.

IPEEC. (2015). *Delivering energy savings in buildings*. Paris, France: IPEEC.

Ji, Y., & Guo, X. (2013). International development of nearly zero energy buildings (In Chinese). *Construction Economy*, 5, 88–92.

Kallis, G., Kiparsky, M., & Norgaard, R. (2009). Collaborative governance and adaptive management: Lessons from California's CALFED water program. *Environmental Science and Policy*, 12(6), 631–643. doi:10.1016/j.envsci.2009.07.002

Khanna, N. Z., Romankiewicz, J., Zhou, N., Feng, W., & Ye, Q. (2014). From platinum to three stars: Comparative analysis of U.S. and China green building rating programs. In *2014 ACEEE summer study on energy efficiency in buildings* (pp. 402–414). Pacific Grove, CA: ACEEE.

Kjaerbye, V. H. (2009). Does energy labelling on residential housing cause energy savings? In *ECEEE 2009 summer study* (pp. 527–537). Retrieved from http://www.eceee.org/library/conference_proceedings/eceee_Summer_Studies/2009/Panel_3/3.068/paper

Kostka, G., & Hobbs, W. (2012). Local energy efficiency policy implementation in China: Bridging the gap between national priorities and local interests. *The China Quarterly*, 211, 765–785. doi:10.1017/S0305741012000860

Liu, N. N., Lo, C. W.-H., Zhan, X., & Wang, W. (2014). Campaign-style enforcement and regulatory compliance. *Public Administration Review*, 75(1), 85–95. doi:10.1111/puar.12285

Lo, K. (2014). China's low-carbon city initiatives: The implementation gap and the limits of the target responsibility system. *Habitat International*, 42, 236–244. doi:10.1016/j.habitatint.2014.01.007

Low Carbon Transition Unit. (2013). *Energy policy toolkit on energy efficiency in new buildings: Experiences from Denmark*.

Lu, D. (2012). *The great urbanization of China*. Singapore: World Scientific Publishing Co. Pte. Ltd.

Lucia, L. Di. (2012). *Public policy and the governance of biofuel systems*. Lund, Sweden: Lund University.

Lund, H. (2010). The implementation of renewable energy systems. Lessons learned from the Danish case. *Energy*, 35(10), 4003–4009. doi:10.1016/j.energy.2010.01.036

Marszal, A. J., Bourrelle, J. S., Nieminen, J., Berggren, B., Gustavsen, A., Heiselberg, P., & Wall, M. (2010). North European understanding of zero energy / emission buildings. In *Zero emission buildings – proceedings of renewable energy conference 2010* (pp. 167–178). Trondheim, Norway: Norwegian Centre for Renewable Energy.

Ministry of Climate Energy and Building of Denmark. (2013). *Energy policy report 2013*. Copenhagen, Denmark: Ministry of Climate Energy and Building of Denmark.

Misuriello, H., Penney, S., Eldridge, M., & Foster, B. (2010). Lessons learned from building energy code compliance and enforcement evaluation studies. In *ACEEE summer study on energy efficiency in buildings* (pp. 245–255). Retrieved from http://aceee.org/files/proceedings/2010/data/papers/2185.pdf

MOHURD. (2014). *Standard for energy consumption of buildings (drafts for comments)*. Beijing, China: Ministry of Housing, Urban and Rural Development of China.

National Bureau of Statistics of China. (2014). *China statistic yearbook 2014.* Retrieved July 2, 2015, from http://data.stats.gov.cn/workspace/index?m=hgndv

National Bureau of Statistics of China. (2015). *Residential energy consumption per capita (In Chinese).* Retrieved August 10, 2015, from http://data.stats.gov.cn/easyquery.htm?cn=C01

New Buildings Institute. (2015). *List of zero energy buildings.* Vancouver, Washington: New Buildings Institute.

NDRC. (2013). *Using price mechanism to promote solar PV industry development (In Chinese).* Retrieved July 31, 2015, from http://www.mlfgw.gov.cn/Article/ShowArticle.asp?ArticleID=428

ODYSSEE. (2015). *Danish households energy consumption per dwelling at normal climate.* Retrieved August 10, 2015, from http://www.indicators.odyssee-mure.eu/online-indicators.html

PG&E. (2012). *California zero net energy buildings cost study.* Davis, CA: PG&E.

Pierre, J., & Peters, B. G. (2000). *Governance, politics and the state.* Hampshire, UK: Palgrave Macmillan.

PRC Bouwcentrum International and Delft University of Technology. (2011). *Screening national building regulations – Denmark.* Delft, The Netherlands: PRC Bouwcentrum International.

Quitzau, M. B., Hoffmann, B., & Elle, M. (2012). Local niche planning and its strategic implications for implementation of energy-efficient technology. *Technological Forecasting and Social Change, 79*(6), 1049–1058. doi:10.1016/j.techfore.2011.11.009

Ran, R. (2015). *Local environmental politics in China: The gap between policy and implementation (In Chinese).* Beijing, China: Central Compilation & Translation Press.

Sartori, I., Napolitano, A., & Voss, K. (2012). Net zero energy buildings: A consistent definition framework. *Energy and Buildings, 48,* 220–232. doi:10.1016/j.enbuild.2012.01.032

Simmons, R. (2015). Constraints on evidence-based policy: Insights from government practices. *Building Research & Information,* (September), 1–13. doi:10.1080/09613218.2015.1002355

Solarplaza. (2013). *Denmark: Minister to speak at solar energy conference.* Retrieved August 3, 2014, from http://thesolarfuture.dk/news/

Sperling, K., Hvelplund, F., & Mathiesen, B. V. (2011). Centralisation and decentralisation in strategic municipal energy planning in Denmark. *Energy Policy, 39*(3), 1338–1351. doi:10.1016/j.enpol.2010.12.006

Statistics Denmark. (2015). *Danish households population.* Retrieved August 10, 2015, from http://www.statbank.dk/BEV21

Szum, C. (2014). Research on very low-energy building operations and management methods. In *US–China Clean Energy Research Center building energy efficiency joint annual meeting.* Berkeley, CA. Retrieved from https://cercbee.lbl.gov/sites/all/files/attachments/Day1-Panel4-LowE-ICF-Carolyn.FINAL_pdf

Treib, O., Bähr, H., & Falkner, G. (2007). Modes of governance: Towards a conceptual clarification. *Journal of European Public Policy, 14*(1), 1–20. doi:10.1080/135017606061071406

UK Green Building Council. (2014). *New build: Domestic and Non-domestic.*

US Census Bureau. (2015). *California resident population 2010–2014.* Retrieved August 10, 2015, from http://factfinder.census.gov/faces/tableservices/jsf/pages/productview.xhtml?pid=PEP_2014_PEPAGESEX&prodType=table

US Department of Energy. (2015). A common definition for zero energy buildings. Retrieved from http://energy.gov/sites/prod/files/2015/09/f26/bto_common_definition_zero_energy_buildings_093015.pdf

US EIA. (2015). *California residential sector energy consumption estimates 1960–2013.* Retrieved August 10, 2015, from http://www.eia.gov/state/seds/data.cfm?incfile=/state/seds/sep_use/res/use_res_CA.html&sid=CA

The World Bank and Development Research Center of the State Council of China. (2014). *Urban China: Toward efficient, inclusive, and sustainable urbanization.* Washington, DC. Retrieved from https://openknowledge.worldbank.org/handle/10986/18865

World Green Building Council. (2013). *Collaborative Policy making case study: Denmark.* Toronto, Canada: World Green Building Council.

Wu, Y., Hou, J., Liu, Y., Li, Y., Xu, K., Zhou, N., & Feng, W. (2015). Research on an energy-efficiency improvement roadmap for commercial buildings in China. In *ECEEE 2015 summer study.* Retrieved from http://proceedings.eceee.org/visabstrakt.php?event=5&doc=6-192-15

Xin, Y., Lu, S., Zhu, N., & Wu, W. (2012). Energy consumption quota of four and five star luxury hotel buildings in Hainan province, China. *Energy and Buildings, 45,* 250–256. doi:10.1016/j.enbuild.2011.11.014

Ye, G., Zuo, Z., & Hu, W. (2013). Nearly-zero carbon/energy building research and practices in Southern China. *Construction Technology (In Chinese), 10,* 44–49. Retrieved from http://211.151.247.143/magazine/Article/KJJS201310018.htm

Young, O. R., Guttman, D., Qi, Y., Bachus, K., Belis, D., Cheng, H., … Zhu, X. (2015). Institutionalized governance processes comparing environmental problem solving in China and the United States. *Global Environmental Change, 31,* 163–173. doi:10.1016/j.gloenvcha.2015.01.010

Yu, S., Evans, M., Delgado, A., & Northwest, P. (2014). Energy code enforcement and compliance evaluation: Comparative lessons learned from the US and China, and opportunities for India importance of code compliance evaluation. In *2014 ACEEE summer study on energy efficiency in buildings* (pp. 415–427). Pacific Grove, CA: ACEEE.

Zero Carbon Hug. (2014). *Zero carbon policy.* Retrieved August 10, 2010, from http://www.zerocarbonhub.org/zero-carbon-policy/zero-carbon-policy

Zhang, S., Xu, W., Jiang, Y., Feng, W., & Sun, D. (2014). Research on demonstration project technology roadmap of zero energy buildings abroad (In Chinese). *HV&AC, 44*(1), 53–60.

Zhang, S., Xu, W., Yiqiang, J., Wei, F., Deyu, S., & Liu, Z. (2013). Research on definition development and main content of zero energy building (In Chinese). *Building Science, 29*(10), 114–120.

Zhang, X. (2015). The current status of passive houses development in China (In Chinese). *Construction Technology, 15,* 16–27.

Zhi, Q., Sun, H., Li, Y., Xu, Y., & Su, J. (2014). China's solar photovoltaic policy: An analysis based on policy instruments. *Applied Energy, 129,* 308–319. doi:10.1016/j.apenergy.2014.05.014

Zhou, X., Ai, Y., & Lian, H. (2012). 'The limit of bureaucratic power in organizations: The case of the Chinese bureaucracy' in rethinking power in organizations, institution, and markets. *Research in the sociology of organizations.* Emerald Group Publishing Ltd. doi:10.1108/S0733-558X(2012)0000034006

Zhou, X., & Lian, H. (2012). Modes of governance in the Chinese bureaucracy: A 'control right' theory (In Chinese). *Journal of Sociological Research, 5,* 69–93.

RESEARCH PAPER

Multilevel governance for building energy conservation in rural China

Kaixun Sha and Shaoyan Wu

Under the city–countryside dual structure, the existing building governance system in China differentiates between urban and rural areas. When updating building regulations and related policies to meet challenges in the built environment, it is essential to develop different strategies for different locations according to local circumstances, requirements and capabilities. Based on two research projects, this article examines the mechanisms and strategies for promoting building energy conservation in rural China from the perspective of economic governance. The challenges and potentials of building energy conservation in rural China are analyzed. The essence of the governance paradigm is briefly reviewed. A three-level analysis framework is developed in which markets, governments and the third party (professionals and others) play complementary roles in regulating stakeholders' behaviour. A key question addressed is how to create a favourable institutional environment in which people are willing to do the right things. Different strategy portfolios are proposed for different levels, including technology strategy, financing strategy, as well as regulations and incentive policies. In conclusion, there is no 'best' but rather the 'most suitable' approach to building governance. In this light, the principle of discriminating alignment and the multilevel analysis approach provides conceptual insights.

Introduction

After 30 years of rapid growth, China has become the world's second largest economy. However, a heavy cost in environmental degradation is being paid for this economic boom. In the building sector, the total completed construction area of 4.23 billion m^2 in 2014 (National Bureau of Statistics of China, 2015a) exhibits both the sector's positive contribution to economic growth and its negative influence on the environment. Challenges in the built environment call for a shift of mindset from purely technical considerations to socio-technical considerations by paying more attention to the interaction between people and technology. Technology is no doubt a major factor, but what is more important is that existing technologies can be effectively applied in practice with the support of a sound governance system.

After three decades of efforts, China has established its basic institutional framework for building energy conservation, including laws, regulations and incentive policies, technological standards and codes, etc. However, due to the emphasis on the gross domestic product (GDP)-first value to promote economic growth (e.g., the opposite of the value of placing people first), the prevailing practice of 'rule of the administrator' (as opposed to the 'rule of law'), uneven regional development and imbalanced relationships between markets, governments and building professionals and their institutions, the current framework for building energy conservation does not work well enough (Sha, 2014; Sha, Song, Qi, & Luo, 2006). What is worse, the framework is basically city-oriented, with little constraint effect on rural areas. As a result, the progress of building energy conservation in the countryside is far behind that in the city. A favourable turn arose in October 2005 when the Chinese government launched its so-called 'new countryside' initiative, creating an opportunity to narrow the gap between urban and rural areas in terms of building energy conservation.

When building regulations and related policies are examined in the context of building energy conservation, externality – the uncompensated impact of one person's actions on the well-being of a bystander – is an inevitable issue, since building-related activities impact negatively not only on various stakeholders but also on uninvolved parties. This is true for the impacts on the present generation as well as on future generations, without, in most cases, adequate compensation being paid.

The issue of internalizing building-related externality is ultimately concerned with tackling the interdependent relationships between relevant parties, individual and organizational, public and private, explicit and implicit (e.g., future generations). In order to make these parties whose interests are not necessarily consistent or even conflicting work together and cooperate for win–win results, it is essential to create a favourable institutional environment in which people are willing to do the right things (building energy conservation) in the right ways. This is bound to be a complex task that possesses the essential properties of organizational complexity: correlated interaction and interdependence between various agents, multiple goals, non-linearity and uncertainty (Ahern, Leavy, & Byrne, 2014; Cleden, 2009; Spering, Wagener, & Funke, 2005), demanding a more flexible mindset (Miller & Hobbs, 2005). Characterized by different forms of flexibility at different levels, the governance paradigm can be regarded as an art of complexity (Jessop, 2003), and is capable of dealing with the above-mentioned interdependent relationships (Müller, Pemsel, & Shao, 2014).

Based on two research projects[1] in Shandong,[2] an eastern province of China, this article will examine the mechanisms and strategies for promoting building energy conservation in rural China from the perspective of economic governance. The study is conducted in two dimensions. In the vertical dimension (y-axis), three geographical levels – so-called 'urban villages', villages in the outskirts and emerging residential quarters in the countryside, and traditional villages – are identified as analytical categories of building governance. In the horizontal dimension (x-axis), governments, markets and the third party (building professionals and their institutions) are taken as complementary approaches to regulate stakeholders' behaviour. The article is organized as follows. The next section introduces the challenges and potentials of building energy conservation in rural China. The third section reviews the literature on the governance paradigm by regarding governance

as an art of complexity. The fourth section briefly introduces the data collection and analysis process. The fifth section develops a three-level analysis framework. The sixth section proposes different strategy portfolios, including technology strategy, financing strategy, as well as regulations and incentive policies. The conclusions section considers the implications of the research findings for new forms of building energy governance in China and identifies directions for further research.

Background

In spite of the steady urbanization progress over the last three decades, more than 45% of China's 1.37 billion population are still living in the countryside. In China, the 'countryside' is not only a regional concept but also a political concept because of the special household registration system, or Hukou[3] in Putonghua (Mandarin Chinese language). Established in 1958, this system artificially divides Chinese citizens into the 'agricultural population' and the 'non-agricultural population', thus creating the dual socio-economic system, or 'one country, two policies for urban and rural areas'.

In addition to the differences between city and countryside, the development in rural areas is also unbalanced. As illustrated in Figure 1, the traditional village in Jining (a), the hometown of Confucius, where a large number of dwellings have a history of more than 100 years, is substantially different from the residential quarter in Qingdao (b), a coastal city of Shandong, in terms of layout, building design, cost, quality, construction process and, ultimately, levels of comfort and building energy consumption. The emerging residential quarter in the countryside near to Jinan (c), the capital city of Shandong, is similar in appearance to the residential quarter in Qingdao (b). However buildings in the former (c) are poorly equipped, without gas and central heating. That is why solar hot water systems as a complementary measure have been widely used in the emerging residential quarter (d).

In this study, the criterion to determine whether a place belongs to the city or the countryside is neither its location nor the occupations of local people, but rather the type of residents' Hukou. A typical example is so-called 'urban villages'. These 'villages' are originally located out of the city. Along with the expansion of the city, their farmland is compulsorily acquired and used for urban development. However, the cost of transforming the 'agricultural population' is so high that the native villagers cannot be simultaneously converted into a 'non-agricultural population'. These 'villages' thus become enclave-like regions – they are actually located in the city, but, in the sense of jurisdiction, still remain assigned to the

(a)

(b)

(c)

(d)

Figure 1 Different types of dwelling in the city and the countryside: (a) a traditional village in Jining, hometown of Confucius, in an average region in Shandong province; (b) a residential quarter in Qingdao, a coastal city of Shandong, which was completed in 2009; (c) an emerging residential quarter in the countryside near Jinan, the capital city of Shandong, which was being developed in 2012; and (d) solar hot water systems being installed in the construction project shown in (c)

'countryside'. Note that only when their land is completely acquired can native villagers be transformed into 'non-agricultural population', and 'urban villages' thus become part of the city.

Under the dual socio-economic system, the existing governance system for buildings (and their energy efficiency) in China is differentiated between urban and rural areas. The biggest challenge facing the countryside is that the national standards for building energy conservation have not yet been incorporated into the scope of compulsory execution in rural areas, *i.e.*, these standards are mandatory for the city but voluntary for the countryside (Ministry of Housing and Urban–Rural Development, 2012). In the absence of compulsory building regulations and proper incentive mechanisms, most dwellings in China's rural areas, either existing or newly built, are characterized by 'three lows and one high' – low cost, low standard, low comfort levels and high levels of energy consumption for space heating and hot water. (There is also a high amount of embodied energy from materials: solid clay bricks and tiles.)

There is a great potential of building energy conservation in rural China. By the end of 2012, per capita living space in the countryside amounts to 37.1 m^2, about 4.6 times as large as that in 1978 when China embarked on the path of gradual free market reform (National Bureau of Statistics of China, 2015a). During the period of the 11th Five-Year Plan (2006–10), the area of newly built farm dwellings is about 700–800 million m^2 each year. However, most newly built farm dwellings are not energy efficient. Therefore, it could be expected that more newly built farm dwellings would result in greater energy consumption. Fortunately, the launch of a 'new countryside' initiative created an opportunity to change this adverse situation. It is also the main motivation of this study.

Governance as an art of complexity

In line with new institutional economics, externalities can sometimes be resolved by private bargaining based on more precisely defined and easily tradable property rights, and a lower transaction cost (Bowles, 2004, pp. 228–229). When private solutions are not found, governments may attempt to solve the problem through either command and control policies or market-based policies (*e.g.*, Pigovian taxes or subsidies). In addition, as a recognition of the plurality of governance mechanisms implied, one's choice in this regard is not restricted to a rigid dichotomy based on market versus hierarchy (Jessop, 2003). The theory of externality has been applied in practice, among which the SO_2 emission allowance scheme and carbon trading scheme are typical examples. Proponents of these schemes argue that it offers the cheapest solutions for tackling climate change. However, the empirical evidence from the European Union Emissions Trading System reveals that the scheme did not achieve its expected goals; in fact, the 'cap' did not create appreciable reductions in the first phase of the programme (2005–08) (Reyes & Gilbertson, 2010). This case indicates that for complex issues such as internalizing the economic costs of climate change, no single alternative will suffice to deal with them. Instead, long-term structural changes should be made in a holistic way, with the functions of markets, governments and the third party being taken into full consideration (Bowles, 2004; Jessop, 2003).

As the complex art of steering multiple agencies, institutions and systems (Jessop, 2003), the governance paradigm is foremost about establishing a new mindset and methods. The rise of governance stems from difficulties of hierarchical coordination by organizations or the state (Miller & Lessard, 2000), which, in turn, is a consequence of increasingly complex business environment characterized by rapidly changing patterns of reciprocal interdependence (Jessop, 1998), and increasingly frequent interactions across all types of pre-established boundaries (Scharpf, 1994, p. 37).

Governance mechanisms refer to processes of institutional, market or network organization through legal, normative, discursive or political processes (Bevir, 2013). In its broadest definition 'good governance' can be thought of as how individuals, groups, organizations, societies and nation-states are held accountable not only for outcomes but also for ethical behaviours (Clegg, Kornberger, & Pitsis, 2011). In the context of organization, governance provides a framework for ethical decision-making and managerial action within an organization that is based on transparency, accountability and defined roles (Müller, 2009). Using the lens of complex problem-solving, the governance approach can be viewed as a distributed coordination mechanism underpinned by a common will of mutual interest, and that operates through what Polanyi (1967, p. 72) calls 'mutual control'. A 'common will' as a consensus among all parties in question is achieved through bottom-up dialogues and negotiations. In a mutual control structure, agents exercise control over each other, based on the twin principles of self-discipline through mutual authority and self-coordination through mutual adjustment (Ahern et al., 2014; Polanyi, 1969).

In spite of diverse and contrary usages, governance is, in essence, 'concerned with creating the conditions for ordered rule and collective action' (Stoker, 1998, p. 155). Under the assumption of bounded rationality, self-interest and opportunistic behaviours, economic governance may be conceptualized as an effort to make the stakeholders with divergent interests work together to realize mutual gains. It implies a favourable institutional environment that gives people, whether individually, in groups or in organizations, the motivation to do the right things in the right way. In essence, it is a framework to shape and influence the attitude of stakeholders rather than directly determining every action of organizational actors (Clegg, Pitsis, Rura-Polley, & Marosszeky, 2002; Müller et al., 2014).

Discriminating alignment is a fundamental principle of economic governance. According to the new institutional economics, social, political, legal and economic institutions may be divided into four categories or levels: (1) social or cultural foundations, or embeddedness; (2) a basic institutional environment; (3) institutions of governance; and (4) short-term resource allocation (Williamson, 2000). At the third level where the purpose is to get the governance structures right, the main task is to define processes and structures, allocate responsibilities, rights and interests, requiring assurance that management is operating effectively and properly within the defined structures (Biesenthal & Wilden, 2014; Too & Weaver, 2014). Transaction cost economics asserts that the supply of a good or service and its governance must be examined simultaneously. The central exercise is to align transactions (which differ in their attributes) with governance structures (market, hierarchy and hybrid, which differ in costs and competencies) in a discriminating way so that an economizing match is achieved (Williamson, 2000). This implies that there is no omnipotent approach to governing economic organizations and activities that are ever-changing in nature; instead, they must be steered in a contingent manner.

Governance is enabled through different forms of flexibility at different levels and thus is a multilevel phenomenon (Biesenthal & Wilden, 2014; Jessop, 2003; Müller et al., 2014). The flexibility changes with the level of governance. Generally speaking, the lower levels require flexibility in the choice of methods and processes, while the higher levels of governance require flexibility in people's mindset and attitudes (Müller et al., 2014). No matter at which level, flexibility implies the compromise between control and freedom (Locatelli, Mancini, & Romano, 2014), requiring dialogue and negotiation between different parties.

In summary, the process of internalizing an externality is a complex task. It demands a more flexible mindset that pays more attention to the complementary roles of governments, markets and the third party in aligning involved parties to work together towards shared goals. Characterized by different forms of flexibility at different levels and the principle of discriminating alignment, governance paradigm is suitable for the task.

Data collection and analysis

Two research projects were undertaken during 2009–10 and 2011–12 respectively. The first concerns the policies and mechanisms for popularizing key technologies used in building energy conservation. The second concerns the management system for building energy conservation in the 'new countryside' initiative. As shown in Table 1, interviews, survey meetings, questionnaires and site investigations were employed to gather data from about 260 and 220 people for the two research projects respectively.

The advantage of interview and survey meeting lies in the cross-fertilization effect and the depth of detail from the interviewees/participants. However, this entails significant time and energy to organize each interview/

meeting. In addition, a possible weakness is missing some information due to the immense multitasking. The strength of questionnaire survey is the standardized answers that make it simple to compile data. However, the limitation of standardized answers may frustrate users or not fully reflect the nuances of their view. The advantage of site investigation is that first-hand information can be obtained. However, it is often difficult for researchers to gain access to a site, particularly a construction site. Besides, the site investigation could not be effective enough without sufficient desk study. In both projects, different methods were chosen in accordance with their appropriateness for the purpose of data collection. For example, in order to acquire the information and documents about related policies and obtain detailed advice on policy and its implementation, several interviews with relevant officials were arranged regardless of the difficulty in doing so.

In the first project, special attention was focused on the factors that hinder available technologies from being effectively applied in practice. A wide range of survey respondents were selected, including developers, research and development (R&D) staff, designers, contractors and subcontractors, property managers, and urban residents. These surveys identified some unfavourable factors, including insufficient incentives, poor performance caused by cutting corners in production and construction, defects in supervision and control, and inadequate penalty for breaches. It was then recognized that popularizing and applying key technologies for building energy conservation depend on, to a great extent, the

Table 1 Main data sources for the two research projects

Interview	*First project* Three interviews with five officials who are in charge of promoting building energy conservation at different levels (province, city and district/county) in Shandong province Eleven interviews with seven developers and nine designers *Second project* Three interviews with six officials who are in charge of land administration, urban–rural planning and the 'new countryside' initiative in Shandong province respectively
Survey meeting	*First project* Three hours meetings with eight research and development (R&D) staff at Shandong Academy of Building Research Three hours meetings with 12 contractors and subcontractors Two hours of meetings with six property managers *Second project* Three hours of meetings with eight heads of villagers' committees
Questionnaire survey	*First project* A total of 240 questionnaires were distributed to urban residents, and 190 valid ones were retrieved *Second project* A total of 160 questionnaires were distributed to farmers living in eastern, middle and western Shandong respectively, and 128 valid ones were retrieved
Site investigation	*First project* Five site investigations including two housing construction sites (Jinan and Qingdao), one urban complex construction sites (Jinan) and two residential quarters (Jinan and Qingdao); more than 20 persons were contacted *Second project* Six site investigations including two 'urban villages' (Jinan and Qingdao), two emerging residential quarters in the countryside (Jinan and Qingdao), and two traditional villages (Jinan and Jining); more than 20 persons were contacted

attitude and behaviour of stakeholders. The job of regulating stakeholders' behaviour can be divided into three categories:

- *Incentives*: the implementation of a reasonable cost-value system would make stakeholders unwilling to behave opportunistically.

- *Difficulty*: the use of strict supervision and restrictions would make stakeholders unable to behave opportunistically.

- *Cost*: the imposition of severe sanctions would make stakeholders dare not behave opportunistically.

In the second project, the emphasis of investigation was placed on the differences of building energy conservation between urban and rural areas, and those between different locations in the countryside. A meeting of heads of villagers' committees was held. Site investigations at different levels were carried out in different regions. A questionnaire survey was conducted through college students who come from the countryside of eastern, middle and western Shandong respectively. When going home on holiday, they were asked to obtain first-hand information from their families. The information includes the status quo of local farm dwellings, perceptions of rural residents about building energy conservation in terms of technologies and policies, their requirement for comfort and the affordability for building energy efficiency technologies.

Existing standards, codes and specifications were collected and analyzed in terms of thermal insulation of building envelope, heating, ventilation, lighting, renewable energy usage, etc., mainly including (1) Chinese industry standard *Design Standard for Energy Conservation of Residential Buildings with Heating Services* (JGJ 26-1986, the criterion of 30% energy saving[4]), (2) Chinese industry standard *Design Standard for Energy Conservation of Residential Buildings in Hot-in-Summer and Cold-in-Winter zones* (JGJ 134-2001, the criterion of 50% energy saving), (3) Shandong industry standard *Design Standard for Energy Conservation of Residential Buildings* (DBJ 14-037-2012, the criterion of 65% energy saving), and (4) *The Catalog of Applicable, Restricted and Forbidden Technologies in Shandong Construction Industry*.

Alternative financing channels were analyzed, including (1) commercial loans such as those from the Agricultural Bank of China, the Agricultural Development Bank and various rural credit cooperatives, which requires considerably high-loan guarantee rates due to greater credit risks in the rural financial system; (2) government subsidies such as those for demonstration projects of energy-efficiency

farmhouses, demonstration projects of renewable energy usage, and renovation of dilapidated buildings; and (3) revenues from land development such as those from 'urban village' transformation and renovation projects of villages in the outskirts or countryside, which could considerably increase the plot ratio, improve infrastructure condition and hence considerably enhance the value of land.

Relevant laws, regulations and government documents were collected and analyzed, including Construction Law (1998), Renewable Energy Law (2006), Regulations on Energy Conservation in Civil Buildings (2008), Special Plan for Building Energy Conservation during the 12th Five-Year Plan (2011), as well as local regulations such as Rules for the Administration of Building Energy Conservation in Shandong Province (2005), Shandong Provincial Plan for Building Energy Conservation during the 12th Five-Year Plan (2011) etc.

Consequently, a 'spectrum' of urban–rural construction modes was developed, as shown in Figure 2. These modes are different in many aspects, including land ownership, the scope of urban–rural planning, financing capacity, infrastructure condition, design and construction process, etc. As for land ownership, the land in the city is owned by the state; while in the countryside it is owned by collectives (*i.e.*, villagers' committees). With regard to the scope of urban–rural planning, the city and 'urban villages' belong to the scope of urban planning; while other regions should be planned in accordance with the regulations and codes of rural planning. These differences necessitate a differentiated system of governance, which can be divided into three levels:

- Level I, or 'urban villages' where a higher standard of building energy conservation is practical. These are 'villages' more in name than in reality, and in many aspects more similar to the city than the countryside.

- Level II, or villages in the outskirts and emerging residential quarters in the countryside. The standard of building energy conservation should be an intermediate one.

- Level III, or traditional villages where the job of building energy conservation is much more difficult than for the other two levels. The expectation for comfort is relatively lower, so a lower energy standard would be acceptable as long as the expectation remains static.

A three-level analysis framework

Based on above literature review and analyses, a three-level analysis framework was developed. As shown in

	urban areas	'urban villages'	villages in outskirts of cities and emerging residential quarters in the countryside	traditional villages
land ownership	owned by the State	owned by collectivities	owned by collectivities	owned by collectivities
scope of urban/rural planning	urban planning	urban planning	rural planning	rural planning

Figure 2 'Spectrum' of urban–rural construction modes

Figure 3, the x-axis is comprised of the three forces as the basic forms of economic governance: markets, governments and the third party. These forces can have roles in regulating stakeholders' behaviour. The market plays the role of 'invisible hand' through price mechanism. For example, the solar hot water system has been widely used in rural China simply because it is economical for farmers. By means of incentive policies, supervision and control, as well as accountability and penalty systems, governments as a 'visible hand' are capable of making people unwilling to, unable to and dare not to behave opportunistically. The third party can be regarded as the 'third hand' to regulate stakeholders' behaviour. This is, in essence, a self-disciplinary mechanism underpinned by reputation and peer pressure. These three forces have their own comparative advantages and limitations. The relationship between them could be a complementary one based upon 'both/and' considerations but not a substitutive one based upon exclusive 'either/or' choices.

The market may be treated as a price-based strangers' society in which there is no place for constraints of social morality (Gauthier, 1986, p. 93). The advantage of the market lies in its efficient disposition of economic resources which 'could be regarded in a well-defined sense as superior to a large class of possible alternative dispositions' (Arrow & Hahn, 1971, p. vi). Note that the above-quoted 'well-defined sense' refers to the market completeness assumptions that do not always hold in real life. So market failure is inevitable, particularly when externality-related issues become the main concern due to self-regulation. Moreover, deeply influenced by the planned economic system, China's construction market is far from mature.

The government may be regarded as a regulation-based hierarchical governance structure. The advantage of the government lies in its capabilities to pass laws, set standards, promulgate bans and enforce regulations. However, when government intervention causes a

governance structure	unwilling to	unable to	dare not
the market (invisible hand)	+ + (price mechanism)	0	0
the government (visible hand)	+ (incentive policies)	+ (supervision and control)	+ + (accountability and penalty)
the third party (the third hand)	+ (reputation)	0	+ (peer-pressure)

making people unwilling to / unable to / dare not behave opportunistically

approach to regulating stakeholders' behavior

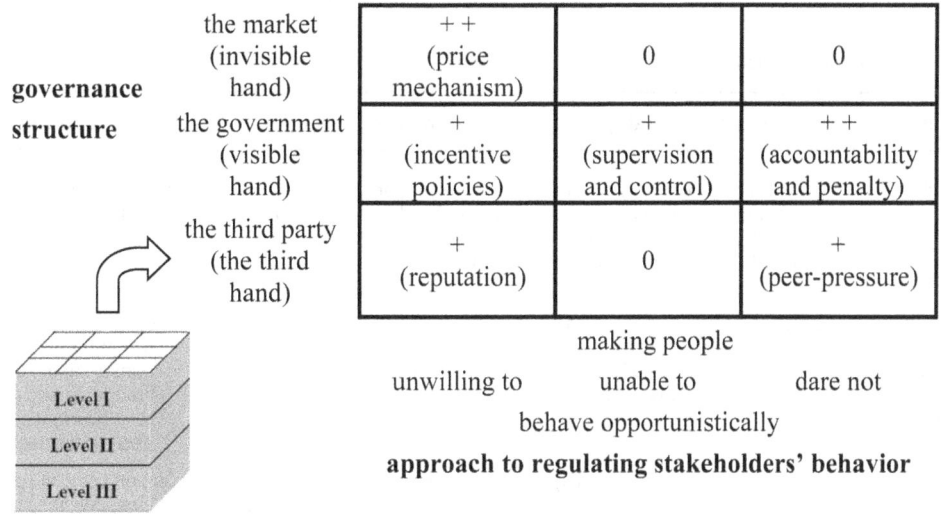

Note: + + = strong; + = semi-strong; 0 = weak.

Figure 3 Matrix of building governance: three forces versus three approaches

more inefficient or unexpected allocation of resources than would occur without that intervention, then government failure occurs. In addition, tenure limitation may make officials and politicians act short-sightedly, devoting themselves to pursuing quick and eye-catching achievements before their next election. In China, the political rivalry between local officials is often a zero-sum game (one participant's gain is exactly another one's loss), which has caused a deep-rooted mindset and practice favouring short- over long-term interests (Sha et al., 2006).

The term of 'the third party' has different meanings in different contexts. In this study, it refers to the professions and professional associations because of their role of the third force augmenting markets and governments. According to the new institutional economics, professionalism can be regarded as a community-based governance structure (Sha, 2013). Its comparative advantage is mainly manifested in a long-term expectation-based incentive mechanism, a reputation-based restraint mechanism, an information-feedback-based cooperation mechanism and a morality cultivation mechanism. These mechanisms are achieved by generating and sharing knowledge, developing good practices, discussing decision-making processes and nuanced judgements, and maintaining peer pressure (Bordass & Leaman, 2013). However, like markets and governments, the third party also has limitations, and sometimes does not work well. The professionalism in China's building sector is still at its fledgling stage of development, which is characterized by following weaknesses: (1) a lack of independence caused by tangled relations between professional associations and government agencies; (2) frequent rent-seeking activities that arise from an inappropriate administrative system for practice qualification and market access; and (3) a tainted impartiality and integrity

within the construction market where the public clients play a predominant role, and professionals are prone to being overwhelmed by the clients (Sha, 2013).

In summary, the analysis framework can be illustrated as three dimensions: analytical level (levels I–III), governance structure (markets, governments and the third party), and approach to regulating stakeholders' behaviour (making people unwilling to, unable to and dare not to behave opportunistically). When developing strategy portfolios to promote building energy conservation in rural China, the essential principle is to make best use of the comparative advantages of markets, governments and the third party in a discriminating way, while avoiding their shortcomings as much as possible.

Developing strategy portfolios

On the basis of the three-level analysis framework, different strategy portfolios are proposed, as illustrated in Table 2.

For level I, it is not only necessary but also feasible to adopt the same technical standard as the city. The criterion of 65% energy saving, or Shandong provincial industry standard, DBJ 14-037-2012, should be satisfied. At level III where farmers' living behaviour and practices are basically traditional, thermal comfort is not a prominent issue. With underdeveloped infrastructure and weaker financing capacity, many advanced technologies could not be effectively applied. Instead, more practical and affordable technologies such as passive solar heating, natural ventilation, solar hot water and marsh gas cooking are more suitable. The baseline here is to forbid completely the use of solid clay brick and tiles in new projects. Taking the Chinese industry standard JGJ 26-1986

Table 2 Three-level strategy portfolio

Analytical level	Technology strategy	Financing strategy	Regulation and incentive policy
Level I: 'Urban villages'	Criterion of 65% energy saving (Shandong provincial industry standard, DBJ 14-037-2012), the same as that of the city	Mainly relying on added land value through either market-driven real estate development or government-led affordable housing programmes	Accountability and penalty system; anti-domicide-related regulations
Level II: Villages in outskirts and emerging residential quarters in the countryside	Criterion of 50% energy saving (see Chinese industry standard, JGJ 134-2001), about 11 years behind the city	Mainly relying on commercial loans, with policy-based loans with favourable rates and special funds as a supplementary channel	Reducing guarantee rates for loans used for building energy conservation; developing relevant standards and codes for town and rural residential quarter planning and farmhouse design
Level III: Traditional villages	Criterion of 30% energy saving (see Chinese industry standard, JGJ 26-1986), about 26 years behind the city	Government subsidies, special funds and interest-free loans as the main financing channel	Ban on using solid clay bricks and tiles; policies in terms of government subsidies, tax breaks and interest-free loans

(enforced in the city in 1986, 26 years earlier than DBJ 14-037-2012) as a reference, the criterion of 30% energy saving should be adopted at level III. As for level II, the technology strategy should be an intermediate one. Taking the Chinese industry standard JGJ 134-2001 (enforced in the city in 2001, 11 years earlier than DBJ 14-037-2012) as a reference, level II should take 50% energy saving as the yardstick.

It was estimated that the 'new countryside' initiative calls for a total fund of 15–20 trillion yuan by 2020, which necessitates a multi-channel financing system (Tang, 2006). With less than 10,500 yuan per capita disposable income that is only 36% as high as that of urban residents (National Bureau of Statistics of China, 2015b), it is clear that farmers' own funds are utterly inadequate for building energy conservation. When raising funds for building energy conservation, level I should give priority to the added land value through either market-driven property (real estate) development or government-led affordable housing programmes. For level II which, in general, has a stronger loan guarantee capability due to its well-run collective business, the financing strategy should be one that relies mainly on commercial loans, with policy-based loans with favourable rates and special funds as a supplementary channel. At level III, which often is the focal area of anti-poverty policies, various government subsidies, special funds and interest-free loans should become the main financing channel.

Along with the proposed technology and financing strategies/channels, the countryside-oriented system of regulations and policies needs development in order to guarantee the implementation of the technology and financing strategies. At level I, existing standards and codes for building energy conservation should be compulsorily executed. This entails the strict supervision from government authorities and the guidance from the third party. Since added land value is adopted as the main financing channel, anti-domicide is thus the major issue that has to be taken into consideration. Domicide refers to the eradication of a home against the will of its inhabitants. This has a series of adverse consequences such as ruthless deprivation (imposed by local officials), land-less farmers and an endless petition process by the dispossessed who seek justice (Sha, 2014; Shao, 2013). The ultimate solution to domicide lies in local governments' self-discipline to prevent its occurrence. At level II, standards and codes for town and rural residential quarter planning and farmhouse design should be established or updated as soon as possible. Guarantee rates for loans used for building energy conservation in emerging residential quarters in the countryside should be reduced. At level III, both constraint and incentive mechanisms are equally important. On the one hand, the ban on using solid-clay bricks and tiles should be effected mainly through accountability and a penalty

system; on the other hand, practical and affordable technologies should be encouraged and promoted through products and services with a higher performance-to-price ratio, and preferential policies in terms of government subsidies, tax breaks and interest-free loans.

Conclusions

China's dual socio-economic system has caused a huge gap between its urban and rural areas. The 'new countryside' initiative provides an opportunity for the countryside to place the affairs of building energy conservation on the agenda, while bringing about great challenges. With inadequate funds, underdeveloped infrastructures and imperfect regulations, standards and codes, the task to promote building energy conservation in the countryside is much more complex than in the city. It should be noted that what the countryside lacks is not the technology. Many technologies exist that are effective in practice but have not yet been applied in rural areas. This is due to either difficulties in some 'hard' aspects such as funding and infrastructures or bottlenecks relating to building regulations and related policies. In order to break these bottlenecks in 'soft' aspects, it is necessary to adopt flexible methods and mindsets that focus on the complementary roles of governments, markets and the third party in aligning involved parties to work together towards shared goals.

This study has provided: (1) the three-level analysis framework that can be regarded as an attempt at applying the rationale of the governance paradigm to the domain of building energy conservation; and (2) different strategy portfolios for different levels in rural China. These strategy portfolios have been submitted to the Shandong Provincial Bureau of Housing and Urban–Rural Development. Some pilot projects have been conducted. It is expected that these strategies could be used to help initiate changes in terms of legislation, regulation, standards and codes, and changes in design and construction practices.

Although this article examines building energy conservation in rural areas of Shandong province, the research findings could have a broader application. The ideal state of building governance should be one where all parties work together towards shared goals that are achieved via compromise between different parties, and/or the trade-offs between various alternative solutions, with each party finding the right niche for itself. However, it is just theoretically realizable due to bounded rationality of diverse participants and changing circumstances. Hence, there is no 'best' one, but rather the 'most suitable' approach to building governance. In this light, the principle of discriminating alignment and multilevel analysis approach may provide some conceptual insight.

It can be seen that as compared with governments and markets, the third party plays only a small role in the above strategy portfolios. In order to make its stunted professional system mature as soon as possible, it is essential for China's building sector to change the imbalanced relationships among markets, governments and professionals.

To some extent, a stronger role of government is beneficial to dealing with some issues such as environment pollution and climate change, since the market, with the aim of maximizing short-term profit, is inherently incompatible with sustainability. In fact, China's government did play a leading role in promoting sustainable construction. However, an overly dominant government could hinder the development of the market and the third party, whereas an unbalanced structure is unfavourable for building governance. Under the present circumstances that are characterized by a powerful government, an immature market and an unfair third party (Sha, 2013), efforts to internalize externalities through private bargaining cannot be successful due to distorted price signals, vaguely defined property rights and higher transaction costs (Bowles, 2004). The key to removing this adverse situation is to make governments at all levels position themselves in the correct position and disentangle their relations with both the market and the third party by establishing a check-and-balance mechanism. This is a difficult task since it would inevitably affect vested interests. However, the job of building energy conservation in rural areas may provide a suitable arena to attempt this. There are two reasons for doing this. First, the influence of the planned economy system on rural areas is relatively weaker – it is for this reason that China's economic reform was initiated in the countryside rather than in the city. Secondly, since the government is responsible to future generations in the domain of building energy conservation, it should have more authority, which might reduce the difficulty in decreasing the 'power distance'[5] (Hofstede, 1980) between the government and other parties. This is an interesting and challenging theme worthy of further exploration in both theory and practice.

Acknowledgement

The authors are grateful for the constructive comments given by the anonymous referees, the guest editors of the special issue, and the editor in chief on previous versions of this paper. They also thank Mr Xiao Li for his English proofreading and corrections. Of course, the responsibility for any errors is solely with the authors.

Disclosure statement

No potential conflict of interest was reported by the authors.

Funding

This work was supported by the National Natural Science Foundation of China [grant number 71202010] and The Energy Foundation China [grant number sd-09-11].

References

Ahern, T., Leavy, B., & Byrne, P. J. (2014). Complex project management as complex problem solving: A distributed knowledge management perspective. *International Journal of Project Management, 32*(8), 1371–1381. doi:10.1016/j.ijproman.2013.06.007.

Arrow, K. J., & Hahn, F. (1971). *General competitive analysis.* San Francisco: Holden-Day.

Bevir, M. (2013). *Governance: A very short introduction.* Oxford, UK: Oxford University Press.

Biesenthal, C., & Wilden, R. (2014). Multi-level project governance: Trends and opportunities. *International Journal of Project Management, 32*(8), 1291–1308. doi:10.1016/j.ijproman.2014.06.005.

Bordass, B., & Leaman, A. (2013). New professionalism: Remedy or fantasy. *Building Research & Information, 41*(1), 1–7. doi:10.1080/09613218.2012.750572.

Bowles, S. (2004). *Microeconomics: Behavior.* Princeton: Princeton University Press, Institutions and Evolution.

Cleden, D. (2009). *Managing project uncertainty.* Farnham, UK: Gower.

Clegg, S. R., Kornberger, M., & Pitsis, T. S. (2011). *Managing and organizations: Introduction to theory and practice* (3rd ed.). London: Sage.

Clegg, S. R., Pitsis, T. S., Rura-Polley, T., & Marosszeky, M. (2002). Governmentality matters: Designing an alliance culture of inter-organizational collaboration for managing projects. *Organization Studies, 23*(3), 317–337. doi:10.1075/aios.9.

Gauthier, D. (1986). *Morals by agreement.* Oxford: Clarendon Press.

Hofstede, G. (1980). *Culture's consequences: International differences in work-related values.* Beverly Hills, CA: Sage.

Jessop, B. (1998). The rise of governance and the risks of failure: The case of economic development. *International Social Science Journal, 50*(155), 29–45. doi:10.1111/1468-2451.00107.

Jessop, B. (2003). *The governance of complexity and the complexity of governance: Preliminary remarks on some problems and limits of economic guidance.* Lancaster: Department of Sociology, Lancaster University. Retrieved June 20, 2015, from http://www.comp.lancs.ac.uk/sociology/papers/Jessop-Governance-of-Complexity.pdf

Locatelli, G., Mancini, M., & Romano, E. (2014). Systems engineering to improve the governance in complex project environments. *International Journal of Project Management, 32*(8), 1395–1410. doi:10.1016/j.ijproman.2013.10.007.

Miller, R., & Hobbs, B. (2005). Governance regimes for large projects. *Project Management Journal, 36*(3), 42–51.

Miller, R., & Lessard, D. R. (2000). *The strategic management of large engineering projects: Shaping institutions.* Cambridge, MA: MIT Press, Risk and Governance.

Ministry of Housing and Urban–Rural Development. (2012). *Special plan for building energy conservation during the 12th Five-Year Plan* (2011–2015). Retrieved June 23,

2015, from http://www.mohurd.gov.cn/zcfg/jsbwj_0/jsbwjjskj/201205/t20120531_210093.html

Müller, R. (2009). *Project governance*. Aldershot, UK: Gower Publishing.

Müller, R., Pemsel, S., & Shao, J. (2014). Organizational enablers for governance and governmentality of projects: A literature review. *International Journal of Project Management, 32*(8), 1309–1320. doi:10.1016/j.ijproman.2014.03.007.

National Bureau of Statistics of China. (2015a). *National data*. Retrieved October 20, 2015, from http://data.stats.gov.cn/easyquery.htm?cn=C01

National Bureau of Statistics of China. (2015b). *Statistical bulletin on national economic and social development in* 2014. Retrieved June 23, 2015, from http://www.stats.gov.cn/tjsj/zxfb/201502/t20150226_685799.html

Polanyi, M. (1967). *The tacit dimension*. Garden City, NY: Doubleday.

Polanyi, M. (1969). Knowing and being: Essays by Michael Polanyi (M. Grene, Ed.). Chicago, IL: University of Chicago Press.

Reyes, O., & Gilbertson, T. (2010). Carbon trading: How it works and why it fails. *Soundings, 45*, 89–100. doi:10.3898/136266210792307050

Scharpf, F. W. (1994). Games real actors could play: Positive and negative coordination in embedded negotiations. *Journal of Theoretical Politics, 6*(1), 27–53. doi:10.1177/0951692894006001002.

Sha, K. X. (2013). Professionalism in China's building sector: An economic governance perspective. *Building Research and Information, 41*(6), 742–751. doi:10.1080/09613218.2013.842459

Sha, K. X. (2014). Destructive construction and constructive conflicts. *Building Research and Information, 42*(3), 391–392. doi:10.1080/09613218.2014.886148.

Sha, K. X., Song, T., Qi, X., & Luo, N. J. (2006). Rethinking China's urbanization: An institutional innovation perspective. *Building Research and Information, 34*(6), 573–583. doi:10.1080/09613210600852730.

Shao, Q. (2013). *Shanghai gone: Domicide and defiance in a Chinese megacity*. Lanham, MD: Rowman & Littlefield.

Spering, M., Wagener, D., & Funke, J. (2005). The role of emotions in complex problem-solving. *Cognition and Emotion, 19*(8), 1252–1261. doi:10.1080/02699930500304886.

Stoker, G. (1998). Governance as theory: Five propositions. *International Social Science Journal, 50*(155), 17–28. doi:10.1111/1468-2451.00106.

Tang, S. N. (2006). *Rethinking unbalanced development of financing market in China's urban and rural areas*. Retrieved May 26, 2015, from http://www.gov.cn/gzdt/2006-07/30/content_349742.htm

Too, E. G., & Weaver, P. (2014). The management of project management: A conceptual framework for project governance. *International Journal of Project Management, 32*(8), 1382–1394. doi:10.1016/j.ijproman.2013.07.006.

Williamson, O. E. (2000). The new institutional economics: Taking stock, looking ahead. *Journal of Economic Literature, 38*(3), 595–613. doi:10.1257/jel.38.3.595.

Endnotes

[1] The full research reports (in Chinese) are available from the authors upon request.

[2] With 95.8 million inhabitants at the 2010 Census, Shandong is the second most populous province in China. Shandong's GDP in 2014 amounts to 5.9 trillion yuan, or US$967 billion, ranking it third in the country.

[3] Under the system, urban residents enjoyed substantially different treatment from rural dwellers in terms of employment, housing, education, medical treatment, taxation, social welfare, etc. Although some reform measures have been tested in some cities since the late 1990s, the *Hukou* system has not yet been abolished.

[4] The criterion of 30% energy saving takes the level of building energy consumption in the early of 1980s as the benchmark. If the level of building energy consumption of a newly built dwelling is not more than 70% of the benchmark, then the criterion of 30% energy saving is satisfied. The criteria of 50% and 65% energy saving can be understood in the same way.

[5] Power distance is the extent to which the less powerful members in organizations accept and expect that power is distributed unequally. Individuals in a society that exhibits a high degree of power distance accept hierarchies in which everyone has a place without the need for justification. Societies with low power distance seek to have equal distribution of power.

The evolution of green leases: towards inter-organizational environmental governance

Kathryn B. Janda, Susan Bright, Julia Patrick, Sara Wilkinson and Timothy J. Dixon

Improving the environmental performance of non-domestic buildings is a complex and 'wicked' problem due to conflicting interests and incentives. This is particularly challenging in tenanted spaces, where landlord and tenant interactions are regulated through leases that traditionally ignore environmental considerations. 'Green leasing' is conceptualized as a form of 'middle-out' inter-organizational environmental governance that operates between organizations, alongside other drivers. This paper investigates how leases are evolving to become 'greener' in the UK and Australia, providing evidence from five varied sources on: (1) UK office and retail leases, (2) UK retail sector energy management, (3) a major UK retailer case study; (4) office leasing in Sydney, and (5) expert interviews on Australian retail leases. With some exceptions, the evidence reveals an increasing trend towards green leases in prime offices in both countries, but not in retail or sub-prime offices. Generally introduced by landlords, adopted green leases contain a variety of ambitions and levels of enforcement. As an evolving form of private–private environmental governance, green leases form a valuable framework for further tenant–landlord cooperation within properties and across portfolios. This increased cohesion could create new opportunities for polycentric governance, particularly at the interface of cities and the property industry.

Introduction

It has been well understood for some decades that energy use in existing buildings is a major source of greenhouse gas (GHG) emissions, and that there is significant potential for energy savings in retrofitting existing buildings (Levine et al., 2007; Ürge-Vorsatz et al., 2012). Reductions in emissions on the scale required to stabilize the global climate cannot be achieved without major change in the patterns of energy use across the entire building stock. Although change at this scale requires modifications in both technologies and organizational practices, research is dominated by technological approaches (Lutzenhiser, 1993, 2014; Schweber & Leiringer, 2012), and towards disciplines and activities that continue this trajectory (Sovacool, 2014). Where social science research does exist, it is skewed towards households (Deline, 2015; Dixon, Deline, McComas, Chambliss, & Hoffmann, 2015; Janda, 2014; Moezzi & Janda, 2014; Taylor & Janda, 2015). Approximately 18% of UK

GHG emissions come from energy use in non-domestic buildings (CSE & ECI, 2012). By one estimate, this rises to 34% if both operational and 'capital' GHG emissions (direct and indirect emissions from construction works, services, materials, transport and products) are included (Green Construction Board, 2013). Yet, research in non-domestic buildings accounts for less than 10% of the end-use energy demand research portfolio in the UK (Hannon, Rhodes, & Skea, 2013; LCICG, 2012). Broadening the understanding of the socio-technical processes and constraints that affect the dynamics of change in non-domestic buildings is of critical national and global importance.

To bolster research in this area, in 2014 the UK Engineering and Physical Sciences Research Council (EPSRC) funded six new projects on energy management in non-domestic buildings. This paper, and part of the research included in it, is largely funded by one of these projects. The project is called WICKED (Working with Infrastructure, Creation of Knowledge, and Energy strategy Development) and is designed to learn from real-world situations, focusing on energy strategy development in the retail sector. The acronym WICKED draws on Rittel and Webber's (1973) conceptualization of complex problems that defy simplistic or straightforward planning responses as 'wicked', or tricky. Improving the building stock needs to address both the technical challenges involved in upgrading the physical infrastructure and also the social challenges of organizational decision-making (Biggart & Lutzenhiser, 2007). Where space is rented, the further challenge is that these strategies need to work within, rather than against or merely alongside, established systems of professional and social practices (Levitt, Heinisz, Scott, & Settle, 2010; Scott, Levitt, & Orr, 2011).

This paper focuses on the governance of rented space in the non-domestic sector, which is a significant and 'wicked' problem. Of particular interest in this area is the 'split incentive' problem between tenants and landlords, where differing property interests and obligations mean that neither party may have sufficient incentive to invest in energy-efficiency upgrades or engage in better energy management. Case studies across both residential and non-residential contexts in Organisation for Economic Co-operation and Development (OECD) countries concluded that split incentives affect up to 90% of the energy used in many major markets (Prindle et al., 2007).

Leases serve a regulatory role in the governance of both domestic and non-domestic rented space, and may contribute to the split incentive problem. Once the content is agreed between the parties, the lease terms constitute the legal basis for resolution of any dispute and operate as a kind of 'local law' binding on them. In this respect the lease is a form of inter-organizational governance for landlords and tenants, fleshing out their relationship and setting practical norms for a specified length of time. In non-domestic settings, the content and style of the landlord–tenant relationship is heavily determined by institutional letting practices, which adopt standardized structural patterns of leasing and lease wording. These lease terms affect what changes can be made to the premises (e.g., improvements) and set out the obligations that parties owe to one another (e.g., in multi-unit properties the landlord's duty to provide maintenance and other services, and the tenant's obligation to pay a service charge towards the landlord's costs). Good environmental governance may require collaboration between landlords and tenants, but standard leasing practices do not provide for, and may even inhibit rather than promote, joint action. In particular, leases do not ordinarily permit tenants to make structural alterations (e.g., energy upgrades) to the premises or entitle landlords to recover the costs of improvements to the tenant's property (e.g., to install energy upgrades). Leases seldom require the parties to share energy data with one another, which can inhibit energy management practices.

Recent developments in 'green leases' and, more broadly, 'green leasing' seek to enable landlords and tenants to meet environmental targets by changing their organizational practices – through the mechanism of the lease – to work more cooperatively. 'Green leases' are built on 'green' clauses within the lease which are designed to account for energy efficiency and other sustainability goals. Green leas*ing* (also referred to as 'best practice' or 'performance' leasing) refers to the environmental processes, engagement and practices adopted by landlords and tenants in relation to the building. Together, these greener practices reflect a change not only to the *wording* of the formal lease document but also to the *relationship* between the landlord and the tenant. Whereas traditional leasehold relationships are frequently characterized as adversarial, distant and distrustful, some greener leasing practices attempt to foster better communication between the parties. Interest in green leases and leasing is growing, and articles on them have been published in a number of countries in the last decade, including the UK (Bright & Dixie, 2014; Hinnells et al., 2008; Langley & Hopkinson, 2009), Australia (Roussac & Bright, 2012; Woodford, 2007), Sweden (Lind, Bonde, & Zalejska-Jonsson, 2014), Singapore (Chua, 2014), the US (Kaplow, 2009; Oberle & Sloboda, 2010), Canada (Sayce, Sundberg, Parnell, & Cowling, 2009), and 20 countries across Europe (Duquesne, 2011).

This paper uses five case studies to investigate the evolving role of 'green leases' in the environmental governance of tenanted non-domestic property in the UK and Australia. The two countries were selected primarily on the accessibility of evidence: green leases were first adopted in Australia and the property markets in both countries support Better Buildings Partnership

(BBP) organizations. These groups promote collaborative efforts across the industry to enhance sustainability, including, but not limited to, green leases. As other papers in this special issue attest (Eisenberg, 2016; Rosenow, Fawcett, Eyre, & Oikonomou, 2016; van der Heijden, 2015 and forthcoming), most building regulations are blind to the differences between owner-occupied buildings and tenanted non-domestic property: they are usually written from the 'top-down' by governments seeking to affect the entire building stock. As this paper will discuss, green leases are a relatively new form of environmental governance negotiated from the 'middle-out' (Janda & Parag, 2013; Parag & Janda, 2014) between landlords and tenants at the property level.

The paper begins with a literature review that seeks insight into the role of inter-organizational negotiations at the property level. A methods section discusses the rationale for using a contextual case study approach, drawing on expertise from industry associations in the UK and Australia, previous research and new qualitative research. The next section provides a context for five case studies, considering general property markets, policy context and leasing practices in the UK and Australia. The case studies illustrate whether and how 'green leases' in the office and retail sectors are becoming more commonplace, as well as the typical forms these leases take. Further discussion and analysis of these findings show that although uptake is growing in both countries, green leases are more commonly adopted in office properties than in retail, and this uptake has been more comprehensively quantified in the Australia than in the UK. As a tool for inter-organizational environmental governance, green leases are mainly implemented by large powerful landlords, with a few exceptions where they are driven by powerful tenants. The concluding section draws together wider implications for opportunities and challenges of green leasing as a tool for the environmental governance of non-domestic buildings and proposes directions for further research.

Context: corporations, inter-organizational governance and property

This section reviews the literature for framing the role of green leases as tools for non-domestic building governance. The literature on corporate environmental governance (CEG) and voluntary environmental programmes (VEPs) provides some guidance, but it also has some gaps. To address more fully the role of leases in tenanted property, the CEG and VEP concepts are augmented with perspectives on 'middle-out' change (Janda & Parag, 2013; Parag & Janda, 2014), 'building communities' (Axon, Bright, Dixon, Janda, & Kolokotroni, 2012), and strategic property management (Edwards & Ellison, 2004).

Corporate environmental governance (CEG) and voluntary environmental programmes (VEPs)

There is a broad literature on corporate social responsibility that addresses various aspects of corporate social, economic and environmental impacts (McWilliams, 2015). Focusing on environmental impacts, specifically, a body of scholarship identifies various factors involved in corporate governance of environmental problems. This work normally focuses on initiatives within firms or across firms of like-kind (*e.g.*, Borck & Coglianese, 2009; Gouldson & Sullivan, 2014; Howard-Grenville, Nash, & Coglianese, 2008; Prakash & Potoski, 2006). This literature shows that an organization's willingness to engage in 'beyond compliance' environmental programmes is shaped by both external conditions (regulation, economic and social) and a range of internal, interacting factors, including management style, organizational culture and organizational structure. In their review of the VEP literature, Borck and Coglianese (2009) note three types of businesses are likely to participate in VEPs: (1) larger businesses, as they have greater resources to participate and may benefit most from recognition, (2) businesses with internal cultures supportive of environmentally friendly behaviour, and (3) businesses that face (or are likely to face) stricter government regulations.

Insofar as green leasing can be considered as a particular form of VEP, this literature provides useful insights into types of external and internal drivers for CEG. However, this literature does not address inter-organizational governance or property-level issues.

Inter-organizational governance: middle-out change

Inter-organizational activities, particularly between dissimilar groups (*e.g.*, landlords and tenants), are often conceptualized as a space where 'intermediaries' serve an important role (Fischer & Guy, 2009; Moss, 2009; Moss, Medd, Guy, & Marvin, 2009). Janda and Parag (2013) and Parag and Janda (2014) augment this literature with new perspectives on 'middle actors', including designers, building professionals and commercial real estate (property) companies. Middle actors have their own agency and capacity to foster innovation from the 'middle out' rather than merely reacting to policy push from the top down or market pull from the bottom up. A middle-out approach recognizes the influence of these actors *upstream* (*e.g.*, to policy-makers), *downstream* (*e.g.*, to customers and clients), and *sideways* (*e.g.*, to other middle actors).

The sideways element of a middle-out approach resonates with Vandenbergh's (2005) attention to private–private contracts, which he views as an important contribution to the regulatory state. He uses the language of 'second-order regulatory agreements' to capture the idea that private–private agreements respond to 'first-order' government regulatory requirements (*e.g.*, standards, taxes and incentives), but also form a new, durable and important form of governance.

This paper conceptualizes landlords and tenants as middle actors in the property market who are able to exert influence sideways through private–private contracts, such as leases.

The role of property

Neither the general CEG literature nor the middle-out perspective consider the role of property, which is critical to both leases and leasing practices. Axon et al. (2012) outline an interdisciplinary 'building communities' approach to reducing energy use in tenanted property. This approach highlights the importance of three levels: (1) the general policy context, (2) the role of organizations, and (3) the level of the building itself, including the particular characteristics of both the premises and the stakeholders. These authors call for more research on the role of leases and their practical efficacy in effecting change. The current paper contributes to this challenge.

Edwards and Ellison's (2004) research on corporate property management suggests that organizations have varying perspectives on the importance of physical property relative to their core business. For some corporates, the building is integral to their core business strategy; for others, it is merely a container in which work happens. For landlords and investors, for example, the building *is* the business. Scholarship focused on the property industry (*e.g.*, Dixon, Ennis-Reynolds, Roberts, & Sims, 2009; Newell, 2008) identifies corporate social responsibility or responsible property investment (UNEP FI, 2011) as a major driver for these organizations. Tenants, however, may only see the value of physical property as an operational asset, not as a physical asset. These different perspectives are at the heart of the tenant–landlord divide. This paper considers whether green leasing can help overcome this split incentive.

Methods

Evidencing change in leasing practices is difficult and complex. The property market is global, but tenanted buildings are located and operated in particular physical, social and political contexts, all of which can affect leasing practices. Each tenanted unit has its own lease, and multi-tenanted buildings have multiple leases. Leases expire at the end of a fixed term, which may be in one year or 99 years. Leases are treated as commercially confidential, and in the UK public records of the contents of agreed leases are not electronically accessible or searchable. Additionally, there is

no internationally standardized method of classifying leases as 'green'.

To address this complex and difficult to access topic, the paper uses a case study approach (Yin, 2009) to draw on a range of empirical evidence and situates these cases within broader market and policy contexts. Using case studies enables international and sectoral comparison and facilitates a mix of methods, units of analysis and case conceptions to capture the complexity inherent in the development of greener leasing practices. As subsystems of practice may influence leasing practices, five diverse cases and methods across countries (UK and Australia), sectors (office and retail), and organizations (landlords and tenants) facilitates further analysis of the landscape of leasing practices. By working across countries and cases, the aim is to triangulate new research within a broader context of existing work to generate additional insights.

Table 1 summarizes the five case studies used, representing both office and retail sectors in Australia and the UK. More detailed information on the methods used in each study is included in subsequent sections. The broader context in which these cases are situated is considered below.

Case context

This section briefly reviews the complex and evolving policy context, property markets and leasing practices in the UK and Australia. The majority of non-domestic property in these countries is rented: 56% in the UK (Property Industry Alliance, 2014) and 70% in Australia (T. Crabb, Head of Research Savills, personal communication, 2015).

The UK context: property market and regulatory environment

The UK commercial property market is worth £683 billion, with London accounting for about 35% of this market (Property Industry Alliance, 2014). The Central London office market is a particularly important sector, with a slight majority of Grade A quality (offices of best quality specification, floor plate efficiency and image; 55% or 307 000 m^2) and with Grade B/C accounting for just 45% (DTZ/Jones Lang LaSalle, 2014, p. 8). Indeed, investment in central London's commercial property market reached £20.5 billion in 2014, marginally below the last investment peak in 2007 when £20.6 billion was traded (PropertyWire, 2015). Additional important markets include Birmingham, Manchester, Leeds, Bristol, Edinburgh, Cardiff and Glasgow (Bilfinger GVA, 2015).

UK building regulations set out specified energy efficiency requirements for new commercial buildings as

Table 1 Case studies on green leases

Country	Case number	Description	Sources
UK	1	Document analysis of leases for 17 'BREEAM' (Building Research Establishment Environmental Assessment Methodology)-rated office buildings and nine (mostly BREEAM-rated) retail premises.	Bright and Dixie (2014)
	2	Twenty interviews of retail property participants; document analysis of company reports and model lease clauses.	Ongoing WICKED research project
	3	Marks and Spencer (M&S) green lease and memorandum of understanding (MoU) experience to date (MoUs in 65 existing stores; green leases in 80 new stores). Methods: document analysis, interviews with M&S staff.	Janda et al. (2015)
Australia	4	Document analysis of leases for 500 plus office buildings in Sydney Central Business District.	Dawson et al. (2014)
	5	Interviews with Sydney Better Buildings Partnership (BBP) and personal communication with Australian property lawyer(s) on office and retail leasing practices.	Ongoing WICKED research project

well as renovations (DECC, 2014). In compliance with the European Union's Energy Efficiency Directive (EP&C, 2012), the UK has a national energy efficiency target to reduce energy consumption by 18% in 2020 relative to the 2007 business-as-usual projection, and a target of 1.5% annual reduction between 2014 and 2020 (DECC, 2014). The Carbon Reduction Commitment (CRC) Energy Efficiency Scheme requires companies that consume over 6000 MWh of electricity to report and buy allowances for their CO_2 emissions (DECC, 2015a). In addition, under the new Energy Saving Opportunities Scheme (ESOS), large organizations are required to carry out an energy audit every four years, measuring and reporting energy use across energy used in buildings, industrial processes and transport (DECC, 2015b). Energy Performance Certificates (EPCs) rate properties based on age, size and fabric of the building and are required when buildings are constructed, let or sold (*e.g.*, DECC, 2014). Display Energy Certificates (DECs) rate properties based on operational energy consumption and are currently required only in public buildings (DECC, 2014). From April 2018 minimum energy efficiency standards (MEES) are being introduced, making it unlawful to let properties that fail to achieve a prescribed MEES, set at EPC rating E (DECC, 2015c). Of these regulations, only MEES recognizes the importance of tenanted commercial property.

In parallel, the voluntary sustainability rating system 'BREEAM' (Building Research Establishment Environmental Assessment Methodology) provides a common standard to enable the assessment and comparison of the environmental impact of buildings. It has been used to certify over 260 000 building assessments across more than 50 countries (BRE, 2014). Under BREEAM 2011, credits were available for green leases, which incentivized owners seeking the highest

rating to negotiate green leases with occupiers. Online commentaries suggest this was 'unpopular', partly because tenants did not want to accept additional obligations, and BREEAM 2014 has removed the green lease credits (Parker, 2014).

The Australian context: property market and regulatory environment

The main property investment market in Australia is worth about AU$280 billion (£130 billion) (Higgins, 2013). Sydney central business district (CBD), with 4.9 million m^2 of total stock, is the largest office market in Australia. Sydney is the primary location for the head offices of most Australian companies; it is also the most sought-after location, highlighted by strong prime property rents and yields. The market is currently comprised of 53% prime space (Premium and A Grade) and 47% secondary grade space (B–D Grades). Other major property markets in Australia include Melbourne, Brisbane and Perth.

The Building Code of Australia Part J sets minimum standards in respect of energy efficiency requirements for new commercial buildings and for refurbishments over a certain level of work (ABCB, 2010). However, unlike the UK, where minimum energy standards were introduced in the early 1980s, energy efficiency only became part of the Australian Building Code in 2006. Subsequently, a much lower proportion of the existing stock has minimum standards of energy efficiency. There is an intention to increase minimum energy standards over time and they were revised upwards in 2010 (ABCB, 2010).

Alongside the mandatory minimum standards, the Australian commercial office and retail property market is characterized by the National Australian

Built Environment Rating (NABERS) system, which has tools for energy and water ratings. In 2010 The Building Energy Efficiency Disclosure Act established the Commercial Building Disclosure Program, which requires energy efficiency information to be provided when commercial office space of 2000 m^2 or more is offered for sale or lease (Australian Government, 2015). These standards are performance ratings (similar DECs in the UK) and are made public in the form of Building Energy Efficiency Certificates (BEECs). The goal is to improve the energy efficiency of Australia's office stock, and also to inform buyers and tenants. The rationale is that buyers and tenants can easily ascertain the level of energy efficiency and many will choose buildings with better standards that align with their leasing practices and/or corporate social responsibility policy. The voluntary Green Star environmental rating tool also sets energy standards that increase in line with the different star ratings. In comparison with the UK (discussed above), Australia has less aggressive CO_2 reduction targets. Whereas European Union countries plan to reduce GHG emissions between 2020 and 2030 by approximately 2.8% per annum, the World Resources Institute estimates that Australia lags behind with a planned 1.8% per annum reduction rate (Gerholdt & Ge, 2015).

Green lease context: UK and Australia

In Australia, the Australian commonwealth and state governments have provided important early leadership for green leases. Under the Energy Efficiency in Government Operations (EEGO), policy standards have been set since 2006 for all new government leases of more than 2000 m^2 through the use of a 'Green Lease Schedule' (GLS) (Woodford, 2007).

In addition, in both the UK and Australia, industry leadership has been crucial to the emergence of greener leasing. The UK Better Buildings Partnership (BBP) and the Sydney BBP were established (in 2007 and 2011 respectively) to work collaboratively with leading landlords 'to develop solutions to improve the sustainability of existing commercial building stock and achieve substantial CO_2 savings' (UK BBP, 2015a). UK BBP members include 'the UK's leading commercial property owners' (UK BBP, 2015a); Sydney BBP members include 'a number of Sydney's leading commercial and public sector landlords' (Sydney BBP, 2015). Both BBPs developed toolkits providing a menu of 'green clauses' (Sydney BBP, 2013; UK BBP, 2013a) that parties can elect to include in leases and that provide a framework 'for sustainable operations and collaboration throughout the life of commercial leases right from the on-set' (Sydney BBP, 2013)

In addition to green leases, the UK BBP has also promoted memoranda of understanding (MoUs) (UK BBP, 2013a, p. 2). MoUs are separate written agreements that capture the intentions between negotiating parties and can be used in conjunction with binding agreements (such as leases). In leasing, MoUs provide a flexible mechanism for enabling collaboration for buildings already let, as they are not legally binding, can be entered into at any stage of the lease and can be updated without amending the lease. By contrast, in Australia, MoUs have not been widely used due to the costs of negotiation and non-binding nature.

Beyond the UK and Sydney BBPs and their members, the Global Real Estate Sustainability Benchmark (GRESB), established in 2009, is an industry-driven assessment that reports annually on the sustainability performance of real estate portfolios around the world (GRESB, 2015). The GRESB survey includes a section on stakeholder and tenant engagement, part of which focuses on the use of green leases and MoUs (Shire & Quispel, 2013).

Case studies: evidence of green leases and leasing

Within the broader context of the markets outlined above, this section discusses a diverse set of case studies in the UK and Australia. In the UK, Case 1 presents a small-scale review of leases in 'green' UK office and retail buildings; Case 2 presents initial findings from the ongoing WICKED research project theme on 'green leasing' practices in the UK retail sector; and Case 3 highlights in more detail the 'green lease' practices of one of the WICKED project participants, a leading UK retailer. In Australia, Case 4 describes a large-scale study carried out by the Sydney BBP into Sydney's commercial office leasing market; and Case 5 provides a brief qualitative assessment of the use of green leases in Australia's retail market.

Case 1: UK office and retail

Using a qualitative document analysis approach, Bright and Dixie (2014) examined the content of 26 UK commercial leases (17 office and nine retail) in detail to develop a categorization of 'green clauses'. The sampling aimed for a mix of locations, landlords and tenants, but most of the leases involved BREEAM-rated properties and were available on the public land register (compulsory for leases more than seven years). Some additional leases were selected by identifying parties with public green commitments (including green lease policies). Sampling was also restricted to leases registered in 2008 or later as green leasing has only emerged since then. This sample group was acknowledged as unrepresentative and was selected in order to increase the chance of identifying green leases within such a relatively small sample.

Bright and Dixie (2014, p. 10) define 'green clauses' as those that are 'designed to facilitate the property being used in a resource efficient manner and which [... take] account of energy efficiency and other sustainability goals and measures'. The authors grouped clauses into 10 categories, such as: producing a sustainability statement or environmental plan; sharing data about energy, water or waste consumption; permitting the landlord to conduct maintenance and other services in an environmentally sensitive manner; and enabling one or both parties (usually the landlord) to make resource efficiency improvements.

In this dataset, despite being deliberately skewed toward green buildings, the authors found that about 40% of the sample (11 leases, five office and six retail) had no discernible green clauses.

Fifteen leases (12 office and three retail) contained one or more green clauses, and these varied significantly in their content, scope and legal commitment. Within the green leases, data-sharing was the most common clause (80%), followed by a clause requiring environmental considerations to be factored into alterations and/or repairs (10/15 leases), and those allowing service provision to take account of environmental considerations (10/15 leases).

Based on their analysis of these leases, Bright and Dixie (2014, p. 19) conclude that the 'flip-side of our findings is that long-term leases are still being entered into for green buildings which pay no attention to environmental issues', and there is 'unlikely to be significant penetration of green lease principles outside green buildings'.

Case 2: WICKED retail research

The WICKED research project investigates energy management strategies in the UK retail sector. In this context, one research theme explores the role of leases and other organizational practices. To date, the researchers have conducted interviews with 20 representatives from 17 different organizations: four property owners (landlords), four retail tenants, four letting and property management companies, four law firms, and one industry organization (the UK BBP). Interviews are supplemented by document analysis of company strategy reports, green lease clauses in company templates and model green lease clauses promoted by industry partnerships (e.g., UK BBP, 2013a).

A number of key findings are emerging. First, interview responses suggest that the uptake of green leasing arrangements in the retail sector remains low, whilst green lease clauses appear more common in the office sector. Green leasing is more likely to be adopted by larger companies (in particular, the BBP members) with public environmental commitments and/or

concern over exposure to environmental regulation, and typically for prime properties. Lawyers for organizations outside this select group of large landlords may have heard of green leases but seldom see green clauses in contracts.

Where green leases are adopted, agreed clauses are typically either (1) quite broad, non-prescriptive and/or non-binding (e.g., provisions that state a general commitment to cooperate to improve environmental performance); or (2) driven by specific regulatory requirements (e.g., clauses dealing with the costs of CRC compliance or the production of EPCs). MEES, BREEAM and GRESB have all been highlighted as drivers for reviewing lease clauses and green lease discussions.

Despite the global interest in green leases in some circles, an interviewee expressed a commonly held view: that leases largely 'sit in a cupboard for the length of the tenancy', except for 'key intervention points' such as alterations or upgrades, when parties (or more probably asset managers) will refer to the formal lease wording (Interview #12).

Although lease clauses, including green ones, appear to have little relevance to day-to-day operations, early adopters point to potential benefits of the process of negotiating green clauses as a platform for discussion between landlords and tenants. The material goal of green leasing is tenant and landlord cooperation around energy and environmental management; a green lease is a piece of paper that symbolizes and supports this effort.

Case 3: WICKED retail research with M&S

Drawing on early evidence from the WICKED research project, a case study of Marks and Spencer (M&S) – a UK-based retailer with an international chain of clothing and food shops – illustrates the leadership role that M&S, the UK BBP and the UK BBP landlords are playing in the roll out of green leases in the UK retail sector. The case study is based on (1) interviews of key M&S staff (over the phone and via e-mail, including the head of Property Plan A, Plan A project managers and an M&S property lawyer); and (2) analysis of public documents and internal M&S documents (e.g., strategy documents and standard lease clauses).

In 2012 M&S announced an 'environmental leasehold policy' – to introduce green clauses through MoUs for existing stores and to include green clauses in new leases – a commitment now incorporated into M&S Plan A 2020 (M&S, 2015, p. 21). The M&S story suggests that strong leadership and concern about climate change are important drivers for the environmental plans that fed into this leasing policy (Vernon, 2007). Plan A, launched in January 2007, sets out

100 commitments to help M&S become 'the world's most sustainable retailer' (M&S, 2015) and is an important part of its brand. The leasing policy emerged from a convergence of drivers: that leases should not undermine Plan A; a desire to control the lease drafting process to create more standardization across the M&S portfolio; an opportunity to save costs through enabling building improvements; and the promotion of green leases by the UK BBP.

Working together, M&S and UK BBP launched an initiative to introduce green MoUs for 70 M&S stores already under lease with BBP landlords (UK BBP, 2013b). This 'buy in' from BBP landlords has meant that the scope of the M&S MoU clauses (broadly based on the BBP green lease toolkit (UK BBP, 2013a)) is broader and more ambitious than the green clauses being used in new M&S leases. Green MoUs with UK BBP landlords have now been introduced for 65 existing stores.

By contrast, for new leases M&S has to negotiate with a much greater diversity of landlords. M&S has developed a standard set of green clauses, informed by the BBP 'Green lease toolkit' (UK BBP, 2013a), with overall a more limited scope than the MoUs. The green clauses include a general commitment to carry out lease obligations with a view to promoting environmental good practice, an agreement to cooperate and use reasonable endeavours to agree and comply with an energy management plan, and an agreement to maintain and share data. Between January 2013 and December 2014 M&S entered around 80 new leases. Early indications are that most of these, other than lease renewals, include green clauses.

Prior to the development of the standard MoU and green lease clauses, M&S reported that 'darker green' clauses (*e.g.*, allowing increases in service charges related to upgrades) were resisted by landlords. Also, the existence of green clauses has provided a framework and incentive for M&S's Plan A project manager to engage with landlords, meeting with them to discuss priorities for cooperation 'under the guise of green leases' (Plan A project manager, 12 January 2015).

M&S's experience suggests that the 'light green' clauses based on the BBP toolkit have proved largely uncontroversial in negotiations, possibly because of the role of BBP in influencing standard industry practice and also M&S's position in the market, where its brand and size add value to landlords' premises. This shows the sideways impact of a large powerful tenant as a 'middle actor' in the property market.

Case 4: Sydney office leasing

In December 2014, the Sydney BBP published the 'BBP Leasing Index,' covering office leasing in Sydney's CBD

(Dawson, Bailey, & Thomas, 2014). This index shows clear evidence of green lease transformation in this market. Whereas in 2009 only 15% all leases signed in Sydney CBD had green clauses, by 2013 over 60% of all leases included green clauses.

The BBP analysed leases from the public register in New South Wales (Thomas & Dawson, 2014), using the Sydney BBP's Model Lease Clauses to define 'green' terms (Sydney BBP, 2013). Of the 7000 commercial office leases in Sydney CBD, approximately 500 were sampled randomly depending on size of tenancy (small, medium and large) and building quality (non-prime and prime grades), with a target sample size of 100 for each segment. Leases were analysed for the presence of one of 22 model lease clauses and a grading system calculated a 'model lease score'. Gradings were based on 'clause breadth' (the number out of the 22 possible) and 'strength' (how binding the clause is, depending on whether dispute resolution process would be triggered by breach). Numerical grades were awarded and averaged to a total model lease score (Thomas & Dawson, 2014).

There was a quadrupling of some form of green leasing between 2008 and 2014, and 27% of the BBP model lease clauses appeared in standard commercial office leases. In prime buildings, over 80% of leases have best-practice leasing in 2013/14, and include 44% of the model lease clauses. Clauses relating to cooperation, management and recycling, waste and consumption were most frequently included. Nearly one-quarter of leases included a clause relating to securing or maintaining an NABERS rating. The next most common clauses relate to information sharing, environmental sustainability (a high-level commitment clause) and waste reduction.

Despite this growth in the numbers of green clauses used, clause strength still lags, indicating that parties agree to collaborative frameworks but hesitate to risk dispute resolution.

This study highlights the importance of landlord and intermediary leadership in green leasing, following the initial government-as-tenant introduction of the GLS. It shows the sideways impact that major landlords have had on standard leasing practices, particularly in prime buildings. Together with the Sydney BBP, major Australian landlords have successfully diffused a new form of governance of the landlord–tenant relationship and environmental practices into the Sydney office market.

Case 5: Australian retail market

Despite the evident adoption of green leases in office properties, qualitative assessments suggest that green leasing is unusual in Australian retail markets. The

Sydney BBP is initiating conversations with retailers but reports that there has been no significant greening of retail leases to date. Although further study is needed, it appears this may be partly due to the strong consumer orientation of state retail legislation and the fact that the retail sector is so highly price sensitive and cost conscious (Professor W. D. Duncan, personal communication, 4 December 2014). Further, in most Australian jurisdictions the cost of capital improvements is not permitted to be passed on to retail tenants as outgoings, although large retail shops are generally excluded from this (Duncan & Christensen, 2014).

Synthesis and discussion

Across the policy and property contexts of the UK and Australia and the five cases presented, this section synthesizes and discusses with reference to the literature (1) what a green lease is and does; (2) who tends to use green leases; and (3) where green leases are present and absent in the market.

What a 'green lease' is and does: potential versus reality

Although there are various sets of green model clauses, there is still no international standard definition of what a 'green lease' *should* be or do. How many clauses and of what kind does a lease need to contain to qualify as 'green'? The Sydney BBP (Case 4) reviewed a large number of green leases and constructed a strong case for green lease market transformation, but even in this study a 'green lease' can contain anything from a single 'light green' clause (*e.g.*, a very general duty to cooperate on environmental matters) to a number of more ambitious 'dark green' clauses (*e.g.*, setting specific environmental rating targets).

A comparison of the UK and Sydney BBP precedent model clauses, as well as green clauses used in Cases 1 and 3, suggests certain core green clauses, typically less ambitious, are promoted through the model clauses and commonly found in agreed leases. These include a general commitment to improve the environmental performance of properties and commitments to cooperate, in particular to share data about environmental performance. The model clauses also all contain similar provisions restricting tenant's rights to make alterations that adversely affect environmental performance, and various provisions to enable the production of energy performance certificates and other ratings (BREEAM in the UK, NABERS and Green Star in Australia). At the same time, the models also contain a number of important differences. In particular, only the Sydney BBP model clauses include the more ambitious provision – arguably key for addressing the 'split incentive' issue – for the landlord to make environmental improvements and recover the cost through service charges. Such ambitious clauses are very rare in the UK, and also meet with firm resistance in Australia although they are successfully negotiated in some leases (Roussac & Bright, 2012). In relation to rent review, the UK BBP requires tenant works to be disregarded but allows landlord works to be taken into account, whilst the Sydney BBP allows NABERS ratings achieved by either party to be taken into account.

In both Australia and the UK, the case studies also show that lease clauses operate at different levels of enforceability. Although lease clauses are, *prima facie*, binding, some lease clauses are intentionally expressed not to be legally binding. For example, Case 1 found that most (but not all) clauses agreeing to agree an environmental plan were 'good faith' obligations and not legally binding. MoUs are likewise non-binding as a matter of law, although in Case 1 there was, exceptionally, reference to a legally binding MoU that would also bind successors. Case 4 shows that binding clauses with 'teeth' – that attach clear remedial consequences to breach – are relatively unusual, and this was also true of leases in Case 1.

Even if binding, many 'green' clauses may be difficult to enforce. Duties to exercise 'reasonable endeavours' and to 'cooperate', lack specificity and it will be hard to prove breach. Additionally, it is unclear whether a promise to 'cooperate' or exercise 'reasonable endeavours' is enforceable as a matter of English contract law.

Interviewees in the WICKED research project (Case 2) commented that parties are unlikely to seek legal enforcement. Indeed, it is not unusual for lease clauses not to be enforced. For example, at the termination of the lease, a Schedule of Dilapidation in the UK, or Make Good in Australia, will be drawn up by the landlord's surveyor, who identifies all repairs that should have been attended to by the tenant(s). Although there is an opportunity to enforce the lease terms, in practice enforcement is dependent on other variables and tenants may be able to walk away without undertaking many of the items agreed to in the lease (Rowling, 2012).

The case studies show 'green leases' in practice to contain a wide variety of ambitions and levels of enforcement. Despite the potential strength of private–private contracts (Vandenbergh, 2005) and the existing role of leases as a 'local law' between two parties, the relatively low level of ambition and enforceability of agreed green leases suggests they have may have more symbolic value than measurable material impact. As a tool for inter-organizational governance, green leases as currently configured have much in common with VEPs, whose impacts are also difficult to quantify (Borck &

Coglianese, 2009). However, interviews with early adopters of green leases (Cases 2 and 3) suggest they facilitate useful conversations about cooperation between tenant and landlord on environmental matters.

Who adopts and uses green leases?

The organizations adopting green leases in the UK and Australia are mainly the same type that are likely to participate in other forms of VEPs: large powerful organizations with sustainability goals, in markets that are facing regulation (Borck & Coglianese, 2009). Corporate responsibility also acts as a major driver for these organizations (Dixon et al., 2009; Newell, 2008).

More specifically, across the UK and Australia, green leases are generally led by the landlord. Almost invariably the landlord's lawyer supplies the draft lease for negotiation. This means that it is usually the landlord's environmental management style that will lead the use of green leases.

There are some exceptions: the Australian government has been an important leader through the required use of the GLS when an Australian government office lease is signed. Case 3 shows M&S as a retail tenant implementing green leasing. What unites these exceptions is the strong market strength of these tenants and their institutional policy commitment to environmental good practices and social responsibility.

The case studies uphold the idea of landlords as middle actors (Janda & Parag, 2013; Parag & Janda, 2014), capable of exerting influence on others. They play this role both sideways (e.g., landlords learning from other landlords, through the mechanism of the BBPs) and downstream (imposing new leasing practices on tenants). In the UK, data from the case studies suggest that only M&S seems to be playing a 'middle actor' role as a private tenant, particularly in the retail sector. In the UK, the impetus for green leases clearly comes from middle actors in industry (mostly landlords) rather than from the top down. The Australian government requires green leases in its rented offices; the UK government does not.

From a 'middle-out' perspective, the 'sideways' and 'downstream' implications of these findings are that green leases may not be an ideal indicator of 'tenant engagement' (e.g., as articulated in and measured by GRESB (GRESB, 2015; Shire & Quispel, 2013). Engagement should be a two-way process, with the tenant and landlord working together. However, if landlords are the more powerful shapers of the 'local law' set by the lease, the nature of the power structure may suggest 'cooptation' (the formalized inclusion of challengers into the authority system that they are challenging (Coy, 2013)) rather than a more equitable

consensus-driven process of 'cooperation' and engagement.

Where are green leases present and absent?

Although green leases are applicable in concept to any rented property, large or small, their diffusion across the market is uneven. The case studies in both the UK and Australia show that green leasing is more prevalent in offices than in the retail sector. Within the office market, the trend seems concentrated in particular geographic locations (e.g., CBDs of major cities). Most of the activity (and research to date) is concentrated in London and Sydney, yet there are important property markets in both the UK and Australia outside these cities. Case 4 also marks a difference in green lease uptake between prime properties (about 80%, representing about half the CBD in Sydney) and sub-prime properties (60% uptake, representing the rest of the tenanted space). Given the evidence available, wider uptake in non-prime properties or amongst smaller landlords and tenants seems unlikely. A 'green lease' might be thought of in the context of letting a modern building designed to high environmental standards, but arguably the need for green leases is even stronger for buildings in the secondary and tertiary markets that have poor environmental credentials.

The case study data to date cannot say why green leases are more widely adopted in the office market than in the retail market. This uneven uptake, however, suggests that the relationship between property and organizational decisions is complex and worthy of study, as Axon et al. (2012) and Edwards and Ellison (2004) suggest.

Conclusions and implications for governance

This paper has argued that green leasing can be conceptualized as a new form of 'middle-out' energy and environmental governance at the property level. It has described the evolution of greener leasing practices drawing on case studies across different countries and sectors. It has highlighted evidence of the uptake of green lease clauses, drivers for this uptake, their content and potential implications for inter-organizational governance. With the exception of the Australian government GLS, there is no government-directed use of green leases. Most green leases or MoUs therefore are 'voluntary' in the sense that the green clauses are additional to the terms ordinarily used within leases and, more generally, go beyond environmental compliance. However, they *could* be legally enforceable, which most other VEPs are not. Top-down regulation provides a backdrop to green leases, but the role of landlords and industry groups, particularly the

BBPs, have been important 'middle-out' influencers. Early adopters of green leases are a mixed group of large organizations, mainly landlord led but with some notable tenant exceptions.

From a larger governance perspective, it is unlikely that green leases are well suited to top-down government-led initiatives such as building regulations or tax incentives (Qian, Fan, & Chan, forthcoming; Zhang et al., forthcoming). Green leases could be made mandatory, but the fact that lease wording is tailored to particular contexts would mean that any mandated clauses would necessarily be broad. Furthermore, monitoring compliance would be difficult given that in the UK, for example, the Land Registry only requires registration of leases that are more than seven years long, so the government itself does not have a complete record of who is leasing what or how these leases are worded. While green leases may be difficult for the regulatory state to embrace, they are an ideal tool for non-state actors, particularly organizations with multiple properties. They can be made mandatory within a portfolio, such as the Australian government requiring them for their rented offices and M&S requiring them for their rented stores. Insofar as 'governance' is not the same as 'government' (Janda & Kwak, 2011), these requirements *are* mandatory, but imposed by one organization upon another rather than by the state across its domain. Considering green leases as a new form of private–private environmental governance could lead to new ways of applying ideas about polycentric governance (Ostrom, 2010) to shared spaces as well as to shared resources.

Polycentric governance is often associated with the management of shared environmental resources in a contiguous space, such as a metropolitan area, a fishery or a forest. This paper has suggested that green leases currently manifest mainly in large multi-site organizations that own or rent spatially distributed spaces, what Axon et al. (2012) call 'fleet' landlords and tenants. How would or do green leases operate as a form of environmental management in a contiguous area, such as a city?

The Tokyo Metropolitan Government and C40 Cities recently surveyed 12 different kinds of energy efficiency initiatives, including green leases, in 16 cities across nine countries (Takagi et al., 2015). This report found programmes promoting green leases in 15 of the 16 studied cities (Takagi et al., 2015, p. 21). Of the 15 cities promoting green leases, however, two-thirds of the programmes (10 cities) were designated as 'partner-led', a category the authors use 'only when no city-led or higher-tier government-led programmes were found' (p 21). Across a matrix of 12 different types of policy elements, the promotion of green leases was the only category where partner-led programmes figured prominently in the analysis.[1] At the city level, these findings support this paper's conclusions that leasing is a mechanism currently led by private entities rather than public ones. As Dixon, Britnell, and Watson (2014) suggest in their study of urban retrofit practices, the commercial property sector may be 'city blind', seeing opportunities for property retrofits in a portfolio that extends across cities rather than fitting neatly within their footprint. Such widely distributed activities challenge the utility of green leases as a policy mechanism for city, regional and national governments. If the geographical level of the policy fits the organizational structure of the industry, however, green leases may be a fitting tool. For example, green leases may be particularly useful in places like Singapore and Tokyo, where strong governments coincide with highly concentrated commercial property activities (Chua, 2014; Nishida & Hua, 2011, forthcoming).

As currently implemented, green leasing is a tool used by businesses, generally with minimal government involvement at the city, regional or national level. In their review of VEPs, Borck and Coglianese (2009) present a typology of three different kinds of VEPs – unilateral, bilateral and public voluntary – depending on the number of participating businesses and on the degree of governmental involvement. Unilateral programmes are led by businesses or industry associations; bilateral programmes are negotiated between government and specific businesses; and governments use public voluntary programmes to recognize achievements beyond mandatory standards. In this typology, green leases are unilateral VEPs in the UK, but they could be characterized as a bilateral VEP in Australia. Governments in other countries could certainly follow Australia's lead and create bilateral programmes, using their own purchasing power to promote the practice. This action would likely only have positive impacts in commercial property markets where government tenancy plays a significant role.

From a legal perspective, leases are at the heart of governance of tenanted space, but from an operational and practical perspective they are less central. The people who manage and occupy space may never see the lease or even have access to it (Roussac & Bright, 2012). The evidence to date in the UK and Australia suggests that green leases may provide a 'necessary but not sufficient' function in tenanted commercial property. Even without 'teeth', green leases encourage inter-organizational negotiations in relation to energy and environmental management. These developments have not flowed directly from top-down governance, VEPs or regulatory measures that are largely (with the exception of MEES in the UK) blind to the joint use of tenanted space. Green leases therefore add a new tool to the box of regulatory and governance tools that 'operate concurrently or in overlapping ways' (Morgan & Yeung, 2007, p. 4) to address the 'wicked' problem of improving the tenanted building stock.

The further study of green leases and leasing could be enhanced by the development of a tiered rating of green leases that develops more nuanced processes and thresholds for environmental ambitions, outcomes and enforceability (*e.g.*, light green, mid-green, dark green), much like BREEAM and NABERS have for green buildings. As much of the green lease work to date is landlord led, genuine forms of tenant engagement and even tenant leadership is another topic for further study. The market may already be moving in this direction. The UK BBP put 'occupier engagement' on its list of priorities (UK BBP, 2015b). The Sydney BBP goes even further, conceptualizing the tenant–landlord relationship as a kind of love match that can lead to happiness (Blundell, 2013). Finally, further quantitative and qualitative research through the WICKED research project should generate firmer insights into the role of green leases across the retail sector as well as their impact on energy management practices of tenants and landlords.

Acknowledgements

The authors would like to acknowledge the contributions of Ben Thomas and Esther Bailey of the Sydney Better Buildings Partnership (BBP) to an earlier version of this paper. They also thank the editor, guest editors and reviewers for their comments and suggestions. This paper is a revision and expansion of a conference paper: Bright et al. (2015). Portions of the research for and preparation of this paper was supported by the UK Engineering and Physical Sciences Research Council under the WICKED research project, grant number EP/ L024357/1. An anonymized summary of interview data from the WICKED project will be made available through ORA, the Oxford University Research Archive (http://ora.ox.ac.uk) on completion of the project.

Disclosure statement

No potential conflict of interest was reported by the authors.

References

ABCB. (2010). BCA section J – Assessment and verification of an alternative solution. Handbook – Non-Mandatory Document. Canberra: Australian Building Codes Board (ABCB). Retrieved from http://www.abcb.gov.au/en/education-events-resources/publications/abcb-handbooks.aspx

Australian Government. (2015). *2015 Commercial building disclosure legal framework*. Canberra: Australian Government. Retrieved September 12 http://cbd.gov.au/overview-of-the-program/legal-framework.

Axon, C. J., Bright, S., Dixon, T., Janda, K. B., & Kolokotroni, M. (2012). Building communities: Reducing energy use in tenanted commercial property. *Building Research & Information*, 40(4), 461–472. doi:10.1080/09613218.2012.680701

Biggart, N. W., & Lutzenhiser, L. (2007). Economic sociology and the social problem of energy inefficiency. *American Behavioral Scientist*, 50(8), 1070–1087. doi:10.1177/0002764207299355

Bilfinger GVA. (2015). *The big nine: Quarterly review of the regional office occupier markets*. London: Bilfinger GVA. Retrieved from http://www.gva.co.uk/uploadedfiles/GVA_UK_Research/2015TheBigNineQ22015.pdf

Blundell, L. (2013). *The tenants & landlords guide to happiness*. Sydney: Fifth Estate. Retrieved September 25, 2015, from http://www.thefifthestate.com.au/tfe_ebooks/e-book-best-practice-and-green-leasing-guide-launched

Borck, J. C., & Coglianese, C. (2009). Voluntary environmental programs: Assessing their effectiveness. *Annual Review of Environment and Resources*, 34(1), 305–324. doi:10.1146/annurev.environ.032908.091450

BRE. (2014). BREEAM UK new construction, non-domestic buildings. Technical Manual: Version: SD5076 – Issue: 0.1. Watford: Building Research Establishment (BRE). Retrieved from http://www.breeam.org/filelibrary/BREEAMUKNC2014Resources/SD5076_DRAFT_BREEAM_UK_New_Construction_2014_Technical_Manual_ISSUE_0.1.pdf

Bright, S., & Dixie, H. (2014). Evidence of green leases in England and Wales. *International Journal of Law in the Built Environment*, 6(1/2), 6–20. doi:10.1108/IJLBE-07-2013-0027

Bright, S., Patrick, J., B. Thomas, K. B. Janda, E. Bailey, T. Dixon, & S. Wilkinson. 2015. The evolution of 'greener' leasing practices in Australia and England. In Proceedings of COBRA, July 8–10, 2015 (Sydney, Australia). Royal Institution of Chartered Surveyors (RICS).

Chua, G. (2014). *New 'green lease' guide launched for commercial landlords and tenants*. Singapore: The Straits Times. Retrieved September 8, 2015, from http://www.straitstimes.com/singapore/environment/new-green-lease-guide-launched-for-commercial-landlords-and-tenants

Coy, P. G. (2013). Co-optation. In *The Wiley-Blackwell encyclopedia of social and political movements* (pp. 280–282). Malden, MA, Oxford: Blackwell Publishing Ltd.

CSE, & ECI. (2012). *What are the factors influencing energy behaviours and decision-making in the non-domestic sector? A rapid evidence assessment*. London: Department of Energy and Climate Change. Retrieved from https://www.gov.uk/government/publications/factors-influencing-energy-behaviours-and-decision-making-in-the-non-domestic-sector-a-rapid-evidence-assessment

Dawson, B., Bailey, E., & Thomas, B. (2014). Progress of best practice leasing: Better buildings partnership leasing index results (Sydney CBD). In *BBP progress of best practice leasing evening*. Sydney, NSW: Sparke Helmore Lawyers. Retrieved from https://drive.google.com/file/d/0B7TfBUAVGwHTR1I3RTdZdnZPLUE/view

DECC. (2014). *UK national energy efficiency action plan*. London: Department of Energy and Climate Change. Retrieved from https://www.gov.uk/government/uploads/system/uploads/attachment_data/file/307993/uk_national_energy_efficiency_action_plan.pdf

DECC. (2015a). *CRC energy efficiency scheme: Qualification and registration*. London: Department of Energy and Climate Change. Retrieved 6 August, from https://www.gov.uk/crc-energy-efficiency-scheme-qualification-and-registration.

DECC. (2015b). *Energy savings opportunity scheme (ESOS)*. London: Department of Energy and Climate Change. Retrieved 6 August, from https://www.gov.uk/guidance/energy-savings-opportunity-scheme-esos.

DECC. (2015c). *Private rented sector minimum energy efficiency standard regulations (Non-domestic)*. URN: 14D/208. London: Department of Energy and Climate Change.

Retrieved from https://www.gov.uk/government/uploads/system/uploads/attachment_data/file/401378/Non_Dom_PRS_Energy_Efficiency_Regulations_-_Gov_Response__FINAL_1_1__04_02_15_.pdf

Deline, M. B. (2015). Energizing organizational research: Advancing the energy field with group concepts and theories. *Energy Research & Social Science, 8*, 207–221. doi:http://dx.doi.org/10.1016/j.erss.2015.06.003

Dixon, G. N., Deline, M. B., McComas, K., Chambliss, L., & Hoffmann, M. (2015). Saving energy at the workplace: The salience of behavioral antecedents and sense of community. *Energy Research & Social Science, 6*, 121–127. doi:http://dx.doi.org/10.1016/j.erss.2015.01.004

Dixon, T., Britnell, J., & Watson, G. B. (2014). 'City-wide' or 'city-blind?' An analysis of emergent retrofit practices in the UK commercial property sector. WP2014/1. Retrofit 2050 Working Paper. Retrieved from http://www.retrofit2050.org.uk/sites/default/files/resources/citywidecityblind.pdf

Dixon, T., Ennis-Reynolds, G., Roberts, C., & Sims, S. (2009). Is there a demand for sustainable offices? An analysis of UK business occupier moves (2006–2008). *Journal of Property Research, 26*(1), 61–85. doi:10.1080/09599910903290052

DTZ/Jones Lang LaSalle. (2014). *City offices: Getting the balance right.* London: City of London Corporation and City Property Association. Retrieved from https://www.cityoflondon.gov.uk/business/commercial-property/city-property-advisory-team/Documents/city_offices_getting_the_balance_right.pdf

Duncan, B., & Christensen, S. (2014). *Commercial leases in Australia* (7th ed). Sydney, Australia: Thomson Reuters, Lawbook.

Duquesne, B. ed. (2011). *CMS study on the use of green lease clauses in Europe.* Version 1.0 ed, e-guide. Europe: CMS.

Edwards, V. M., & Ellison, L. (2004). *Corporate property management: Aligning real estate With business strategy.* Oxford, UK: Blackwell Science.

Eisenberg, D. (2016). Transforming building regulatory systems to address climate change. *Building Research & Information.* doi:10.1080/09613218.2016.1126943

EP&C. (2012). *Directive 2012/27/EU of the European parliament and of the council on energy efficiency, amending directives 2009/125/EC and 2010/30/EU and repealing directives 2004/8/EC and 2006/32/EC.* Brussels: European Parliament and Council (EP&C). Retrieved from http://eur-lex.europa.eu/legal-content/EN/TXT/?qid = 1399375464230&uri = CELEX:32012L0027

Fischer, J., & Guy, S. (2009). Re-interpreting regulations: Architects as intermediaries for low-carbon buildings. *Urban Studies, 46*(12), 2577–2594. doi:10.1177/0042098009344228

Gerholdt, R., & Ge, M. (2015). *Australia offers lackluster 2030 climate target.* Washington, DC: World Resources Institute. Retrieved October 28, 2015, from http://www.wri.org/blog/2015/08/australia-offers-lackluster-2030-climate-target

Gouldson, A., & Sullivan, R. (2014). Understanding the governance of corporations: an examination of the factors shaping UK supermarket strategies on climate change. *Environment and Planning A, 46*(12), 2972–2990. doi:10.1068/a130134p

Green Construction Board. (2013). *Low carbon routemap for the UK built environment.* London: The Green Construction Board. Retrieved from http://www.greenconstructionboard.org/otherdocs/Routemap final report 05032013.pdf

GRESB. (2015). *About GRESB*: Global real estate sustainability benchmark. Retrieved September 24, from https://www.gresb.com/about

Hannon, M., Rhodes, A., & Skea, J. (2013). *Summary of workshop on energy in the home and workplace.* Working Document. London: Centre for Environmental Policy. Retrieved from https://workspace.imperial.ac.uk/rcukenergystrategy/Public/reports/Expert Workshop Reports/Energy in the Home and Workplace Working Document Final.pdf

van der Heijden, J. (2015). Voluntary programmes for building retrofits: Opportunities, performance and challenges. *Building Research & Information, 43*(2), 170–184. doi:10.1080/09613218.2014.959319

van der Heijden, J. (forthcoming 2016). The new governance for low-carbon buildings: Mapping, exploring, interrogating. *Building Research & Information.*

Higgins, D. (2013). *Australian commercial property investment market: Styles, performance and funding September 2013.* Melbourne: Australian Centre for Financial Studies. Retrieved from http://www.australiancentre.com.au/News/australian-commercial-property-investment-market-styles-performance-and-funding

Hinnells, M., S. Bright, A. Langley, L. Woodford, P. Schiellerup, & T. Bosteels. (2008). The greening of commercial leases. *Journal of Property Investment & Finance, 26*(6), 541–551. doi:10.1108/14635780810908389

Howard-Grenville, J., Nash, J., & Coglianese, C. (2008). Constructing the license to operate: Internal factors and their influence on corporate environmental decisions. *Law & Policy, 30*(1), 73–107. doi:10.1111/j.1467-9930.2008.00270.x

Janda, K. B. (2014). Building communities and social potential: Between and beyond organisations and individuals in commercial properties. *Energy Policy, 67*(April), 48–55. .1016/j.enpol.2013.06.024

Janda, K. B., & Parag, Y. (2013). A middle-out approach for improving energy performance in buildings. *Building Research & Information, 41*(1), 39–50. doi:10.1080/09613218.2013.743396

Janda, K. B., Patrick, J., Granell, R., Bright, S., Wallom, D., & Layberry, R. (2015). A WICKED approach to retail sector energy management. In Proceedings of *ECEEE Summer Study,* 1–6 June 2015 (Toulon/Hyères, France). Vol. 1 – Foundations of Future Energy Policy, pp. 185–195. Stockholm, Sweden: European Council for an Energy-Efficient Economy.

Janda, K. F., & Kwak, J.-Y. (2011). *Party systems and country governance.* London and New York: Routledge.

Kaplow, S. D. (2009). Does a green building need a green lease. *University of Baltimore Law Review, 38*(3), 375–409.

Langley, A., & Hopkinson, L. (2009). *Greening the commercial property sector: A guide for developing and implementing best practice through the UK leasing process: Good practice guide.* Cardiff: Welsh School of Architecture. Retrieved from http://www.greenleases-uk.com

LCICG. (2012). *Technology innovation needs assessment (TINA): Non-domestic buildings summary report.* London: Low Carbon Innovation Coordination Group (LCICG). Retrieved from http://www.lowcarboninnovation.co.uk/working_together/technology_focus_areas/nondomestic_buildings/

Levine, M., Ürge-Vorsatz, D., Blok, K., Geng, L., Harvey, D., Lang, S., . . . Yoshino, H. (2007). Residential and commercial buildings. In *Climate change 2007: Mitigation contribution of working group III to the fourth assessment report of the intergovernmental panel on climate change,* edited by B. Metz, O. R. Davidson, P. R. Bosch, R. Dave, and L. A. Meyer (pp. 387–446). Cambridge, United Kingdom and New York, USA: Cambridge University Press.

Levitt, R. E., W. Heinisz, W. R. Scott, & D. Settle. (2010). Governance challenges of infrastructure delivery: The case for socio-economic governance approaches. In *Proceedings of Construction Research Congress 2010: Innovation for Reshaping Construction Practice* (pp. 757–767). Reston, VA: American Society of Civil Engineers.

Lind, H., Bonde, M., & Zalejska-Jonsson, A. (2014). The economics of green buildings. In J. Metzger, & A. R. Olsson (Eds.), *Sustainable Stockholm: Exploring urban sustainability in Europe's greenest city* (pp. 129–146). New York: Routledge.

Lutzenhiser, L. (1993). Social and behavioral aspects of energy Use. *Annual Review of Energy and the Environment, 18*, 247–289. doi:10.1146/annurev.eg.18.110193.001335

Lutzenhiser, L. (2014). Through the energy efficiency looking glass: Rethinking assumptions of human behavior. *Energy Research and Social Science, 1*(March), 141–151. doi:http://dx.doi.org/10.1016/j.erss.2014.03.011

McWilliams, A. (2015). Corporate social responsibility. In *Wiley encyclopedia of management* (pp. 1–4). Chichester, West Sussex: John Wiley & Sons, Ltd.

Moezzi, M., & Janda, K. B. (2014). From 'if only' to 'social potential' in schemes to reduce building energy use. *Energy Research and Social Science, 1*(March), 30–40. doi:http://dx.doi.org/10.1016/j.erss.2014.03.014

Morgan, B., & Yeung, K. (2007). *An introduction to law and regulation: text and materials, Law in context.* Cambridge: Cambridge University Press.

Moss, T. (2009). Intermediaries and the governance of sociotechnical networks in transition. *Environment and Planning A, 41*(6), 1480–95. doi:10.1068/a4116

Moss, T., Medd, W., Guy, S., & Marvin, S. (2009). Organising water: The hidden role of intermediary work. *Water Alternatives, 2*(1), 16–33.

M&S. (2015). *M&S plan A report 2015.* London: Marks & Spencer. Retrieved from http://planareport.marksandspencer.com

Newell, G. (2008). The strategic significance of environmental sustainability by Australian-listed property trusts. *Journal of Property Investment & Finance, 26*(6), 522–540. doi:10.1108/14635780810908370

Nishida, Y., & Hua, Y. (2011). Motivating stakeholders to deliver change: Tokyo's cap-and-trade program. *Building Research & Information, 39*(5), 518–533. doi:10.1080/09613218.2011.596419

Nishida, Y., & Hua, Y. (forthcoming). Alternative building emission governance: Outcomes from the Tokyo cap-and-trade program. *Building Research & Information.*

Oberle, K., & Sloboda, M. (2010). The importance of 'greening' your commercial lease. *Real Estate Issues, 35*(1), 32–41.

Ostrom, E. (2010). Beyond markets and States: Polycentric governance of complex economic systems. *American Economic Review, 100*(3), 641–672. doi:10.1257/aer.100.3.641

Parag, Y., & Janda, K. B. (2014). More than filler: Middle actors and socio-technical change in the energy system from the 'middle-out'. *Energy Research and Social Science, 3*(September), 102–112. doi:http://dx.doi.org/10.1016/j.erss.2014.03.014

Parker, J. (2014). *BREEAM 2014.* Retrieved 6 August, from http://www.flattconsulting.com/Articles/breeam-2014.html

Prakash, A., & Potoski, M. (2006). *The voluntary environmentalists: Green clubs, ISO 14001, and voluntary environmental regulations.* New York, NY: Cambridge University Press.

Prindle, W. R., Sathaye, J., Murtishaw, S., Crossley, D., Watt, G., Hughes, J., … Joosen, S. (2007). *Quantifying the effects of market failures in the end-use of energy.* ACEEE Report Number E071. Paris: International Energy Agency. Retrieved from http://psb.vermont.gov/sites/psb/files/publications/Reports to legislature/EEU/Act89Sec29Report/Thermal Efficiency Report VEIC Appendix A.pdf

Property Industry Alliance. (2014). *Property data report 2014.* London: Property Industry Alliance. Retrieved from http://www.bpf.org.uk/sites/default/files/resources/BPF-PIA-Property-Data-Report-2014-are_0.pdf

PropertyWire. (2015). *Investment in London commercial property market close to last peak in 2007.* London: PropertyWire.com. Retrieved November 17, 2015, from http://www.propertywire.com/news/europe/london-commercial-property-investment-2015012110060.html

Qian, Q., Fan, K., & Chan, E. (forthcoming). Regulatory incentives for green building: Gross floor area concessions. *Building Research & Information.*

Rittel, H. W. J., & Webber, M. M. (1973). Dilemmas in a general theory of planning. *Policy Sciences, 4*(1973), 155–169. doi:10.1007/BF01405730

Rosenow, J., Fawcett, T., Eyre, N., & Oikonomou, V. (2016). Energy efficiency and the policy mix. *Building Research & Information.* doi:10.1080/09613218.2016.1138803

Roussac, A. C., & Bright, S. (2012). Improving environmental performance through innovative commercial leasing: An Australian case study. *International Journal of Law in the Built Environment, 4*(1), 6–22. doi:10.1108/17561451211211714

Rowling, J. (2012). Can dilapidations be improved? *Journal of Building Survey, Appraisal & Valuation, 1*(2), 106–113.

Sayce, S., Sundberg, A., Parnell, P., & Cowling, E. (2009). Greening leases: Do tenants in the United Kingdom want green leases? *Journal of Retail and Leisure Property, 8*(4), 273–284. doi:10.1057/rlp.2009.13

Schweber, L., & Leiringer, R. (2012). Beyond the technical: A snapshot of energy and buildings research. *Building Research & Information, 40*(4), 481–492. doi:10.1080/09613218.2012.675713

Scott, W. R., Levitt, R. E., & Orr, R. J. (2011). *Global projects: Institutional and political challenges.* Cambridge, UK: Cambridge University Press.

Shire, P., & Quispel, E. (2013). Benchmarking on track. *Real Estate*, May/June (2013), 48.

Sovacool, B. K. (2014). What are we doing here? Analyzing fifteen years of energy scholarship and proposing a social science research agenda. *Energy Research and Social Science, 1*(1), 1–29. doi:10.1016/j.erss.2014.02.003

Sydney BBP. (2013). *BBP releases model lease clauses.* Sydney: Sydney Better Buildings Partnership. Retrieved February 3, 2015, from http://www.sydneybetterbuildings.com.au/2013/bbp-releases-model-lease-clauses/

Sydney BBP. (2015). *Members.* Sydney: Sydney Better Buildings Partnership. Retrieved September 24, from http://www.sydneybetterbuildings.com.au/members/

Takagi, T., Sprigings, Z., Nishida, Y., Graham, P., Horie, R., Lawrence, S., … Miclea, C. P. (2015). *Urban efficiency: A global survey of building energy efficiency policies in cities.* Tokyo, Japan: Tokyo Metropolitan Government and C40 Cities. Retrieved from https://www.kankyo.metro.tokyo.jp/en/int/attachement/Full_Report.pdf

Taylor, M., & Janda, K. B. (2015). New directions for energy and behaviour: Whither organizational research? In Proceedings of *ECEEE Summer Study*, 1–6 June 2015 (Toulon/Hyères, France). Vol. 9 – Dynamics of Consumption, pp. 2243–2253. Stockholm, Sweden: European Council for an Energy-Efficient Economy.

Thomas, B., & Dawson, B. (2014). *A method for analysing and assessing the collaborative potential and environmental commitment of commercial leases.* Sydney, NSW: Better Buildings Partnership.

UK BBP. (2013a). *BBP green lease toolkit* [website]. London: Better Buildings Partnership. Retrieved January 14, 2015, from http://www.betterbuildingspartnership.co.uk/download/bbp-gltk-2013.pdf

UK BBP. (2013b). *M&S & BBP unveil green lease collaboration* [Website]. London: Better Buildings Partnership. Retrieved from January 13, 2015, from http://www.betterbuildingspartnership.co.uk/media/news/details/82/mands-and-bbp-unveil-green-lease-collaboration

UK BBP. (2015a). *BBP – Aims and objectives* [Website]. London: Better Buildings Partnership. Retrieved from September 22, 2015, from http://www.betterbuildingspartnership.co.uk/about-us/aims-and-objectives

UK BBP. (2015b). *BBP – Our priorities: occupier engagement* [Website]. London: Better Buildings Partnership. Retrieved September 22, 2015 from http://www.

betterbuildingspartnership.co.uk/our-priorities/occupier-engagement

UNEP FI. (2011). *Implementing responsible property investment strategies*. Geneva: United Nations Environmental Programme Finance Initiative. Retrieved from http://www.unepfi.org/fileadmin/documents/responsible_property_toolkit4.pdf

Ürge-Vorsatz, D., Eyre, N., Graham, P., Harvey, D., Hertwich, E., Jiang, Y., … Novikova, A. (2012). Chapter 10 – Energy end-use: Building. In *Global energy assessment – Toward a sustainable future* (pp. 649–760). Cambridge, UK and New York, NY, USA and the International Institute for Applied Systems Analysis, Laxenburg, Austria: Cambridge University Press.

Vandenbergh, M. P. (2005). The private life of public Law. *Columbia Law Review, 105*(7), 2029–2096. doi:10.2307/4099486

Vernon, P. (2007). Saint Stuart – The man who is turning M&S green. *The Guardian*, 15 April 2007. Retrieved January 13, 2015, from http://www.theguardian.com/environment/2007/apr/15/fashion.ethicalliving

Woodford, L. (2007). The green lease schedule. In Proceedings of *ECEEE Summer Study*, (Colle Sur Loop, France), pp. 547–556. Stockholm: European Council for an Energy-Efficient Economy.

Yin, R. K. (2009). *Case study research: Design and methods* (4th ed). Newbury Park, CA: Sage.

Zhang, J., Zhou, N., Hinge, A., Feng, W., & Zhang, S. (forthcoming). Effective governance strategies to achieve zero-energy buildings in China. *Building Research & Information.*

Endnote

[1]For example, only two other policy categories noted the existence of partner-led programmes. Reporting and benchmarking was partner-led in only one city; green building and energy rating was partner-led in two cities. Green lease programmes were partner-led in 10 cities.

RESEARCH PAPER

Governance of heritage buildings: Australian regulatory barriers to adaptive reuse

Sheila Conejos, Craig Langston, Edwin H. W. Chan and Michael Y. L. Chew

The resilience and capacity of historic buildings to adapt plays a vital role in mitigating climate change through adaptive reuse. The adaptive reuse of buildings is a practical substitute to demolition and has substantial economic, environmental and social benefits. However, tensions exist between the retention of heritage buildings and conformance with regulatory requirements (e.g. energy efficiency to reduce greenhouse gas emissions, disability access, etc.). This raises questions about whether regulatory systems can embrace both green building technologies and heritage conservation principles. This paper examines the challenges/barriers to successful adaptive reuse projects in Australia using a qualitative approach that involves multiple case studies and in-depth interviews with industry experts coupled with field observation and building plan appraisals. The findings show that compliance to codes/regulations and current design requirements are the major challenges encountered in undertaking adaptive reuse projects. The underlying parameters of the identified challenges will serve as an initiative for formulating prospective regulations that address changing building use, encourage the integration of modern technologies and inhibit unnecessary building demolition for future global climate protection.

Introduction

Adaptively reusing existing buildings provides a significant opportunity to address climate change by reducing energy use while simultaneously improving the building's environmental performance over their entire life cycles. Existing buildings have embodied energy and original qualities that cannot be surpassed by demolition and new construction in terms of their environmental, social and cultural contributions. The practice of adaptive reuse is not new. It extends the useful life cycle of an asset and the preservation of resources involved in its construction.

The global implementation of adaptive reuse can significantly contribute to sustainability (Ball & Ball, 1999; Department of Environment & Heritage (DEH), 2004; Douglas, 2006; Gregory, 2004; Rabun & Kelso, 2009; Remøy, & van der Voordt, 2007; UNEP, 2007; Velthuis, & Spennemann, 2007; Wilkinson & Reed, 2008) in terms of economic rewards (Browne, 2006; Bullen & Love, 2011; Cooper & Cooper, 2001), environmental benefits (Gorse & Highfield, 2009; UNEP, 2009), and socio-cultural gains (English Heritage, 1997, 2008, 2013; UNESCO, 2007, 2009). However, according to Shipley et al. (2006) and Ellison, Sayce and Smith (2007), reusing buildings according to sustainability standards entails higher costs in comparison with conventional adaptive reuse undertakings. The higher initial reuse cost is due to mechanical and electrical engineering systems (Siddiqi, 2006). Additionally, the integration of green features is harder to apply on heritage-listed buildings than non-heritage-listed buildings due to the minimum intervention approach stipulated in conservation guidelines.

The aim of this paper is to determine the various challenges encountered in undertaking adaptive reuse and, through critical examination of these challenges, identify the underlying parameters that will guide the formulation of prospective regulations to encourage heritage building sustainability and avoid unnecessary demolition. This research focuses only on the adaptive reuse practices for heritage-listed buildings, as described in the Burra Charter (Marquis-Kyle & Walker, 1994). Heritage buildings often have a certain flair for quality and timelessness that contributes to community value and culture that, in many cases, sets them apart from more contemporary buildings. Conserving these heritage buildings to make them useful in the long-term is a top priority and demands great responsibility from designers, developers and policy-makers to manage their sustainability.

Adaptive reuse and green adaptive reuse

Adaptive reuse is essential when the building's function becomes obsolete and reinventing its purpose is the only way to preserve and maintain its intrinsic heritage value. The essence of the adaptive reuse of heritage buildings is the restoration of the building's value to a place or community while ensuring its future usefulness and contribution. Adaptive reuse is a strategy that extends the building's physical and social functions by giving the building a new purpose while conserving its historic and cultural significance (DEH, 2004; Latham, 2000; Wilkinson et al., 2009). It is often a more successful and cheaper alternative to demolition and new construction (Ball, 1999, 2002; Brand, 1994; Cooper & Cooper, 2001; DEH, 2004; Douglas, 2002; Gregory, 2004; Kohler & Hassler, 2002; Latham, 2000; Pearce, 2004; Petersen, 2002; de Valence, 2004).

With the development of new technologies, specialist knowledge is required in order to incorporate these technologies into existing buildings, particularly those that are heritage-listed (Kincaid & Kincaid, 2000). The blending of new technologies in heritage buildings is significant for the future of architecture (Snyder, 2005). Fournier and Zimnicki (2005) developed adaptive reuse guidelines that integrate the goals of sustainable design and historic preservation. This leads to the concept of 'eco-vernacular' (Dittmark, 2008) or 'green adaptive reuse' (Langston, 2009), which integrates green principles and technologies into the adaptation of historic buildings to improve performance while preserving heritage and cultural values.

Barriers to adaptive reuse

Adaptive reuse has many benefits and opportunities; however, it does carry with it several obstacles especially when it pertains to heritage buildings. Table 1 shows a list of identified barriers to adaptive reuse with their underpinning literature.

Heritage conservation in the Australian context

In an Australian context, heritage buildings are built heritage that is distinct from natural, indigenous and historic heritage (Department of Environment, 2015). This includes the buildings erected following settlement by European and other migrants and dwellings with designated cultural or heritage significance (Australia ICOMOS, 1999; Marquis-Kyle & Walker, 1994). The heritage significance of a place is determined according to its association during the formative period in Australia's national history. This includes the aboriginal cultural landscape, the convict landscape where the colonial frontier was expanded by convicts working for free settlers, the 19th-century European colonial expansion, the gold rush era and the commercial and industrial development after the Second World War (Australia ICOMOS, 1999; Marquis-Kyle & Walker, 1994).

Conservation philosophy and practice

Australia uses the Burra Charter to conserve and manage places of cultural significance and also adopts the Venice Charter principles to suit local Australian requirements (Marquis-Kyle & Walker, 1994) for conserving and managing places of cultural significance. In addition to following the Burra and Venice charters, a conservation management plan, which is globally recognized and has been widely adopted in Australia, New Zealand and the UK (Kayan & Kayan, 2015), must be undertaken by developers prior to the refurbishment or adaptive reuse of heritage buildings that involve heritage management and historic assessment. In 2008, the adaptation guidelines published by the NSW Department of Planning and RAIA (2008) were considered best practice for the conservation and adaptation of heritage items of either local or state significance. These adaptation guidelines are used when assessing development applications for adaptation projects. Local councils use the guidelines for the same purpose. The seven principles for the adaptation of historic buildings and sites for new functions are summarized as follows:

- Identify and understand the heritage significance of the place and its fabric.

- Provide an appropriate new use that is compatible with the place's heritage significance.

Table 1 List of barriers to adaptive reuse

Barrier	Brief description	References
(1) Building codes and regulations/legal constraints	Compliance with current building codes, regulations, conservation guidelines, licensing and planning requirements	Bruce et al. (2015), Bullen and Love (2011), Cooper (2001), Douglas (2006), Shipley et al. (2006)
(2) Physical restrictions	Restrictions due to existing floor layouts, number of columns/walls and structural system layouts	Bruce et al. (2015), Bullen and Love (2011), Cox (2004), Reyers and Mansfield (2001)
(3) High remediation costs and construction delays	Contamination due to the use of hazardous materials in buildings that causes additional costs and time delays	Bruce et al. (2015), Bullen and Love (2011), Wilkinson et al. (2009)
(4) Availability of materials and lack of skilled tradesmen	Compatibility of new materials with existing materials, as well as the availability of local expertise and tradesmen capable of implementing conservation works	Cox (2004), Bullen and Love (2011), Douglas (2006), Remoy and van der Voordt (2007), Reyers and Mansfield (2001)
(5) Complexity and technical difficulties	Refurbishment techniques, technical installations and innovative solutions for the adaptive reuse of heritage buildings	Ball and Ball (1999), Bruce et al. (2015), Bullen and Love (2011), El Kerdany (2002), Kronenburg (2007), Shipley et al. (2006)
(6) Economic considerations	Direct and indirect cost considerations in terms of the conservation requirements for the adaptation of heritage buildings	Cox (2004), Douglas (2006), O'Donnell (2004), Reyers and Mansfield (2001), Shipley et al. (2006), Yung and Chan (2012), Wang and Zeng (2010)
(7) Social considerations	Pertains to the intangible non-economic values considered to maintain the community's daily life (e.g. a sense of attachment to the place)	Bond (2011), DEH (2004), Jonas (2006), Yung and Chan (2012)
(8) Inaccuracy of information and drawings	Lack of accurate information and drawings for heritage buildings (includes defects or dimensional and material inconsistencies)	Cox (2004), Remoy and van der Voordt (2007), Reyers and Mansfield (2001)
(9) Limited response to sustainability agenda	Limited support from building owners and commercial property markets in updating buildings to sustainability standards	Ellison and Sayce (2007), O'Donnell (2004), Pivo and McNamara (2005)
(10) Maintenance	High maintenance and repair costs due to physical deterioration and defects	Bullen and Love (2011), O'Donnell (2004), Remoy and van der Voordt (2007)
(11) Classification change	Scope and classification changes of buildings that need building code and zoning compliance	Bullen and Love (2011), Cox (2004), Langston et al. (2007), Reyers and Mansfield (2001)
(12) Inertia of production and development criteria	Different production and developmental criteria of cities pose challenges to urban regeneration or redevelopment approaches	Bromley et al. (2005), Bullen and Love (2011)
(13) Commercial risk and uncertainty	Lengthy and difficult renovation or reuse often leads to reduced profit margins	Bruce et al. (2015), Bullen and Love (2011), Shipley et al. (2006),
(14) Financial and technical perceptions	Notion that demolition is the only way to get a reasonable profit since adaptive reuse is seen as too expensive	Bruce et al. (2015), Bullen and Love (2011), Shipley et al. (2006), Yung and Chan (2012)

- Minimize the effect on the place's heritage significance by determining the appropriate level of change to be applied to its significant fabric.

- Ensure the principle of reversibility is applied for the place's conservation in the future.

- Protect the association among the settings and conserve important views that contribute to the place's significance.

- Establish a sustainable management and feasibility scheme for the heritage place that includes fund sourcing and heritage agreements for its future preservation.

- Publicize and translate the place's heritage significance as a meaningful and important part of the building adaptation project.

As well as buildings, monuments, gardens, landscapes and archaeological sites, Australia also conserves areas and regions that have heritage value. The Environment Protection and Biodiversity Conservation (EPBC) Act 1999 and the Commonwealth and National heritage lists identify places of outstanding or significant heritage to the nation through a set of criteria. The commonwealth, state and territory, and local governments (councils) handle the implementation of heritage policies according to the national, state and local heritage significance levels respectively (Department of Environment, 2015; Yu, 2008).

Australia's existing legislation encourages both government and community engagement in matters concerning heritage conservation. The existing state and territory legislation allows most local planning authorities to relax planning and building requirements to encourage the use or conservation of a heritage site (*e.g.* transfer of development rights). To encourage the public to share the government's burden of heritage conservation, many tax concessions and a variety of economic and planning incentives are provided and the active participation of the private sector is sought. The funding sources are government appropriations, grants and/or loans, revolving funds and private sector investment (Department of Environment, 2015; Yu, 2008).

Adaptive reuse of historic buildings

The Burra Charter states that adaptation is acceptable only where the adaptation has minimal impact on the cultural significance of the place (Article 21.1), while minimal changes to the significant fabric should take place after considering alternatives (Article 21.2). In Australia, adaptive reuse is another form of conserving heritage buildings (Bromley, Tallon, & Thomas, 2005). This process conserves the building's heritage significance when its original use becomes obsolete (DEH, 2004; Douglas, 2006; NSW Department of Planning and RAIA, 2008). Most Australian states legislate policies for the adaptive reuse of heritage buildings to guarantee that the process will have a minimal effect on the building's heritage values.

Smart and innovative green technologies

The ability of green building and its smart features to improve building performance and energy efficiency has led architects to integrate these features into new buildings. However, it is more difficult to integrate these features successfully in heritage buildings to improve their performance and energy efficiency than in new buildings (Bruce et al., 2015; DEH, 2004). Table 2 shows some of the smart and innovative green features used in new green buildings that can also be integrated into heritage buildings.

Building regulations and/or codes

Australia has a centralized approach to building code administration. The Australian Building Codes Board (ABCB) develops, maintains and administers the

Table 2 Some smart and innovative green technologies

Categories	Green features
Energy efficiency	Double-/low-e glazing; automatic blinds/louvres; natural ventilation, *e.g.* by layout optimization/wind catchers; high-performance mechanical and electrical systems (AHU, chiller, cooling tower, pump lift, escalator, etc.); high-performance lighting, *e.g.* LED, T5 with high-efficiency ballast and control system; and photovoltaic cells, *e.g.* solar panels
Water efficiency	Rain sensors/rainwater harvesting; storm water management system; and greywater recycling
Environmental protection	Vertical greenery; roof/sky garden/green roof; and carbon-neutral/low embodied energy/recyclable materials
Indoor environmental quality	CO/CO_2 monitoring to activate actions, *e.g.* AHU fresh air intake; and low VOC paints/adhesive
Other green features and innovation	Waste food biodigester; sustainable construction/demolition technology; design using less material, *e.g.* flat slab design, glass, steel; and waste segregation/sorting, *e.g.* double rubbish chute/recyclable

Source: Adapted from Chew (2016).
Note: AHU = air handling unit; VOC = volatile organic compound.

National Construction Code (NCC) on behalf of the commonwealth and state and territory governments. Under intergovernmental agreement, while state and territorial governments have some control over certain codes and regulations, the NCC prevails over any conflicting codes and regulations that do not pertain to resource management and is mandatory for all of Australia (Australian Building Codes Board, 2010). Existing building are required to conform to the Building Code of Australia (BCA) if a change of use occurs and/or if upgrading work to more than 50% of the building is proposed. The BCA places an extremely high level of importance on fire protection, safety and disabled access requirements. The main considerations in complying with the BCA are to do so in such a manner that the essential character and appearance of a building is preserved. For the adaptive reuse of existing buildings (both listed and non-listed), references to the 'Fire and Heritage' and the 'Improving Access to Heritage Buildings' guidelines supplements the BCA and Burra Charter in the adaptive reuse process.

Since 2000, the BCA has been influential in bringing energy efficiency standards to the residential and commercial sector of the building industry. The Green Star Green Building Council of Australia (GBCA) standard ensures existing and new buildings conform to energy-efficient building designs. The National Australian Built Environment Rating System (NABERS) helped boost the energy efficiency retrofits in Australia and was an effective marketing tool for commercial office building use while complementing other Australian rating tools (Judson et al., 2014). However,

> rating tools frequently do not acknowledge the sustainability benefits of conserving existing buildings or recognize the embodied energy inherent in these structures. Additionally, they do not consider the contribution that the inherent quality of materials makes to the life cycle of a structure or the cultural value of a building to the community.
>
> (Australian Government, 2016, p. 769)

Heritage and sustainability

The 2011 Australian State of the Environment Report states that:

> the green building agenda places significant pressure on heritage buildings. Sustainability legislation aims to achieve immediate greenhouse gas savings by increasing efficiencies in heating, cooling and ventilation, saving water and minimizing waste; hence, it measures only the operational efficiencies of buildings. The imbalance in the current rating tools is being addressed by a number of organizations across a number of different spheres. These organizations include the GBCA, which is developing a rating tool for existing buildings; the Australian Tax Office, which is proposing a green investment tax incentive for retrofitting; and RMIT University and Heritage Victoria, which are researching the embodied energy of various heritage building typologies. CSIRO [Australian Commonwealth Scientific and Research Organization] is developing the Australian Life Cycle Inventory materials database for eventual incorporation into the Nationwide House Energy Rating Scheme, and the Property Council of Australia run seminars on retrofitting existing buildings.
>
> (Australian Government, 2016, p. 769)

Redden (2014) notes there has been little discussion concerning BCA requirements for energy use or environmental sustainability relating to the heritage fabric. To incorporate adaptive reuse into conservation practices requires heritage professionals to be up-skilled and educated about current Australian standards for ecologically sustainable design.

Methods

This research uses a qualitative approach that involves multiple case studies and in-depth interviews with experts coupled with field observation and building plan appraisals. The study starts with a literature review to establish the relevant background knowledge and provide the conceptual framework that allows for the identification of the list of barriers in the adaptive reuse of heritage buildings (Table 1). The case study approach was used in this study since it is considered an effective approach in researching a variety of adaptive reuse issues (Conejos et al., 2014; HSMO, 1987; Langston et al., 2013; NSW Department of Planning & RAIA, 2008; Park, 1998; Snyder, 2005; Wang et al., 2010).

Selection of case studies

The selected case studies were taken from the New Uses for Heritage Places: Guidelines for the Adaptation of Historic Buildings and Sites publication of the NSW Department of Planning and RAIA (2008), which features 11 adaptive reuse projects in New South Wales (NSW). These adaptive reuse Australian case studies highlight the successful merging of modern design and technologies in heritage buildings. Among 20,000 heritage-listed buildings in NSW, the 11 case studies from NSW are those chosen and advertised since they represent different types of use and illustrate the recent adaptation guidelines in practice. These successful award-winning projects were heritage buildings with different typologies and have undergone different types of conversions or adaptive reuse, allowing the study to get varied results. These case studies were selected due to the accessibility of data and the experts' willingness to contribute their valuable and specific technical knowledge in designing the selected adaptive reuse case studies. While it is expected that the results of the study will have global application, this aspect of the research is not tested. Furthermore, issues of geographic location and climate have also been removed from the study given the case studies largely reside in one Australian state.

The selected case studies are award-winning projects that illustrate the successful merging of modern design and technologies with respect to the building's historic fabric. Table 3 summarizes the selected 11 case studies and mentions some of the practical innovations undertaken to comply with current building codes and regulations and the design brief.

Experts' interviews

A purposeful sampling approach was used in this study with the intention to revisit the professionals directly involved in the selected case studies. Since this is a retrospective study, it required expert

Table 3 Selected adaptive reuse case studies

Case study	Statement of significance	Design approaches
(1) Units at Egan Street, Newtown, NSW	Representative example of a 1920s' light industrial development that makes a positive aesthetic contribution to the O'Connell Town Estate Conservation Area	In achieving the new design requirement, the designers were able to lift up the roof and timber trusses for additional headroom without causing any negative effect to the adjacent properties while using the existing grid to divide the three apartments and office at the front. The original line shafting and other industrial archaeology items were retained
(2) Babworth House, Darling Point, NSW	Diverse mix of Classical Revival, Arts and Crafts, and Art Nouveau styles designed by Morrow and de Putron in 1912 and commissioned by Sir Samuel Horden	This majestic 93-room stately home was converted into five high-end apartments with an underground swimming pool, individual lifts and parking area. The designers were able to keep the external integrity of the house as intact as possible amidst the new design requirements by excavating the whole house and supporting it with new beams
(3) Tocal Visitor Centre, Tocal, NSW	Integral part of the Tocal Homestead precinct that exemplifies the historic character of an Australian rural shed that was built after 1907	Adaptively reused into a multipurpose visitor centre with a seating capacity of 100, a mini-theatre, exhibition areas and modern amenities. The challenges that the designers encountered were focused on the materials, as they had to find the correct type of timber and have it air-dried for a year. It took three years to complete the project
(4) Toxteth Church and Hall, Glebe, NSW	Late 19th-century church and hall that is an important part of the Toxteth Estate conservation area with a Victorian streetscape	In adapting the church and hall into a modern four-bedroom residence, the designers built a mezzanine floor fastened in the church interior. A new lightweight structure of steel and glass that clipped inside the church was also provided to ensure reversibility in the future. A state-of-the-art rainwater and water storage system was installed that acts as a heat bank and provides dual plumbing for tank and main water supply
(5) Bushells Building, The Rocks, NSW	State Heritage-listed landmark designed by renowned architects H. E. Ross and Rowe in 1923. Its significance lies in its industrial character and its historical association with the Bushells Company, once synonymous with Australia's cultural identity	This factory building was converted into modern offices, with a number of significant artefacts incorporated in the new use. Contamination issues were prevalent. New additions included lifts, air-conditioning, fire services and stairs, and acoustic separation in an expressed timber structure with timber floors. Insertions included false ceilings, raised floors, glass floor bridges going across to meet the old building with light wells and small breakout areas that are an exemplar of that time
(6) Sydney Harbour Federation Trust Offices, Georges Heights, NSW	Three First World War buildings that were part of a 1915 Army Auxiliary hospital and are considered rare survivals	When reused as the headquarters for the Sydney Harbour Federation Trust, the hospital wards were turned into offices, a staff lounge and a bar. The offices and historic barracks are linked by courtyards and verandas. Partition walls were removed to create an open-plan area, while green roofs and solar panels were added. Asbestos sheeting on roofs and internal linings were removed and engineering works complied with current standards
(7) Sully's Emporium, Broken Hill, NSW	State Heritage-listed landmark designed by Adelaide architect George Abbott in 1889	Adaptively reused into a regional art gallery. The installation of air-conditioning was a major design challenge and expense due to viability issues and the location of ducts. Lifts were added at the rear of the complex, and careful joinery work reused original components to meet the correct safety height of the damaged balustrade
(8) The Mint, Sydney, NSW	Constructed in 1811–16 and in 1854 for the Royal Mint and located in the most important civic precinct of Sydney	Adaptively reused as the Historic Houses Trust (HHT) head office. Challenges included major contamination issues due to asbestos products, lead paints, and the presence of mercury contamination and hydrocarbons. Additionally, an archaeological programme was undertaken and the brief for current requirements was articulated before the project started. New contemporary additions such as the new building wings and mini-theatre were incorporated within the existing complex
(9) Forum Health and Wellness Centre, Newcastle, NSW	State Heritage-listed building that is an integral part of the Honeysuckle urban regeneration area	Adaptively reused as the University of Newcastle sports health and wellness centre. The infilling of clear glass in the mezzanine structure does not touch the external walls and maintains a sense of space, while achieving acoustic separation. The new additions such as the changing areas were clearly discrete from the heritage building

(Continued)

Table 3 Continued.

Case study	Statement of significance	Design approaches
(10) George Patterson House, Sydney, NSW	Queen Anne revival-style building built between 1892 and 1895 for Holdsworth MacPherson and Co. It was considered the grandest emporium of its period	A new building was inserted beside the old existing heritage structure. The designers utilized the atrium that was damaged by fire by restoring the courtyard with a glass roof over it. The existing heritage elements such as the pilasters and elaborate mouldings were retained and services were fitted around them. The cast-iron columns and the elaborate pressed metal ceilings were retained. The original surviving spaces were used as the basis and inspiration for the interior design, finishes and furnishings, turning it into luxurious New York-style warehouse apartments
(11) Prince Henry, Little Bay, NSW	This aboriginal place of social, historic, architectural, scientific and aesthetic significance has been associated with the treatment of infectious diseases since 1881	The Prince Henry redevelopment project provides a model for redevelopment in Australia since it embodies an environmentally sustainable design redevelopment approach

Sources: Updated based on NSW Department of Planning & RAIA (2008) and Conejos (2013).

participants who were willing to discuss their past projects. The participants are considered a rich source for data collection and are professional consultants responsible for key design decisions and have all been involved in successful adaptive reuse projects. Yin (1994) suggests that the unit of analysis should be the focus of the study to avoid the collección of unnecessary data. The units of analyses in this study are the key experts/professional consultants involved in the 11 successful adaptive reuse case studies in NSW. They are the heads of the consultant-based team of each adaptive reuse project and are directly involved in heritage conservation, adaptive reuse and green building projects across Australia.

These experts belonged to the middle to upper level of management, handling small to large-scale organizations and have up to 50 years of experience in the fields of adaptive reuse, design and construction.

Table 4 Profile of experts

Code	Gender	Office location	Position/influence	Years of experience	Company size
E1	Male	NSW	Principal architect/head of the consultant-based team	Fewer than 20	Small
E2	Male	NSW	Principal architect/head of the consultant-based team	Over 50	Large
E3	Male	Canberra	Principal architect/head of the consultant-based team	Over 50	Small
E4	Male	NSW	Principal architect/head of the consultant-based team	31–40	Large
E5	Male	NSW	Principal architect/head of the consultant-based team	Over 50	Large
E6	Female	NSW	Project architect/firm's associate designer	20–30	Medium
E7	Male	NSW	Project manager/firm's associate director	20–30	Large
E8	Male	NSW	Heritage architect and project manager/firm's associate director	31–40	Medium
E9	Male	NSW	Quantity surveyor/firm's associate director	31–40	Large
E10	Male	NSW	Principal architect/head of the consultant-based team	20–30	Medium
E11	Male	NSW	Architect and project manager/head of the consultant-based team	20–30	Large
E12	Female	NSW	Architect and project manager/firm's associate designer	20–30	Large
E13	Male	NSW	Architect and project manager/firm's associate designer	20–30	Small
E14	Male	NSW	Structural engineer/managing owner	20–30	Medium

Note: NSW = New South Wales.

Most of the experts interviewed are the principal architects and act as representative of their organizations due to their influence and position (ownership/stakeholder). A systematizing expert interview was used in this study, where the expert is involved in presenting solutions and decision-making structures as well as aids in reconstructing the expert's special objective knowledge in a specific field (Van Audenhove, 2007). The interviews lasted for one hour and mostly occurred at the experts' offices. The interviews were audio recorded and with the expert's consent, were transcribed and coded using NVivo 10 software. NVivo is used in the data analysis since it is an effective and efficient software choice for linking contextual data in many ways. Its tools can also distinguish patterns and spot relationships that facilitate iterative coding during data analysis (Kolenic, 2013). Table 4 shows the experts' profiles.

Interview questionnaire

A semi-structured open-ended interview questionnaire guide was prepared to minimize bias and encourage respondents to reflect and expound on their experiences. The study's reliability is enhanced through the consistency of interview questions asked to all key experts. An interview protocol was used to steer the conversation to the issues pertinent to the investigation and allow the conversations to flow freely to encourage the experts to tell their stories. The interview protocol was divided into three parts:

- Background and history of the project, before and after adaptive reuse:

  - Provide a brief background of the project from its existing use to its newly adaptive use or building function.

  - What were the major decisions/events that lead to its reuse?

  - What were the major considerations before undertaking the project?

  - Were there any latent conditions (*e.g.* physical conditions on, underlying or adjacent to the site that could not be identified by the contractor by reasonable observations or investigations of the site or the site information provided in the tender documentation at the time that the tender for the works was being prepared)?

- Design and technical considerations in undertaking adaptive reuse:

  - What were the impediments encountered during the design process?

  - How were modern and green design features (if any) incorporated or blended into the existing heritage building? What were the structural and utility challenges encountered?

  - Were there any legal and building code considerations?

- Design process and interventions to improve building performance:

  - What were the design principles and criteria applied or implemented in the project?

  - How were the design consultations conducted with stakeholders and how often?

  - How were adaptive reuse strategies identified or applied to the project?

  - What do you think are the critical factors that affected the success of the adaptive reuse projects?

Coding of themes

The collected data and their interpretations were scrutinized, synthesized and a generalization was constructed through the use of the NVivo software. The interview results were grouped in a particular theme based on a direct content analysis method to determine the parameters that guided the formulating prospective regulations in this study. Hickey and Kipping (cited in Hsieh, Shannon, & Hsieh, 2005) state that direct content analysis offers a more structured guide than the conventional process, where the ranking based on frequency codes can be used (Curtis et al., 2001). The identified list of barriers to adaptive reuse based from the literature review (Table 1) was used to guide the coding query. The coding query relies heavily on the themes extracted from the texts and wherein a search by nodes and attributes can be done, or a combination of both. An advanced coding query was used in the data analysis in order to search for queries with a combination of nodes by using Boolean operators such as 'AND', 'OR' to help build up the query criteria via clicking the 'Define More Criteria' section of the window. Queries provide a systematic way of discovering patterns that may not be obvious when dealing with vast amounts of qualitative data.

The coding query results were gathered from all sources of nodes created in NVivo. They were organized in two ways. Firstly, according to the case study profile based on comprehensive documentation (*e.g.* approved building plans and maps, literature, news

Table 5 Major challenges experienced by experts undertaking adaptive reuse

Categories (challenges)	Parameters (themes)		Experts (*n* = 14)
(1) Compliance with codes and regulations	High remediation cost and construction delays	Experts 7 and 11: asbestos in the building and loose fibre asbestos in the ground that was used to insulate hot water pipes lead to high contamination and remediation costs	10
	Availability of materials and lack of skilled tradesmen	Expert 2: 'The slates on the roof originally came from Wales in the UK [. . . we] reroofed the house in the same way as the original house.' Experts 8 and 9 added that 'there is a lack of skilled tradesmen in the industry [. . .] there's a limited market for them in Australia' in terms of the availability of local tradesmen	7
	Building codes and regulations/legal constraints	Expert 15: 'Adaptive reuse is difficult since codes change and sometimes this can be an obstacle such as the seismic code in 1994.' Expert 2: 'In terms of project compliance, the ceilings, the walls and the floors were the only new material introduced to make sure those modern codes and controls were adhered to'	7
	Acoustics/noise control	Expert 12 related that after completing the case study project, they still could not prevent external sound from the surroundings from going inside the mall	7
	Fire safety	Expert 5: 'The building wasn't fire-rated, being constructed in the 19th century, so the council required [the designers] to upgrade the building in a way to make it safe for use by the public but also in a way which respects the heritage significance of the buildings.' Expert 1: 'The thing that we could have improved was the common corridor, although this perhaps would be in some conflict with fire regulations'	5
	Disability access	Expert 14 explained that the 'adaptive reuse of new buildings requires access to new service for disability'. Expert 8 revealed that the building standards are prescriptive. For instance, 'the disability access code specifies a 50-millimetre tubular handrail, which [the designers] didn't follow but instead designed flat handrails to make it look sympathetic to the heritage building'	5
(2) Current design requirements	Modern technologies	Expert 11: 'The project's master plan incorporates green design and tackles heritage policies.' Expert 2: 'The principles are to design buildings that have a contemporary feel about them, that have a use of appropriate modern materials that are sustainable, sensitive to their future energy demands and true to ESD principles of design.' Expert 5: 'It can be a challenge to introduce solar panels to historic buildings'	14
	Physical restrictions	Expert 8: 'Fireplaces can be used for modern services and they even put the air-conditioning into the chimneys.' Expert 14: 'Opportunity to provide access to replace mechanical features, hydraulics, buildings extra capacity must be considered for future servicing and insulation'	10
	Complexity and technical difficulties	Expert 1 needed to get a head height for a good floor space ratio and solved the problem by 'floating the floor above to be able to get the slope we wanted [. . .] to give [the suitable] head height for the upper floor'. Expert 5 'had to find a way to solve the problem of uneven floors and find a way to incorporate the redesigned building with other buildings, [by installing] glass floor bridges [. . .] going across to meet the old buildings [. . .] by putting a false floor, we were able to level the uneven floors'	4
	Inaccuracy of information and drawings	Expert 15: the adaptive reuse process requires good documentation and availability of as-built plans. Expert 4 suggests the layering of the different period of building styles on the plan for reference. Expert 8 recommends that a good measured drawing and survey of the old building is essential	3

clippings, articles and public reports) concerning the 11 case studies and the experts interview transcriptions, and secondly, according to the aim of this paper, which is to identify the barriers to adaptive reuse of heritage buildings based on the validation between the gathered literature reviews and the in-depth expert interviews of the selected professionals involved in the case studies. These were pattern coded. The issue of possible bias is low in this case since triangulation of gathered data from different sources was found to be consistent.

Discussion

Ten parameters (themes) that are specific to the challenges encountered by experts in undertaking adaptive reuse were identified and established in this study. Out of the list of fourteen parameters (themes) based on the underpinning literature, only six parameters (themes) matched the experts interview results, while four parameters (themes) were specifically created into new nodes as they received a high frequency of answers from the interviews. The six matching parameters are 'High remediation cost and construction delays'; 'Availability of materials and lack of skilled tradesmen'; 'Building codes and regulations/legal constraints'; 'Physical restrictions'; 'Complexity and technical difficulties'; and 'Inaccuracy of Information and Drawings'. The four new parameters (themes) are 'Acoustics', 'Fire Safety', 'Disability Access' and 'Modern Technologies'. These parameters were grouped into two categories (challenges) namely: 'Compliance to codes and regulations' and 'Current design requirements' (Table 5).

Summary of the findings

The experts unanimously stated that compliance to codes and regulations, and compatibility to design requirements were the greatest challenges they encountered in undertaking adaptive reuse projects. The experts unanimously stated that compliance to codes and regulations, and compatibility to design requirements were the greatest challenges they encountered in undertaking adaptive reuse projects. The challenges experienced by the experts in adaptively reusing heritage buildings are:

- Onsite contamination and remediation costs, which led to delays in time and a high cost of implementation. All hazardous materials had to be removed before the development of the site could proceed.

- Adhering to the Burra Charter, which requires them to retain the significance of the place, take into account the building's history, apply minimum changes while ensuring that any change is reversible.

- Soundproofing problems in heritage buildings need solutions in order to control or manage noise inside the buildings.

- Fire rating and codes were non-existent during the 1900s when many of the heritage buildings were constructed; hence, there is a need to integrate fire safety features to blend in with the existing heritage building elements.

- Meeting disability access standards, which are prescriptive and not always appropriate for retaining the intrinsic character of heritage buildings.

- Accurate and good documentation is necessary for ease in undertaking adaptive reuse process.

- Modern technologies are necessary to upgrade the heritage buildings, although their use creates challenges for designers in finding approaches that do not limit their innovation but allow the blending of new technologies into the existing building elements.

Towards a green adaptive reuse protocol

Based on the above findings and discussions, it is possible to develop a conceptual framework to develop a protocol that better governs the green adaptive reuse of heritage buildings. The introduction of green adaptive reuse protocol for the governance of heritage buildings is a proactive strategy that will encourage the seamless integration of modern green technologies in heritage buildings. It is anticipated that it will aid in boosting developers' commitment to undergo adaptive reuse projects especially with heritage buildings and encourage stakeholder participation in the entire green adaptive reuse process. Table 6 shows the conceptual framework of the green adaptive reuse protocol with the prospective regulatory approaches to meet the corresponding challenges. The framework relies on the overall Australian regulatory setting rather than the local regulations from the different states.

Conclusions and future research

Preserving the world's cultural heritage is one of the many benefits of sustainability as it retains the historical identity of places and buildings. The reuse of buildings will result in less land consumption and no waste disposal. This is an effective way to reduce the built environment's ecological footprint. The research in this paper confirmed that experts attached importance to adaptive reuse and upgrading heritage buildings through the integration of modern technologies. However, heritage and modern technologies are difficult to integrate under the current regulatory framework. These apparent challenges can be resolved through the application of a protocol that better governs the adaptive reuse of heritage buildings.

Evidence was presented on successful adaptive reuse practices in Australia. The findings reveal that even successful adaptive reuse projects meet significant challenges and threats prior to the delivery of these projects. Two types of challenges were identified for the adaptive reuse of heritage buildings: 'Compliance to codes and regulations' and 'Current design

Table 6 Conceptual framework green adaptive reuse protocol

Challenges (themes)	Parameters	Australian context	Prospective regulatory approaches
(1) Compliance with codes and regulations	High remediation cost and construction delays	Non-existent regulation on this aspect	Incentives to lessen the financial burdens on building owners caused by remediation and construction delays
	Availability of materials and lack of skilled tradesmen	Training opportunities for heritage professionals in Australia is lacking (Redden, 2014)	Capacity-building and financial incentives to promote skills and professionalism will strengthen local materials, skills and craftsmanship, and the marketability of skilled heritage tradesmen
	Building codes and regulations/legal constraints	Building Code of Australia (BCA) compliance for new building work associated with existing buildings is mandatory for all designers to consider (Australian Building Codes Board, 2010)	Exemptions from normal zoning requirements and funding schemes to encourage the patronage of green technologies appropriate for heritage building's upgrading
	Acoustics (*e.g.* noise control)	BCA has a guideline written to support the sound insulation of buildings (Australian Building Codes Board, 2004). However, there are no provisions pertaining to heritage building concerns	Incentives to promote technological and smart innovations for the sound insulation of heritage buildings
	Fire safety	The 'Fire and Heritage' guidelines are mandatory	Funding support from suppliers, etc. to embed fire safety features harmoniously in the heritage buildings
	Disability access	The 'Improving Access to Heritage Buildings' guidelines are mandatory	Relaxation of disability access standards to be more sensitive to the historic character of the building
(2) Current design requirements	Modern technologies	Weak guidelines for adapting historic structures to meet ESD standards and the high cost of introducing new products to the Australian market (Redden, 2014)	Incentives to promote the integration of technological and smart innovations into heritage buildings (*e.g.* tax reduction for new green technologies products)
	Physical restrictions	Physical characteristics of heritage buildings may pose difficulties in achieving a high energy rating and may affect in competing in the current rating system (Australian Government, 2016)	Funding to promote the integration of technological and smart innovations into heritage buildings via training and the development of a manual to aid practitioners in dealing green adaptive reuse issues
	Complexity and technical difficulties	Mainstream focus on integrating heritage conservation and ESD, as well as on preservation technologies and opportunities is lacking (Redden, 2014)	Incentives to promote the integration of technological and smart innovations into heritage buildings via research and development
	Inaccuracy of information and drawings	Non-existent regulation on this aspect	Incorporating as-built plans and historical layering of plans using architectural and building software as part of the building clearance process. University students should take part in greening heritage buildings

In addition, public awareness and participation is important and necessary in sponsoring the adoption of a green adaptive reuse protocol

Note: ESD = ecologically sustainable development; BCA = Building Code of Australia.

requirements'. The detailed findings point to the need of a green adaptive reuse protocol to encourage the adaptive reuse of heritage buildings whilst applying practical innovations to make these heritage buildings on par with their modern counterparts. Through practical and smart innovations, contemporary and sophisticated green technologies can sometimes be embedded in heritage buildings in unobtrusive or invisible methods. These findings have critical implications for policy and practice that are beyond the scope of this research. Further exploration can be done by conducting a comprehensive analysis of the regulatory barriers from views of other stakeholders (other than architects) and conducted in different states in Australia.

The next step is to develop a green adaptive reuse protocol that will provide prospective regulations that support each identified parameter in this research. This will assist industry practices and public policy by providing clarity on the factors involved in decision-making and allow potential for relaxation of some requirements or alternative solutions. As the most influential stakeholder for the implementation of this proposed protocol, the government can improve legislation to reduce the conflicts between adaptation, conservation and regulatory conformance. It is in the interest of wider civil society to manage heritage buildings in a socially and environmentally beneficial manner.

Disclosure statement

No potential conflict of interest was reported by the authors.

References

Australia ICOMOS. (1999). The Burra charter: The Australia ICOMOS charter for places of cultural Significance. retrieved from http://australia.icomos.org/wp-content/uploads/BURRA_CHARTER.pdf.

Australian Building Codes Board. (2004). Sound insulation handbook 2004. Retrieved from www.abcb.gov.au/.../Handbooks/2004_sound_insulation-buildings.ashx.

Australian Building Codes Board. (2010) Applying energy efficiency provisions to new building work associated with existing Class 2 to 9 buildings. Retrieved from http://www.abcb.gov.au/~/media/Files/Download%20Documents/Education%20and%20Training/Handbooks/2010_applying_ee_provisions_to_new_building_work_associated_with_existing_class_2-9_buildings.ashx?la=en

Australian Government. (2016). Australian state of the environment report 2011. Retrieved from https://www.environment.gov.au/science/soe/2011-report/contents

Ball, R. (1999). Developers, regeneration and sustainability issues in the reuse of vacant industrial buildings. *Building Research & Information*, 27(3), 140–148. doi:10.1080/096132199369480

Ball, R. (2002). Re use potential and vacant industrial premises: Revisiting the regeneration issue in Stoke-on-Trent. *Journal of Property Research*, 19(2), 93–110. doi:10.1080/09599910210125223

Bond, C. (2011). *Adaptive reuse: Explaining collaborations within a complex process*. Retrieved from https://scholarsbank.uoregon.edu/xmlui/bitstream/handle/1794/11680/Bond%20Adaptive%20Reuse%20final_cbond.pdf?sequence=1

Brand, S. (1994). *How buildings learn: What happens after they're built*. New York: Viking Penguin.

Bromley, R. D. F., Tallon, A. R., & Thomas, C. J. (2005). City centre regeneration through residential development: Contributing to sustainability. *Urban Studies*, 42(13), 2407–2429. doi:10.1080/00420980500379537

Browne, L. A. (2006). Regenerate: Reusing a Landmark Building to Economically Bolster Urban Revitalization (Unpublished master's thesis). University of Cincinnati, Ohio.

Bruce, T., Zuo, J., Rameezdeen, R. and Pullen, S. (2015). Factors influencing the retrofitting of existing office buildings using Adelaide, South Australia as a case study. *Structural Survey*, 33 (2), 150–166. doi:10.1108/SS-05-2014-0019

Bullen, P. A., & Love, P. E. D. (2011). Factors influencing the adaptive re-use of buildings. *Journal of Engineering, Design and Technology*, 9(1), 32–46. doi:10.1108/17260531111121459

Chew, M. Y. L. (2016). *Maintainability of facilities: Green FM for building professionals* (2nd ed.). Singapore: World Scientific.

Conejos, S. (2013). Designing for future building adaptive reuse (Unpublished doctoral thesis). Bond University, Gold Coast, Australia.

Conejos, S., Yung, E. H. K., & Chan, E. H. W. (2014). Evaluation of urban sustainability and adaptive reuse of built heritage areas: A case study on conservation in Hong Kong's CBD. *Journal of Design Research*, 12 (4), 260–279. doi:10.1504/JDR.2014.065843

Cooper, I. (2001). Post-occupancy evaluation – Where are you? *Building Research & Information*, 29(2), 158–163. doi:10.1080/09613210010016820

Cox, F. (2004). *The engineering and management of retrofit projects in process industries*. Loughborough: European Construction Institute.

Curtis, J. R., Wenrich, M. D., Carline, J. D., Shannon, S. E., Ambrozy, D. M., & Ramsey, P. G. (2001). Understanding physicians' skills at providing end-of-life care: Perspectives of patients, families, and health care workers. *Journal of General Internal Medicine*, 16, 41–49. Retrieved from www.ncbi.nlm.nih.gov/pubmed/11251749

Department of Environment. (2015). The Australian government department of environment. Retrieved from http://www.environment.gov.au/heritage

Department of Environment and Heritage (DEH). (2004). *Adaptive reuse: Preserving our past, building our future*. Commonwealth of Australia, ACT: Author. Retrieved from www.environment.gov.au/resource/adaptive-reuse-preserving-our-past-building-our-future

Dittmark, H. (2008). Continuity and context in urbanism and architecture: The honesty of a living tradition. *Conservation Bulletin*, 59(Autumn), 7–9. Retrieved from www.english-heritage.org.uk/content/publications/publicationsNew/conservation-bulletin/conservation-bulletin-59/conbull59pp1-13. pdf

Douglas, J. (2002). *Building adaptation*. Oxford: Butterworth Heinemann.

Douglas, J. (2006). *Building adaptation* (2nd ed.). Oxford: Elsevier Ltd.

El Kerdany, D. (2002). Innovation and conservation: Case studies in intervening with valuable buildings. Proceedings from UIA Int. Conf. WPAHR on Architecture and Heritage as a Paradigm for Knowledge and Development. Alexandria, Egypt: Biblioliotheca.

Ellison, L. and Sayce, S. (2007). Assessing sustainability in the existing commercial property stock: Establishing sustainability criteria relevant for the commercial property investment sector. *Property Management*, 25(3), 287–304. doi:10.1108/02637470710753648

Ellison, L., Sayce, S., & Smith, J. (2007). Socially responsible property investment: Quantifying the relationship between sustainability and investment property worth. *Journal of Property Research*, 24(3), 191–219. doi:10.1080/09599910701599266

English Heritage. (1997). *Sustaining the historic environment*. London: Author.

English Heritage. (2008). Climate change and the historic environment. Retrieved from www.english-heritage.org.uk/content/publications/publicationsNew/guidelines-standards/climate-change-and-the-historic-environment/climate-change.pdf

English Heritage. (2013). The English Heritage. Retrieved from http://www.english-heritage.org.uk

Fournier, D., & Zimnicki, K. (2004) *Integrating sustainable design principles into the adaptive reuse of historical properties*. Washington, DC: U.S. Army Corps of Engineers.

Gorse, C., & Highfield, D. (2009). *Refurbishment and Upgrading of Buildings* (2nd ed.). New York: Spon Press.

Gregory, J. (2004). *New Ways for Older Housing. New South Wales Department of Housing Rehabilitation*. Retrieved from www/housing.nsw.gov.au/rehab.htm

HMSO. (1987). *Town and country planning (use classes) order 1987*. London: HMSO.

Hsieh, H. F., & Shannon, S.E. (2005). Three approaches to qualitative content analysis. *Qualitative Health Research*, 15(9), 1277–1288. doi:10.1177/1049732305276687

Jonas, D. (2006). Heritage Works: The reuse of historic buildings in regeneration. A toolkit of good practice. Retrieved from www.english-heritage.org.uk/.../heritageworks.pdf

Judson, E. P., Iyer-Raniga, U., and Horne, R. (2014). Greening heritage housing: Understanding homeowners' renovation practices in Australia. *Journal of Housing and the Built Environment*, 29, 61–78. doi:10.1007/s10901-013-9340-y

Kayan, B. A. (2015). Conservation plan and "green maintenance" from sustainable repair perspectives. *Smart and Sustainable Built Environment*, 4(1), 25–44. doi:10.1108/SASBE-08-2014-0042

Kincaid, D. (2000). Adaptability potentials for buildings and infrastructure in sustainable cities. *Facilities*, 18(3/4), 155–161. doi:10.1108/02632770010315724

Kohler, N., & Hassler, U. (2002). The building stock as a research object. *Building Research & Information*, 30(4), 226–236. doi:10.1080/09613210110102238

Kolenic, G. (2013). *Introduction to NVivo 10*. Retrieved from http://sitemaker.umich.edu/gkolenic/files/nvivo10_winter2013.pdf

Kronenburg, R. (2007). *Flexible: Architecture that responds to change*. Great Britain: Laurence King.

Langston, C. (2009). Green adaptive reuse: Issues and strategies for the built environment. Retrieved from ibrarian.net/navon/paper/Green_Adaptive_Reuse__Issues_and_strategies_for_t.pdf?paperid=17624822.

Langston, C., Wong, F. K. W., Hui, E. C. M. and Shen, L. Y. (2007). Strategic assessment of building adaptive reuse opportunities in Hong Kong. *Building and Environment*, 43(10), 1709–1718. doi:10.1016/j.buildenv.2007.10.017

Langston, C., Yung, E. H. K., & Chan, E. H. W. (2013). The application of ARP modelling to adaptive reuse projects in Hong Kong. *Habitat International*, 40(4), 233–243. doi:10.1016/j.habitatint.2013.05.002

Latham, D. (2000). *Creative reuse of buildings*. Shaftesbury: Donhead.

Marquis-Kyle, P., & Walker, M. (1994). *The illustrated Burra Charter: Making good decisions about the care of important places*. Sydney: Australia ICOMOS with the assistance of the Australian Heritage Commission.

NSW Department of Planning & RAIA. (2008). *New uses for heritage places: Guidelines for the adaptation of historic buildings and sites*. Sydney: Heritage Council of New South Wales and the Royal Australian Institute of Architects (RAIA). Retrieved from www.environment.nsw.gov.au/resources/heritagebranch/heritage/NewUsesforHeritagePlaces.pdf

O'Donnell, C. (2004). Getting serious about green dollars. *Property Australia*, 18(4), 1–2.

Park, S. C. (1998). Sustainable design and historic preservation. *CRM*, 2, 13–6. Retrieved from tusculum.sbc.edu/toolkit/toolkit_pdfs/Park,Sharon_SustainableDesignHP.pdf

Pearce, A. R. (2004). Rehabilitation as a strategy to increase the sustainability of the built environment. Retrieved from http://maven.gtri.gatech.edu/sfi/resources/pdf

Petersen, E. H. (2002, September). An LCA based assessment tool for the building industry. 3rd International Conference on Sustainable Building, BEAT, Oslo.

Pivo, G., and McNamara, P. (2005). Responsible property investing. *International Real Estate Review*, 8(1), 26–42.

Rabun, S., & Kelso, R. (2009). *Building evaluation for adaptive reuse and preservation*. New Jersey: John Wiley and Sons, Inc.

Redden, R. (2014). *Greening historic buildings: A study of Heritage protection and environmental sustainability*. Melbourne: International Specialised Skills Institute. Retrieved

from http://www.cpsisc.com.au/resources/CPSISC/ISSI%20Research%20papers/REDDEN-Report-FINAL-LowRes.pdf

Remoy, H. T., & van der Voordt, T. J. M. (2007). A new life: Conversion of vacant office buildings into housing. *Facilities*, 25(3/4), 88–103. doi:10.1108/02632770710729683

Reyers, J., and Mansfield, J. (2001). The assessment of risk in conservation refurbishment projects. *Structural Survey*, 19(5), 238–244. doi:10.1108/02630800110412480

Shipley, R., Utz, S., & Parsons, M. (2006). Does adaptive reuse pay? A study of the business of building renovation in Ontario, Canada. *International Journal of Heritage Studies*, 12(6), 505–520. doi:10.1080/13527250600940181

Siddiqi, K. (2006). Benchmarking adaptive reuse: A case study of Georgia. *Environmental Technology and Management*, 6 (3/4), 346–61. doi:10.1504/IJETM.2006.009005

Snyder, G. H. (2005). Sustainability through adaptive reuse: The conversion of industrial building (Unpublished master's thesis). University of Cincinnati, Ohio.

UNEP. (2007). *Buildings and climate change status, challenges and opportunities*. Nairobi: UNEP publications.

UNEP. (2009). *Buildings and climate change: Summary for decision-makers*. Nairobi: UNEP publications.

UNESCO. (2007). *Asia conserved: Lessons learned from the UNESCO Asia-Pacific Heritage awards for culture Heritage conservation (2004–2004)*. Bangkok: Author.

UNESCO. (2009). *Hoi an protocols for best conservation practice in Asia: Professional guidelines for assuring and preserving the authenticity of Heritage sites in the context of the cultures of Asia*. Bangkok: Author.

de Valence, G. (2004). The construction sector system approach. CIB Publication No. 293, Rotterdam: CIB.

Van Audenhove, L. (2007). Expert interviews and interview Techniques For Policy Analysis. Retrieved from www.ies.be/files/060313%20Interviews_VanAudenhove.pdf

Velthius, K., & Spennemann, D. (2007). The future of defunct religious buildings: Dutch approaches to their adaptive reuse. *Cultural Trends*, 16, 43–66. doi:10.1080/09548960601106979

Wang, H.-J. & Zeng, Z.-T. (2010). A multi-objective decision-making process for reuse selection of historic buildings. *Expert Systems with Applications*, 37(2), 1241–1249. doi:10.1016/j.eswa.2009.06.034

Wilkinson, S., & Reed, R. (2008, January). The business case for incorporating sustainability in office buildings: The adaptive reuse of existing buildings. Proceedings of the 14th Annual Pacific Rim Real Estate Society Conference. Pacific Rim Real Estate Society, Kuala Lumpur.

Wilkinson, S., Reed, R., & Kimberley, J. (2009). Using building adaptation to deliver sustainability in Australia. *Structural Survey*, 27, 46–61. doi:10.1108/02630800910941683

Yin, R. K. (1994). *Case study research: Design and methods* (2nd ed.). California: Sage.

Yu, M. (2008). Built heritage conservation policy in selected places. Retrieved from www.legco.gov.hk/yr07-08/english/sec/library/0708rp10-e.pdf

Yung, E. H. K., & Chan, E. H. W. (2012). Critical social sustainability factors in urban conservation: The case of the central police station compound in Hong Kong. *Facilities*, 30(9/10), 396–416. doi:10.1108/02632771211235224

Regulatory incentives for green buildings: gross floor area concessions

Queena K. Qian, Ke Fan and Edwin H. W. Chan

Incentive schemes formed by regulatory or administrative instruments are measures to promote green building (GB) and increase the motivation of developers to meet higher standards. The hidden costs to different stakeholders during the GB transaction are often ignored. Understanding these hidden transaction costs (TCs) helps appraise the costs and benefits of GB and policy effectiveness. The example of a gross floor area (GFA) concession scheme is used systematically to explore and understand the fundamental issues of TCs' typology and chronology in the GB development process. The GFA concession scheme is a popular incentive due to its indirect compensation to developers by allowing additional floor area without expenditure by government to implement GBs. A TCs' framework is used critically to review and evaluate the costs and benefits of the GFA concession scheme. Its particular implementation in both Hong Kong and Singapore is explored. Hong Kong is used as a case study, complemented with in-depth expert interviews on GFA concession in Hong Kong. The key contribution is to establish the parameters for estimating the optimum GFA bonus that could both motivate various stakeholders and minimize the negative impacts on the built environment in future.

Introduction

Buildings account for over 40% of global energy consumption and one-third of greenhouse gas (GHG) emissions (UNEP, 2007). In 1950, 30% of the world's population resided in urban areas. By 2014, this rose to 54%; and it is projected to be 66% by 2050 (UNDESA, 2014). Taking China as an example, with rapid urbanization, building energy consumption has steadily increased due to increased affluence in living (Cai, Wu, Zhong, & Ren, 2009; World Bank, 2005). Worldwide, the way the new buildings are conceived and built will be decisive for influencing the level and pattern of energy consumption, therefore they will affect the climate and natural environment.

Green building (GB) is the practice of creating and using more resource-efficient models of construction, renovation, operation, maintenance and demolition.

GB brings together a vast array of practices and techniques to reduce the impacts of buildings on energy consumption, environment and human health. Energy-saving measures such as solar photovoltaics, sun-shading devices, low-emissivity glass, energy-efficient air-conditioning systems, and building-space planning and orientation have become the common design considerations for GB. Sophisticated technologies have been developed ready for GB implementation through good management and policy support. For example, Leadership in Energy and Environmental Design (LEED)-certified buildings can lower the operating costs by 8–9% compared with regular buildings, and these savings can pay for higher initial costs in the relatively shorter payback period (Cole, 2015).

Reports of past attempts to save energy (WBCSD, 2009) agree that with the technology then available, the energy-efficiency level could be increased by 40%, yet this did not happen and even today does not happen. It is assumed that underlying reasons prevent or inhibit the GB market from realizing this implementation and this raises the question of whether the current incentives are appropriate and sufficient to act as drivers for the take-up of energy-efficient buildings. The objectives for instituting GB incentives (Ocampo, 2011; Qian, Chan, & Choy, 2013) are:

- to correct for external costs

- to supply information

- to reduce investors' risk in a new technology

- to accelerate the pace of adoption of efficient technologies

To fulfil these objectives, the incentives should be attractive to business and also to be administratively easy for government to implement. A disregard of the role of hidden transaction costs (TCs) affects the economic effectiveness of policy implementation and market efficiency (Simmons, 2015).

One example of an economic incentive is gross floor area (GFA) concessions. This has been widely used worldwide to motivate the private sector to provide public amenities in exchange for additional GFA (Tang & Tang, 1999). This mandates GB construction to some extent, and has achieved success in promoting GB. The design of the GFA concession scheme was connected with development control measures, such as sustainable building design guidelines (SBDGs) (Hong Kong), GB labelling programmes (Hong Kong, Singapore, US, etc.), and government land sale conditions (Hong Kong and Singapore), etc. Recently, the idea has also been applied as an effective instrument for long-term practices defined to promote an aspect of the public good – typically by creating the incentives and conditions for the market to bear the costs. For example, the affordable housing programme in the US, Australia and UK, renewable energy of buildings in New Zealand, Japan, France and US (Paetz & Pinto-Delas, 2007), and provision of new parks and plazas in New York.

The GFA concession scheme can be particularly beneficial for high-rise, dense cities with high land prices and property/rent prices, *e.g.* Hong Kong and Singapore. More saleable GFA means more profits for developers. In a high-rent place, a density bonus could be easily used to encourage developers (Küçükmehmetoğlu & Büyükgöz, 2013). In Hong Kong, due to the limited provision of development land each year (maximum 50 ha), a higher building density arising from a GFA concession can increase the profitability of a project, which appeals to local developers (Fan, Qian, & Chan, 2015; Liu &

Lau, 2013; Qian, 2010). Therefore, in Hong Kong, the GFA concession scheme has motivated developers actively to commit to GB investment, and the registered GB has increased almost one-third within one year since the scheme was launched in 2011 (Liu & Lau, 2013).

On the other hand, the additional GFA bonus could cause negative impacts to the built environment, particularly in a dense city (*e.g.* Hong Kong). Research has shown that an excessive GFA concession resulting in the increase of building bulk and height has brought the negative impacts (*e.g.* lack of daylight, views and air ventilation problems, *i.e.*, canyon effect), which reduces the effectiveness of air ventilation and strengthens the concentration of pollutant at a pedestrian level (Council for Sustainability Development, 2010; Fan et al., 2015; Feiock, Tavares, & Lubell, 2008). The drawback could be arguably compensated to a certain extent by the fact that the GFA incentive requires the building to meet the sustainable building guidelines and GB certification. These require the building to have separating gaps, a setback from street, permeability in building block, open green areas and other GB features that contribute to 'environmental friendliness' and the reduction of energy use. However, the balancing point between the environmental advantages and disadvantages is not clear. Therefore, when designing this economic instrument, a government must carefully decide how much GFA bonus could be granted. For example, in Hong Kong, a 10% cap of GFA concession was set to reduce its impacts. However, in Singapore, the maximum bonus of GFA is only 2%. The key research question that arises is: what is the optimum GFA bonus that could motivate developers and also minimize the negative impacts on the built environment? What is the appropriate amount of GFA bonus with a GB market progressing towards maturity? By employing an incentive scheme, a government's intention is to compensate developers for extra cost of GB, rather than give them more profits, as the cost is eventually borne by government and taxpayers/GB buyers. This implies a need to understand what the hidden costs (TCs) are and what extra works are caused by the GFA concession scheme to the key stakeholders. Without this information, it is impossible to predict whether the optimum GFA bonus will motivate the stakeholders and leverage the market system. It will also be impossible to design the incentive and fair distribution among the participating stakeholders without knowing their costs and benefits (actual and hidden). Therefore, the aim of the current research presented in this paper is to explore the extra work during the property (real estate) development process due to the commitment to the GFA concession scheme, with due consideration of TCs.

The research design of this study is summarized in Figure 1. First, the GFA concession schemes are

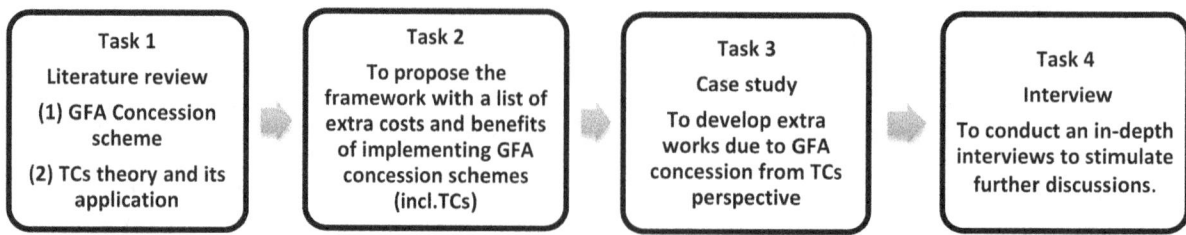

Figure 1 Research design

reviewed, using the bonus GFA concession schemes in Hong Kong and Singapore to provide an overview of the scheme (Task 1 in Figure 1). Then TCs theory is explained and applied to it to examine the hidden TCs and extra costs and benefits of committing to the GFA concession (Task 1). Both the tangible and intangible costs/benefits are mapped for the key stakeholders. A comprehensive analytical framework is provided (Task 2) that forms a basis for an empirical study in Hong Kong (Tasks 3 and 4). The Hong Kong case study is used to collect a list of extra works entailed by the incentive scheme during the real estate development process using the RIBA Plan of Work. A comparison is made between GB and traditional building (Task 3). The list of the extra tasks is verified by in-depth expert interviews. Finally, the interview results are complemented to verify the theoretical framework from the views of practitioners (Task 4 in Figure 1).

Background

GFA concession scheme

A variation in terminologies exists in different regions that share a similar meaning and intent, *e.g.* GFA concession, GFA incentive scheme, DB (density bonus), and FAR (floor area ratio) bonus. The DB and FAR bonus are used in North American, Japan, France, etc. (Paetz & Pinto-Delas, 2007). Density bonus, one of the most attractive incentives for developers in the US, refers to increasing allowable density by FAR or increasing allowable height for LEED-certified buildings (Abair, 2008; Miller, Spivey, & Florance, 2008). In Hong Kong, the GFA concession scheme refers to the floor area of certain green features and extra bonus floor area could be discounted from the total GFA calculation under the building regulations. In Hong Kong, this entails the prerequisite requirement of obtaining the BEAM Plus certification, and meeting some specific SBDGs (Building Department, 2011; Council for Sustainable Development, 2010). The Green Mark (GM) GFA incentive scheme in Singapore refers to buildings that attain the Platinum and Goldplus GM standards, which are then entitled to additional GFA. The principle underpinning these schemes is to encourage GB by granting additional bonus GFA to the particular site.

GFA concession scheme in Singapore

In 2009, Singapore implemented its GM GFA incentive scheme to reward the developers who acquire the certificates of GM Goldplus and Platinum. The Singapore Building Construction Authority (BCA) report showed the total number of registered GBs labelled with GM was 17 in 2005 and the number increased to 1696 by 2012 (BCA, 2014). The GFA concession scheme in Singapore has four levels of GB ratings, which makes the incentive scheme more flexible. It links the GFA concession scheme to the development control system. In Singapore, the GM Platinum or Goldplus certification are tied to the land sale conditions in certain new development sites. Generally, these sites are located within the new strategic growth areas, such as the Marina Bay and Downtown Core. By regulating land sale conditions, GB is a prerequisite for urban development

GFA concession scheme in Hong Kong

The GFA concession scheme in Hong Kong has four levels according to the GB ratings as well, which allows some flexibility. Similar to Singapore, it links to the development control system. However, in Hong Kong, GB design is stipulated in joint practice notes and SBDGs that are developed under building regulations. According to the GFA concession scheme, if there is no BEAM Plus certification, only a few green features that are mandatory in building regulations could receive any GFA concession. Therefore, BEAM Plus certification becomes mandatory if developers want to ensure all the green features can receive a GFA concession. The Hong Kong GFA concession scheme has been implemented since 2011. The total number of registered BEAM Plus projects increased from 225 (before the implementation of the GFA concession to 416 by 2015 (HKGBC, 2015).

Differences between Singapore and Hong Kong

Key differences between the Singapore and Hong Kong GFA concession schemes are that Hong Kong aims to promote sustainable building design and green features that were formulated according to its unique built environment. GFA concession will be granted only if the green features required under the SBDGs can comply with the minimum level of BEAM Plus certification. In contrast, Singapore's GM incentive scheme promotes the attainment of higher tiers of GM building. Only projects certified with GM Goldplus or above can acquire GFA bonus. Also, the methods to calculate GFA concession are different. In Singapore, the GM GFA is sensitive to land value with the total GFA regulated in the master plan and green premium. If the project is located in a city centre that has a high land value, then the GM GFA will be less than that of the same project located in suburban areas. This calculation restricts the GFA bonus in high land-value areas that usually have high density in order to reduce negative impacts of increased density on the surroundings. In Hong Kong there are three types of GFA concession: exempted GFA, disregarded GFA and GFA bonus, subject to the building features. Some building features, *e.g.* those deemed beneficial to the community, are not set with a cap of GFA concession. This in turn encourages developers to provide as much as possible. However, green features and amenity features (*e.g.* balcony and utility platform[1]) are subject to the cap of 10% GFA concession.

Other differences are:

- Singapore implemented incentive schemes to promote higher-tier GB ratings. A higher bonus is given to higher-tier GB ratings in Singapore. In Hong Kong, the GFA concession scheme does not distinguish the rating levels of BEAM Plus, as long as the project meets the minimum rating level.

- Singapore's GM GFA incentive schemes have a strong emphasis on energy efficiency in order to achieve the reward. This entails professionals spending additional time working on energy efficiency.

- Singapore has special financial incentives for architects and engineers to get paid for their additional efforts and time spent on GB. This explicitly recognizes the importance of design stage. In contrast, the reward from Hong Kong's GFA concession scheme is only targeted at the developers.

Transaction costs theoretical application

The term of 'transaction cost' was proposed by Arrow (1969). TCs were defined as the costs of running an economic system, including exclusion costs and costs of communication (*e.g.* supplying and learning terms where transactions would be undertaken), and the costs of disequilibrium. The difference between TCs and production costs is that TCs varied with the modes of resource allocation while production costs relied on the technology and tastes, and would not change with economic systems (Arrow, 1969). Arrow's opinion linked TCs with institutions, which

was supported by Cheung (1987) who claimed TCs were essentially institutional costs and North (1990b) who stated that TCs are the sources of power for social, economic and political institutions. North (1990a) claimed that TCs were the costs of measuring and enforcing agreements. Measurement costs are those of measuring the valuable attributes (*e.g.* colour, size, durability, robustness, performance, etc.) of what are being exchanged, while enforcement costs are those of protecting and enforcing agreements. Williamson (1985) further developed the concept of TCs. TCs comprised *ex ante* and *ex post* that the former occurred in drafting and negotiating agreements, while the latter included setup and the costs of running governance structure. TCs are equivalent to friction force in physical systems (Williamson, 1985). Similarly, Matthews (1986) stated that TCs comprised *ex ante* and *ex post* that were the costs of arranging contract, and monitoring and implementing it respectively. A more recent study stated that TCs are the costs relevant to search and information, policing and enforcement, as well as bargaining and decision-making processes. The exchange process was regarded as the major source of TCs (Furubotn & Richter, 2005).

From a transaction cost economics (TCE) perspective, incentive schemes can be deemed as a governance structure shaping transactions among the key stakeholders (Finon & Perez, 2007). TCs occur during the process of incentive implementation to both the government and the involved parties who are willing to take part in the incentives, which will eventually be borne by the government and citizens (tax payers) (Andersen & Sprenger, 2000; Qian, 2012; Qian, Chan, Visscher, & Lehmann, 2015a). With regard to the GFA concession scheme, TCE sheds light on the implicit contractual relationship between the policy-maker and the real estate developers, given the extra costs and market uncertainty caused by committing to the GFA concession scheme and GB. The TCs will not only decrease the effectiveness of the incentive scheme itself but also may decrease the desire of stakeholders to participate in the (voluntary) GFA concession scheme or GB. Economists argue that the compliance cost of incentives is more cost-effective as they allow the stakeholders the flexibility to seek innovative and cost-saving solutions.

To date, no comprehensive study exists on the application of TCs analysis to GFA concession scheme. Therefore, this study conducted a wide-range review of TCs associated with energy efficiency, GB and environmental policy to identify the possible TCs in the process of GFA concession scheme implementation (Table 1). The list of possible TCs will be verified in the interviews presented below.

Table 1 Transaction costs associated with energy-efficiency and green building promotion, and environmental policy implementation

Transaction cost items	Mundaca, Mansoz, Neij, and Timilsina (2013)	Hein and Blok (1995)	Dudek and Wiener (1996)	Coggan, Whitten, and Bennett (2010)	McCann, Colby, Easter, Kasterine, and Kuperan (2005)	LBNL (2007)	Michaelowa and Jotzo (2005)	Ofei-Mensah and Bennett (2013)	Singh (2009)	Hagemann, Prager, and Bartke (2015)	Joas and Flachsland (2014)
Cost of information searching	×	×	×	×	×	×	×	×	×	×	×
Research cost				×		×		×			
Decision-making cost		×									
Implementation cost			×	×							
Negotiation cost	×		×			×	×	×	×		
Project documentation/ administration cost				×	×		×	×			×
Approval cost			×			×	×		×		
Validation cost							×		×		
Registration cost							×				
Monitoring and verification cost	×	×	×	×	×	×	×	×	×		×
Certification cost							×		×		
Enforcement cost			×	×	×		×	×	×	×	×
Trading cost	×							×			
Transfer cost							×				
Insurance cost			×			×					
Coordination cost										×	

176

Costs and benefits of the GFA concession
Transaction costs of implementing GFA concession

Williamson (1985) further developed this theory of TCs with three dimensions, namely asset specificity (AS), frequency and uncertainty, which are the three determinants of TCs in the GFA concession application. AS refers to the durable investment to support the particular transactions. Coggan et al. (2010) and Rørstad, Vatn, and Kvakkestad (2007) stated that AS influences TCs via information collection, administration, contracting, monitoring and implementation.

In the current study, AS refers to the specific investment for the particular transaction in the process of implementing the GFA concession scheme as well as achieving GB. For example, developers may need to undertake extra work and deploy extra facilities or staff to get access to capital, GFA concession/GB design, and government approval, etc. (Chai & Yeo, 2012). The administrative burden will increase due to the need for documentation and additional contracts. Moreover, the SBDG in Hong Kong provide instructions for building design in different sizes and shapes of sites. Thus site-specific building design must be performed by architects to fulfil the requirements of the SBDG. Also, specialized GB knowledge and equipment are needed to acquire BEAM Plus registration or certification. These lead to the non-standard contracting and building design scheme as well as TCs in exchange. In the event of a shortage of human and technical capacity, the TCs would rise for the information searching cost and contract negotiation cost (Mundaca et al., 2013). Coggan et al. (2010) also stated that non-standard contracts would generate more TCs than the standard ones.

Frequency (F) refers to how often stakeholders performed a particular practice with the GFA concession scheme. The more often this happens with one particular practice/transaction, the less extra costs will arise due to the economies of scale. Frequency would affect TCs because when the transactions are recurrent. TCs could be reduced through repeating transactions that reduce learning costs and create a fixed, appropriate contract that reduces the efforts involving the collection of information (Nilsson, 2009). In this study, frequency means the how frequently stakeholders participate in the GFA concession scheme. Apparently, participating frequency is higher when the GFA concession scheme is mandatory.

Uncertainty (U) refers to market, policy and economic uncertainty in this context. For example, this can include: GB assessment uncertainty, technique uncertainty, incentive expiration after a period of time, economic downturn, etc. Technique uncertainties include the uncertainty about the extent of efficiency improvement (Chai & Yeo, 2012). The impacts of uncertainty on TCs are conditional and only if there is AS, can the degree of uncertainty influence TCs (Williamson, 1981). Uncertainties contribute to TCs via information collection, clarifying and negotiating contracts, monitoring and enforcement actions (Coggan et al., 2010). In this study, uncertainty exists particularly in the process of application of GFA concessions and BEAM Plus registration or certification.

This study also aims to map both the tangible and intangible costs/benefits among the key stakeholders in real estate development projects, which include developers (D), contractors (C), professionals (P) and government (G). Table 2 provides the list of TCs (with the three dimensions of AS, F and U) assigned to the stakeholders due to the GFA concession scheme.

Actual costs and benefits

In general, GB requires comparably higher initial costs and extra risks to deliver compared with traditional buildings. Some stakeholders will therefore decide to avoid voluntarily entering the GB market. Yu and Tu (2011) stated that GM buildings require a range of 1–3% extra cost compared with non-GM buildings in Singapore. BCA (2015) stated that the cost premium for GM Platinum, and Goldplus are \$123\$/m^2 and \$97\$/m^2 in the residential sector. Kats, Alevantis, Berman, Mills, and Perlman (2003) claimed 0.66% extra cost for LEED certification, 2.11% for Silver, 1.82% for Gold and 6.50% for Platinum in the US. Davis Langdon (2007) suggested 3–5% greater cost for five star and 6% for six star in Australia where a Green Star rating system is employed. In Hong Kong, under the Hong Kong Building Energy Assessment Methods (BEAM Plus), the cost premiums for Silver, Gold and Platinum building are 0.8%, 1.3% and 3.2% respectively (Burnett, Chau, Lee, & Edmunds, 2008). It is widely acknowledged that a cost premium of GB exists and varies according to the level of GB ratings.

In terms of financial benefits to the developers, Fuerst and McAllister (2008) claimed that GBs have a price premium of 10% and 31% for GB certified by Energy Star and LEED respectively, if the market reflect its value. Miller et al. (2008) suggested 9.94% price premium for LEED and 5.76% for Energy Star per square foot. Yu and Tu (2011) stated that GM buildings do have price premium that increases according to the levels of the GM ratings. Burnett et al. (2008) studied the financial benefit of GB to end-users, such as a reduced sewage charge. Based on the literature review, Table 3 summarizes the actual costs and benefits of committing the GFA concession scheme among the different stakeholders.

Hidden benefits

Hidden benefits include improved health and productivity, reduction in demands for water and

Table 2 Barriers causing transaction costs (TCs) to the stakeholders due to the gross floor area (GFA) concession scheme

TCs items	AS	F	U	D	C	P	G
Costs of access to capital and budget (Chai & Yeo, 2012) • Internal constraints on the budget • Problems with external financing	×			×			
BEAM Plus certification • Evidence that credit has been achieved • Time pending for assessment of results Costs of calculating costs and benefits of different efficiency levels	×			×			
Opportunity cost: priority over GB/GFA (Qian et al., 2015a) • Lack of time priority (Chai & Yeo, 2012) • Attraction (business/policy) of counterparts over GB/GFA • High sunk costs	×		×	×			×
Opportunity costs of planning incentives • Plan alternative incentive programmes, e.g. promote the provision of affordable housing, public open space.	×		×				×
New implementation process (Meacham, 2010) • Barriers of internal organization • Additional testing and inspection in construction	×			×		×	
Contracting organization (Walker & Chau, 1999) • Information gathering • Contract involve more elements (new contract) with different (often more) stakeholders	×			×		×	
Cooperation and working relationships between the more parties within the specific project (Love, Niedzweicki, Bullen, & Edwards, 2011)	×			×	×	×	
Delays may occur due to the capability of contractors to implement the specifications relating to GB technology.			×	×	×		
Additional working time without reward • Additional coordinating and analytic effort • High-standard design details (BEAM Plus) • Demonstrating in a credible way that a new building will reduce prospective energy costs	×			×	×	×	
Administrative burden (Ahn & Pearce, 2007) • Paperwork • Documentation and photographic evidence (BEAM Plus) • Assessment costs/certification costs • Additional contracts	×			×		×	×
Monitoring costs and report on construction activities, material use.	×			×		×	
Cost of technology innovation and management innovation (Qi, Shen, Zeng, & Jorge, 2010) • Managing equipment and materials • Negotiate with suppliers • Lack of information, awareness or expertise to achieve sustainable measure • Barriers of internal structure and interaction	×		×	×	×	×	
New process, green technology and working methods (Häkkinen & Belloni, 2011) • Possible risks and unforeseen costs due to the new adoption • Unfamiliar techniques/lack of experience • Speed of arrival of new technology • Inadequate, untested or unreliable sustainable materials, products or systems • Misunderstanding of green technological operations • Limited availability and accessibility of green technology, suppliers and lack of quality and performance information (Chai & Yeo, 2012) • The extent of efficiency improvements of new technologies • Concern over the high price of new technology deployed now will drop drastically in short due to the market economy of scale • Information and learning costs of green technology and energy efficiency measure (Chai & Yeo, 2012)	×	×	×	×	×	×	
Costs of training the workforce with specialized knowledge of GB/GFA • Employing suitably qualified person to report and document • Learning costs of the professionals and key stakeholders	×			×	×	×	

(continued)

Table 2 Continued

TCs items	AS	F	U	D	C	P	G

Unfamiliarity of the design team and contractors with GB methods
    Lack of manufacturer and supplier support
    Credibility and reliability of new suppliers and subcontractors
    Costs of identifying project, new networks/supply chain (Walker & Chau, 1999)
    Mistrust and dispute resolution costs and time
		×	×	×	×	×	

Costs of increased congestion (Kayden, 1978)		×	×				×
Increased building bulk and height • Influence air ventilation, visual effects, city image		×	×				×

Note: AS = asset specificity; F = frequency; U = uncertainty to the key stakeholders: D = developer; C = contractor; P = professionals; G = government.

Table 3 List of actual costs and benefits of committing the gross floor area (GFA) concession scheme

Stakeholders	Actual costs	Actual benefits

Developers	More construction cost due to risk in a longer construction time, new construction methods and new GB technologies (Rehm & Ade, 2013)
	  • Increased architectural and engineering design time (Kats et al., 2003)
	
	Costs of GB certification
	  • Assessment cost
	  • Survey costCertification cost about HKD75 000–150 000 depending on the project scale and complexity (Burnett et al., 2008)
	Additional or increased consultant fee (Häkkinen & Belloni, 2011)
	  • Higher cost for green appliance design and energy-saving material at design stage
	  • The design fee rises from around 9–10.5% of total cost (Larsson & Clark, 2000)
		GFA concession bonus
		
		Higher market selling price (Hebb, Hamilton, & Hachigian, 2010)
		Costs saving from efficient use of materials
		  • Reduction of material use through modular design (off-site prefabrication, lean construction methods), reuse of building elements
		  • Improved material management and on-site sorting

Government	Professional training – continuous professional development (CPD) course	Tax revenues derived from the extra floor area (Kayden, 1978)
		  • Tax from additional housing units transactions
		  • Tax from extra construction activities

Contractor	More construction cost due to longer construction time Increased architectural and engineering design time (Kats et al., 2003)	Material saving
End-users	Higher property price	Operational cost saving (quantity depends on building performance) • Energy and water saving (Kats et al., 2003) Higher property value (resale)

electricity infrastructure (Burnett et al., 2008). Isa, Rahman, Sipan, and Hwa (2013) stated that developers can improve their corporate image by developing GB. A contractor's future competitiveness could be improved with the GB development and practice. However, the location and affordability and aspects such as culture, individual preference etc., still dominate buyers' considerations, especially in a residential sector where consumers are uncertain about GB performance and lack GB awareness (Burnett et al., 2008). It is possible that in an immature GB market with a lack of awareness from the public, hidden benefits of GB cannot be fully taken into account for decision-making by developers as well as consumers. Through this literature review, the hidden benefits to the stakeholders due to GFA concession scheme are encapsulated in Table 4.

Table 4 Hidden benefits to the stakeholders due to the gross floor area (GFA) concession scheme

Hidden (invisible) benefits to the stakeholders	D	G	P	C	E
Good company reputation/profile, status, market power, job satisfaction, rewards, personal development (Isa et al., 2013)	×		×	×	×
Future business competitiveness over the long-term	×	×			
Extra GFA bonus to sell more and gain more profits	×				
Energy efficiency and environmental protection can help GB sell quicker (Bartlett & Howard 2000)	×				
Reduction in construction pollution (BEAM Plus) • Reduction of pollution, resource depletion, energy and waste consumption (Addae-Dapaah, Hiang, & Sharon, 2009)		×			
Reduced demands on infrastructure (Pearce, DuBose, & Bosch, 2007), public water treatment, electricity demands, and landfill (Kats et al., 2003)		×			
(National) savings of healthcare (Pivo & McNamara, 2005) • Reduced respiratory infections, allergies and asthma • Decrease demand for healthcare facilities • Enhanced occupant productivity and health (Kats et al., 2003) • Reduced healthcare cost		×			×
Create more job opportunities		×	×	×	×
Improved working efficiency and social productivity • Increased economic activities, e.g., activity associated with bonus GFA (Kayden, 1978) • Green premium increase construction spending • Stimulate more consumers spend more in the long-term, due to the savings from energy bills • Higher interest paid to bank on construction loans (Kats et al., 2003)		×			×
Support from company to take training course (Ahn & Pearce, 2007), i.e., professional certificate			×		
Obtaining and using new professional skills (Ahn & Pearce, 2007) • Serving new technology • BEAM Pro • Life-cycle cost of GB • GB design process • Familiar with GB standard • Knowledgeable about low environmental impacts materials	×		×	×	×
Better quality of life from, for example, sky/podium garden, wider corridor, quality indoor environment, natural light and ventilation (Hebb et al., 2010), better site plan and design, fewer carbon emissions, etc. (Kats et al., 2003)		×			×
New knowledge and skills about green construction (Qian, Chan, & Bin Khalid, 2015b) • Basic knowledge and concepts of green construction and management • GB rating system • General knowledge of sustainability in the built environment • GB materials and method	×		×	×	

Note: D = developer; C = contractor; P = professionals; G = government; E = end-users; AS = asset specificity; F = frequency; U = uncertainty.

Methods

GFA concession in Hong Kong

Hong Kong aims to promote sustainable building design and green features that were formulated according to its unique built environment.[2] After building plan approval and before consent to commence building works, developers need to provide provisional assessment result of BEAM Plus rating to the buildings department. A GFA concession will be granted only if the green features are provided as required under the SBDGs and supported with the minimum level of BEAM Plus certification.

With due reference to the TCs barriers to stakeholders in Table 2, the framework used (Qian et al., 2015b) to ascertain the TCs in delivering GB projects is extended here to consider extra work (considering TCs) due to committing to the GFA concession scheme during the real estate development process. The RIBA Plan of Work (2007/08) provides the structure to consider all stages of a development project in order to identify the extra tasks involved with the GFA concession scheme (when compared with its traditional counterpart). The extra work to meet the GFA concession scheme will incur additional costs. Therefore, interviews were designed to collect additional information and verify the review results.

Expert interviews

In order to support the case study, expert interviews were conducted to understand the GFA concession practice in Hong Kong. Based on the literature review of costs and benefits (including hidden and TCs) shown in Tables 3 and 4, as well as the extra work/TCs in Table 5 due to GFA concession in the RIBA plan, the interview questions were developed. In-depth expert interviews were conducted with 10 experienced senior industry practitioners, *i.e.*, architects, surveyors, green consultants, developers and professors, to gain practical insights. The aim was to validate the identified list of costs and benefits, and provide explanation for each item of costs and benefits and the effectiveness of the GFA. The profile of interviewees is shown in Table 6. All are at the management level and actively involved with the GFA concession scheme and GB practice, with a minimum of 10 years' experience in the building industry, and a wide knowledge of surveying, urban planning, law, finance and accounting, etc. Some of the interviewees are also authorized persons (AP) who are qualified to perform the duties and roles in accordance with buildings ordinance. They have a good overview the costs and benefits due to participating in the GFA concession scheme in practice. The decision to use 10 experienced experts who have been actively involved in implementing the GFA concession scheme in Hong Kong will yield insightful, highly relevant and more convincing views than a massive survey of people without necessary expertise and hands-on experience.

The structured interviews were designed to discuss the extra costs and benefits due to participating in the GFA concession scheme. The interview questions were divided into three parts:

- the actual costs and benefits (Table 3)

- the TCs (Table 2)

- the hidden benefits (Table 4)

The interviewees were encouraged to share their views beyond this framework, which is believed to be essential to capture any novel factors. The discussion also included the relevant background knowledge that is not shown in the website or publications, and the future perspective of GFA concession scheme in their views. The comprehensive views of the interviewees in market practice of the GFA concession scheme help to verify and complement the theoretical framework of this paper from practical perspectives.

Results and discussion
Increased construction costs and land cost

As already noted above, various studies have estimated the increased cost for different GB schemes to vary between 0.6% and 6.5% This may be offset by increased market value, market share or reduced risk.

The uncertainties caused by GFA concession due to the complex design to be approved by the government, directly affects the estimation of the developers' profits, especially at the land bidding stage. Developers in Hong Kong have to estimate the possible GFA concession granted and decide the maximum land cost they could afford. Therefore, the GFA concession scheme causes land prices in Hong Kong to be increased, which in turn decreases developers' expected profits. This also happened in New York City. When the New York City government provided a density bonus for developers constructing moderate-cost housing, the land cost increased as well. In this context, the planning agency began to calculate the profits of each specific GB project and control the bonus accordingly (Johnston, Schwartz, Wandesforde-Smith, & Caplan, 1989), which inevitably generated administration costs.

In Singapore, the land cost is prescribed and this reduces the uncertainties of total costs. The granted GFA concession could be exactly calculated through the prescribed formula, in which GFA is calculated in reverse accordance with land price, *i.e.* a higher land price will lower the GFA bonus. The GFA bonus will be given less to the project in the city centre where land is usually expensive compared with the same project located in the suburban area. Thus, this calculation method of the GFA concession potentially reduces the negative impacts of extra GFA on the built environment. It also largely reduces the uncertainties to participate in the GFA concession scheme and results in decreased corresponding TCs, such as information searching costs and research costs.

Uncertainty results in higher negotiation and approval costs

The interviewees mentioned that BEAM Plus assessment largely depends on the variation between assessors, which leads to the unexpected or inconsistent results. The evaluation can have different subjective measurements that may give the same/similar projects different ratings, depending on who is the assessor and the assessor's measurement approach. For example, the BEAM Plus assessor may have a biased view due to conflict of interests of working in a rival firm. The BEAM Plus assessment process is perceived by the interview participants to lack transparency and consistency. Generally, there is 20–25% rejection

Table 5 Extra work arising from the gross floor area (GFA) concession scheme mapped onto the RIBA 2007 Outline Plan of Work

	RIBA Stage of Work	Tasks to be done for traditional projects	Extra work with TCs incurred (in concern of GB/GFA concession scheme)
Briefing	A: Inception	Set up client organization for briefing. Consider requirements, appoint architect. *Developer's key actions:* Identify opportunities (property/need/use/idea); assemble co-developer; identify and review information; identify seed money; evaluate investment climate	(1) Appoint an authorized person and involve special stakeholders for proposing the GFA concession options (2) Set up extra organization for briefing on GB in terms of granting GFA concession, *e.g.*, new offices, new staff (3) Consider extra GB-related market and policy requirements: market study in GB; policy study in GB
	B: Feasibility	Carry out studies of user requirements, site conditions, planning, design and cost, etc., as necessary to reach decisions. *Developer's key actions:* preliminary market analysis (community/supply/competitive); assemble technical team; identify potential users; consider alternative site; preliminary financial plan; formal analysis (site/building/market/design/financial/appraisal); investment threshold; legal issues; public participation; review available information; review objectives	(4) Need JV or co-developer for such special project? (5) Undertake extra studies of market requirements, potential and expectation on GB (considering local community need/supply/competitiveness) (6) Extra GB planning, design and cost, etc., in terms of GFA concession, as necessary to reach decisions (7) Extra effort to identify potential users (8) Study the extra financial risk (9) Consideration of extra legal liability risk of the GB product in terms of GFA concession (10) More careful review of available information on GB (11) Decision process for determining the level of BEAM Plus certification (12) Identification process(-es) for specialist expertise and assembling the project team (13) Registration process for the BEAM Plus (14) Others
Sketch plans	C: Outline proposals	Develop the brief further. Carry out studies on user requirements, technical problems, planning, design and costs as necessary to reach decisions *Developer's key actions:* obtain control of the land/property; preliminary plans and specifications; negotiation with government for approval	(1) Extra work about site planning and design, building material and equipment selection to reach the requirements of the expected GFA concession approval (2) Special user requirement study (3) Explore special technical solutions
	D: Scheme design	Final development of the brief, full design of the project by architect, preliminary design by engineers, preparation cost plan and full explanatory report. Submission of proposals for all approvals	(4) Special concept/design that need negotiation with government for approval (5) Design leading to non-efficiency use of floor area (6) Special cost study for using new design features (7) Submission to HKGBC for BEAM Plus provisional assessment (8) Supplementing additional information if required by HKGBC (9) Inadequate information will cause the delay of process of BEAM Plus registration (10) Examination of sustainable building design and green features design and their integration into the building design (11) Discussion of the different design scenarios more with clients and reach agreement with cost consideration (12) Submission of application for GFA modification to Building Department with supporting documents (13) Others.

(continued)

Table 5 Continued

	RIBA Stage of Work	Tasks to be done for traditional projects	Extra work with TCs incurred (in concern of GB/GFA concession scheme)
Working drawings	E: Detail design	Full design of every part and component of the building by collaboration of all concerned. Complete cost checking of designs. *Developer's key actions:* finalize plans and specifications; revise financial projections; financial negotiations (mortgage/loan/ construction loan); tax consideration	(1) Financial negotiations for new design feature (consideration of mortgage/Loan/construction loan) (2) Limited number of available contractors with expertise reduces competition (3) Learning cost of new professional qualification and skills who are lack of experience in GB and GFA concession scheme (4) Risk of rejection by HKGBC or building department and need for resubmission (5) Appointment of contractor with endorsement of an authorized person (6) Inadequate information will cause the delay of process of BEAM Plus registration (7) In granting modification of or exemption from the provision of the building ordinance, conditions may be imposed by the building authority (8) Negotiation with government for special building plan approval (9) Others
	F: Production information	Preparation of final production information, *i.e.*, drawings, schedules and specifications	
	G: Bills of quantities	Preparation of bill of quantities and tender documents.	
	H: Tender action	Compile a list of tenders; issue tender documents; check and open tenders	
Site operation	J: Project planning	Notify acceptance of tender; check all contract document are in order; brief all project personnel of the project requirement and procedure for administer the project ; check approvals and site condition to ensure the project can be carried out on site *Developer's key actions:* acquire property; select construction company; marketing and leasing; initial financing; assemble construction management team; tenant involvement	(1) Extra effort to brief all project personnel of the project requirement and procedure for administering the project (2) Special promotion strategy and materials for marketing and leasing (3) Additional consideration of tenant for GB products (4) Extra requirement on testing and commissioning of service installations to obtain green labelling etc. (5) Special effort to prepare maintenance manual (6) Extra fee for certificates involving green items (7) Extra administration: paperwork and time for preparation, and BEAM Plus application (8) Documentation and photographic evidence to apply for BEAM Plus (9) Cost of paperwork related to the application of BEAM Plus (10) Speed of arrival of new technology (11) Uncertainty of efficiency improvements of new technologies (12) Cost of managing and monitoring, *e.g.* to monitor and report on construction activities, like material use; to manage equipment and materials (13) Cost of negotiation, *e.g.* negotiate with suppliers (14) Speed of arrival of new technology (15) Submission of materials for project assessment of BEAM Plus to HKGBC for final assessment within six months of the date of issuance of the occupation permit by the building authority (16) Supplementing additional information if required by HKGBC (17) Inadequate information will cause the delay of process (18) Risk of rejection by HKGBC and need resubmission (19) Others
	K: Operations on site	Setting out the building on site; site meetings; supervision and site visits; financial monitoring of each construction stages; testing and commissioning of service installations; prepare maintenance manual	
	L: Completion	Check works ready for completion; hand-over inspection; rectify defects; final inspection and final certificate *Developer's key actions:* inspection; certificate of occupancy; permission to sell/rent	

(continued)

Table 5 Continued

	RIBA Stage of Work	Tasks to be done for traditional projects	Extra work with TCs incurred (in concern of GB/GFA concession scheme)
Feedback and maintenance	M: Feedback	Analysis of job records. Inspections of competed buildings. Studies of building in use. *Developer's key actions*: prepare property management plan; revise marketing plan; oversee marketing or leasing	(1) Special property skill requirement for property management plan (2) Special strategy and materials for overseeing marketing or leasing (3) Operation: keep building running effectively and under good repair (4) Set up and manage ownership entity (5) More green items require special care for property maintenance and improvement (6) Easy to sell or rent property (7) Management of additional guarantee certificates (8) Others
	N: Maintenance	*Developer's key actions*: set up and manage ownership entity; property improvement; property disposition; closing ownership entity	

184

Table 6 Profile of interviewees

Profession	Qualification and position
Architect	Authorized person; more than 20 years' working experience; director of architectural firm
	Registered architect; chairman of architectural firm
	Authorized person; Hong Kong Institute of Architects fellow member
	Senior architect; working in leading architecture firm for five years in Hong Kong; all projects the architect has joined are green buildings
	Manager, working in leading architecture firm that all the projects it did are green buildings
Developer	Chief executive officer (CEO) in one of leading real estate development firms in Hong Kong
Surveyor	Green building professional, environmental officer working in leading construction firm. Familiar with LEED and BEAM Plus
	Authorized person; project director of consultancy firm
	Director of consultancy firm
Professor	Full professor; over 10 years' working experience in project management and building control

risk to all applications. Developers whose applications are rejected have to negotiate or resubmit. This process increases the risk and leads to 20–30% extra work. This explains why developers usually prefer to aim for attaining a lower standard (BEAM Plus at Bronze level) and not a higher level as it is perceived to contain a higher level of risk.

Similarly, uncertainties also exist in the process of GFA concession application. If there are some special designs, the Hong Kong Buildings Department will hold a conference to discuss the decision of GFA concession of special design. Architects have to negotiate and convince the municipal government to accept their design with strong evidence of environmental benefits. Negotiation between design teams and developers or contractors can also generate TCs due to the complex requirements for building design. In Singapore there was also an increased number of meetings with green specialists (Hwang & Ng, 2013). The misinterpretation of clients' requests by the design team is a vital element that negatively influences the project schedule (Hwang, Zhao, & Tan, 2015). This reflects that stakeholders have not developed a standard procedure of cooperation and tacit agreement, which usually takes much time to build.

In order to be granted a GFA concession, BEAM Plus registration and assessment are required, together with the additional administration fee to be paid by developers. An approval cost arises when the transactions must be approved by government. It may result in the delay of transaction completion and impose modifications. Consultants need to prepare supporting documents for BEAM Plus registration/certification and GFA concession application. Additional information may be required may cause the delay of processing. In granting a modification or exemption from the provision of the buildings ordinance,

conditions may be imposed by the buildings authority. If there are special designs, architects have to prepare relevant documents in detail to support the application of the GFA concession. It is worth noting that in some US cities and counties the certification and building permit fees are reduced as incentives to promote GB (Olubunmi, Xia, & Skitmore, 2016).

Market uncertainty and GB property value
According to the interviews, there is an apparent inconsistency about the perceptions of market value of GB. The majority of the interviewed Hong Kong experts agree that in Hong Kong GBs generally do not attain a higher market value than their counterparts, and that the GFA concession scheme is the main driver causing developers to construct GBs. Some disagree by acknowledging that GB does indeed enhance the building value but the actual enhanced value depends on many factors. Green features and energy efficiency are not the main considerations of residents. For office buildings, some international firms may prefer a GB-labelled office which gives GB a competitive market advantage to the traditional buildings. However, there is little difference in rental or sale prices between the levels (Bronze Silver, Gold and Platinum) of BEAM Plus ratings. The interviewees stated that the GFA concession scheme does not help improve the building quality much, which is the main reason why the general public is reluctant to pay more.

After developers bid on the land and determine the building design scheme, many other factors need consideration. For example, they would project the market price for a certain period in future (*e.g.* three years), and then decide to provide the extent of facilities to acquire the GFA concession (*e.g.* car park, podium garden and green features). Normally, the estimated property price that depends on the location (*i.e.* the uses and price of comparable property in the area), economic situation, time, development cost, net floor area and the standard of development would determine the design and green feature provisions. A clear conflict exists between the GFA concession and market price. This indicates that the uncertainty of the GFA concession and property market arise more research costs.

Extra workload to professionals

All the interviewees agree that GFA concession is the main factor attracting developers to participate. However, to acquire a 10% GFA concession is difficult, and may not be cost-effective. Indeed, a few interviewees questioned whether some of the building features getting GFA concessions are really environmentally friendly. They suggest that government should distribute GFA concessions to a wider array of quality features.

The searching cost refers to the cost of collecting information. In this study, consultants collect specialized information of GB such as the performance of green equipment and GB design information. Developers usually commission experienced architects and GB consultants because these consultants' experience largely affects the amount of GFA concession granted to developers and the assessment results of BEAM Plus. According to the interviews, there is 20–25% risk of obtaining unexpected results, depending on the consultants' level of experience. This is also supported by Coggan, Buitelaar, Whitten, and Bennett (2013) and Ducos, Dupraz, and Bonnieux (2009) that past experience could improve the ability of decision-making and influence TCs, because experienced professionals spend less time and effort collecting and processing information. The searching cost accounts for the more time and money spent in the implementation process. However, two interviewees mentioned that there is a shortage of experienced consultants, which indicates that the GB market still has much room to develop.

Costs associated with negotiation, communication and approval

The validation cost includes the cost of review and revision of project document by operational parties. When the results of application of BEAM Plus registration/certification or GFA concession are not satisfactory, a review and revision of the application has to be conducted to reach the relevant requirements. This results in extra time and effort expended by consultants, and it will be reflected in the consultancy fee. The same situations also occur in Singapore such that mistakes or delays in preparing design documents significantly affect the schedule performance of GB (Hwang et al., 2015).

High monitoring costs and verification costs

The monitoring cost is the cost of monitoring policy compliance, contract implementation and the outcome. Site monitoring and reporting on the execution of the instructions have to be conducted to provide evidence for BEAM Plus certification. Some interviewees mentioned that contractors have to monitor and work longer – this cost would be reflected in the total construction cost. Likewise, in Singapore, monitoring the project progress by consultants is ranked fourth out of 36 significant factors that affect schedule performance in GB projects (Hwang et al., 2015).

Verification cost refers to the cost to verify the effectiveness of green materials or equipment. Three interviewees stated that the information provided by suppliers on the effectiveness of green materials or equipment may not be complete. Hence, the consultants have to undertake research or testing to verify the effectiveness. The replacement of material and equipment is common if there is lack of information before procurement. Green materials have to be tested by an accredited laboratory to ensure its effectiveness (Lam, Chan, Poon, Chau, & Chun, 2010). If the green specification could be specified in the contract, the verification costs could be reduced.

Benefits

- *Reputation*
  Developing GB could gain a reputation for developers, but this is not the main reason for GB development. For the developers who only achieve the BEAM Plus registration, participating in the GFA concession scheme is perceived as not enhancing their reputation or may even negatively influence their reputation. Some residents do not acknowledge the utility of concession features and regard them merely as developers' instruments to acquire extra GFA and make more money.

- *Faster sales*
  Only two interviewees stated GBs are comparatively easier to sell than the traditional ones. On the contrary, eight interviewees stated that there is no difference between GB and non-GB because this is not the main consideration of buyers. Further, they suggested that Hong Kong people do not have a strong awareness of 'green' or a willingness to choose 'green' products, supported by Chan, Qian, and Lam (2009).

- *More flexible design for environmental benefits*
  Though there are inconsistent views about the

flexible approach to design, some opine that the building regulations tend to restrict design innovation, while the GFA concession incentive scheme provides designers more flexibility to innovate and design for environmental reasons. The GFA concession encourages architects to give greater consideration to users and the general public. For example, without the GFA concession scheme, a sunshade device would be calculated on total coverage, and developers may support the design. However, with GFA concession scheme, they could apply for GFA exemption, the buildings department will permit the concession as long as the innovative design is justified. On the other hand, three interviewees pointed out that the GFA concession scheme restricts building design and discourages innovation due to the cap of 10% GFA concession. Even if architects could negotiate with the Hong Kong Buildings Department to apply for a concession of innovative design, it is still perceived as a risk and not worth trying. Overall, it is felt that the GFA concession scheme is designed to encourage innovation and provide design flexibility for architects, but the potential risks and uncertainties prevent architects actually doing it.

- *Job opportunities*
Over half the interviewees mentioned that the GFA concession scheme created more job opportunities. One interviewee specifically stated that his/her architect firm has employed an extra 20% of employees to do BEAM Plus projects. There are new job positions created by the GFA concession scheme, including green professionals, environmental consultants, green material/equipment supplier, BEAM Plus assessor and energy simulation consultant.

- *Energy and water efficiency benefits*
Some interviewees claimed that the new technologies are not cost-effective because of high upfront costs and low energy and water savings. Opposing views endorsed the energy and water-efficiency benefits because government could save the cost of energy and water infrastructure expansion. In short, it seems that GBs do generate energy and water efficiency benefits for the public, but developers have to bear the upfront costs that may be even more than the lifecycle savings. That is why some countries and regions have provided subsidies to compensate developers.

- *Increased land price*
Through indirectly providing financial aid by GFA bonus, the government can save money or even generate more money through the incentive scheme. For example, the GFA concession scheme in Hong Kong makes developers willing to pay more for

the land, as discussed above. Since all land transactions bring levies/duties to the government, the increased land price becomes government's additional income, while developers' actual benefits are less than the profits of 10% GFA concession due to the increased land cost.

Mandatory or voluntary GFA concession scheme
Hong Kong and Singapore have each integrated their voluntary scheme with the regulatory system in a different way. Therefore, different costs and benefits are generated and distributed. Whether the GFA concession scheme is mandatory or voluntary would affect the frequency of implementing the scheme. Apparently, a mandatory scheme has a high participation rate and could reduce TCs soon, as more experience of participating is quickly accumulated. In Singapore, the GFA concession scheme is connected with the land sale conditions in designated areas. This forces developers to participate in the GFA concession scheme and allows them to gain experience. In Hong Kong, BEAM Plus registration is a mandatory requirement for obtaining the GFA concession, which helps to increase the frequency of GB construction. Both Singapore and Hong Kong take the built environment into consideration when they integrate the GB labelling scheme into their respective regulatory system. In Singapore, the features and needs of the built environment were addressed by connecting the GFA concession scheme with the land sale conditions, which make GFA concession scheme mandatory for all sites. Key development areas that usually have a high density are forced to construct GB. In contrast, the GB labelling scheme and building design guidelines in Hong Kong were integrated to facilitate GB registration with the promotion of a specific requirement of GB design tailored for the unique built environment. The concession scheme is free for developers to adopt by complying with a combination of administrative and regulatory requirements administered jointly by self-governing non-governmental organizations (NGOs) and government authorities.

Conclusions
After the GFA concession scheme is implemented, as in Hong Kong and Singapore, GB becomes a popular market practice. It is time to revisit the rationale underpinning the GFA concession scheme to assess the costs and benefits to industry and society. The clear and certain requirements in the GFA concession scheme reduce the approval and communication costs. The GFA concession scheme in Hong Kong encourages innovative designs, but it requires more time and effort from government, developers and architects in the approval process. This paper provides a theoretical

framework to analyze the extra costs and benefits for stakeholders participating in the GFA concession scheme. Extra costs comprise additional construction costs, administration costs, consultancy fees and financing costs, etc., together with TC considerations. Benefits consist of GFA concession, enhanced property value as well as the hidden benefits such as future business competitiveness, etc. A theoretical framework was established to test and explain the effectiveness of a concession scheme, which provides the foundation to find the optimum GFA concession in future studies.

Acknowledgements

The authors acknowledge the assistance of Dr Paul W. Fox on an earlier draft of this paper.

Funding

This paper was prepared with the support of a research fund from the Construction Industry Council (CIC) of the Hong Kong SAR Government [project account number K-ZJJR] and a RGC research grant from Hong Kong Polytechnic University. The first author is thankful for the Delft Technology Fellowship (2014) for its generous research support.

References

Abair, J. W. (2008). Green buildings: When it means to be green and the evolution of green building laws. *Urban Law, 40*, 623–632.

Addae-Dapaah, K., Hiang, L. K., & Sharon, N. Y. S. (2009). Sustainability of sustainable real property development. *Journal of Sustainable Real Estate, 1*(1), 203–225.

Ahn, Y. H., & Pearce, A. R. (2007). Green construction: Contractor experiences, expectations, and perceptions. *Journal of Green Building, 2*(3), 106–122. doi:10.3992/jgb.2.3.106

Andersen, M. S., & Sprenger, R.-U. (Eds.). (2000). *Market-based instruments for environmental management*. Cheltenham: Edward Elgar.

Arrow, K. (1969). *The limits of organization, the Fels lectures on public policy analysis*. New York: Norton.

Building and Construction Authority, Singapore (BCA). (2014). Retrieved January 16, 2016, from https://www.bca.gov.sg/greenmark/others/sg_green_buildings_tropics.pdf

Building and Construction Authority, Singapore (BCA). (2015). Retrieved January 16, 2016, from http://www.bca.gov.sg/greenmark/others/Green_Premium_Rates_(updated_Mar_2016).pdf

Building Department, Hong Kong. (2011). *New practice notes to foster a quality and sustainable built environment*. Retrieved January 16, 2016, from http://www.info.gov.hk/gia/general/201101/31/P201101310250.htm

Burnett, J., Chau, C., Lee, W., & Edmunds, K. (2008). Costs and financial benefits of undertaking green building assessments: Summary report of the CII-HK research project.

Cai, W., Wu, Y., Zhong, Y., & Ren, H. (2009). China building energy consumption: Situation, challenges and corresponding measures. *Energy Policy, 37*(6), 2054–2059. doi:10.1016/j.enpol.2008.11.037

Chai, K., & Yeo, C. (2012). Overcoming energy efficiency barriers through systems approach—A conceptual framework. *Energy Policy, 46*, 460–472. doi:10.1016/j.enpol.2012.04.012

Chan, E. H. W., Qian, Q. K., & Lam, P. T. I. (2009). The market for green building in developed Asian cities—the perspectives of building designers. *Energy Policy, 37*(8), 3061–3070. doi:10.1016/j.enpol.2009.03.057

Cheung, S. N. (1987). Economic organization and transaction costs. *The New Palgrave: A Dictionary of Economics, 2*, 55. Retrieved from http://www.dictionaryofeconomics.com/article?id=pde1987_X000658

Coggan, A., Buitelaar, E., Whitten, S., & Bennett, J. (2013). Factors that influence transaction costs in development offsets: Who bears what and why? *Ecological Economics, 88*, 222–231. doi:10.1016/j.ecolecon.2012.12.007

Coggan, A., Whitten, S. M., & Bennett, J. (2010). Influences of transaction costs in environmental policy. *Ecological Economics, 69*(9), 1777–1784. doi:10.1016/j.ecolecon.2010.04.015

Cole, R. (2015). *Understanding regenerative design challenging the orthodoxy of current green building practice*. Retrieved January 16, 2016, from https://www.reminetwork.com/articles/understanding-regenerative-design/

Council for Sustainable Development. (2010). *Building design to foster a quality and sustainable built environment*. Hong Kong: Council for Sustainable Development, Hong Kong Government.

Davis Langdon Australia. (2007). *The cost and benefit for achieving green buildings*. Retrieved September 14, 2016, from http://www.aecom.com/What+We+Do/Program+Cost+Consultancy/Davis+Langdon

Ducos, G., Dupraz, P., & Bonnieux, F. (2009). Agri-environment contract adoption under fixed and variable compliance costs. *Journal of Environmental Planning and Management, 52*(5), 669–687. doi:10.1080/09640560902958248

Dudek, D. J., & Wiener, J. B. (1996). *Joint implementation, transaction costs, and climate change*. Paris: OECD.

Fan, K., Qian, Q. K., Chan, E. H. W. (2015). *Floor area concession incentives as planning instruments to promote green building: A critical review of International practices*. Smart and Sustainable Built Environments (SASBE) conference 9–11 Dec, 2015, Pretoria, South Africa.

Feiock, R. C., Tavares, A. F., & Lubell, M. (2008). Policy instrument choices for growth management and land use regulation. *Policy Studies Journal, 36*(3), 461–480. doi:10.1111/j.1541-0072.2008.00277.x

Finon, D., & Perez, Y. (2007). The social efficiency of instruments of promotion of renewable energies: A transaction-cost perspective. *Ecological Economics, 62*(1), 77–92. doi:10.1016/j.ecolecon.2006.05.011

Fuerst, F., & McAllister, P. (2008). Green noise or green value? Measuring the price effects of environmental certification in commercial buildings. Working Papers in School of Real Estate & Planning, University of Reading, Reading, UK.

Furubotn, E. G., & Richter, R. (2005). *Institutions and economic theory: The contribution of the new institutional economics*. Ann Arbor: University of Michigan Press.

Hagemann, N., Prager, K., & Bartke, S. (2015). Costs of implementing agricultural soil protection policies—Insights from two German cases. *Journal of Environmental Policy & Planning*, 1–17. (ahead-of-print).

Häkkinen, T., & Belloni, K. (2011). Barriers and drivers for sustainable building. *Building Research & Information, 39*(3), 239–255. doi:10.1080/09613218.2011.561948

Hebb, T., Hamilton, A., & Hachigian, H. (2010). Responsible property investment in Canada: Factoring both environmental and social impacts in the Canadian real estate market. *Journal of Business Ethics, 92*(1), 99–115. doi:10.1007/s10551-010-0636-5

Hein, L. G., & Blok, K. (1995). *Transaction costs of energy efficiency improvement*. Paper presented at the Energy Efficiency Challenge for Europe. ECEEE summer study.

Hong Kong Green Building Council (HKGBC). (2015). Retrieved January 16, 2016, from https://www.hkgbc.org.hk/eng/BEAM PlusStatistics.aspx

Hwang, B.-G., & Ng, W. J. (2013). Project management knowledge and skills for green construction: Overcoming challenges. *International Journal of Project Management, 31*(2), 272–284. doi:10.1016/j.ijproman.2012.05.004

Hwang, B.-G., Zhao, X., & Tan, L. L. G. (2015). Green building projects: Schedule performance, influential factors and solutions. *Engineering, Construction and Architectural Management, 22*(3), 327–346. doi:10.1108/ECAM-07-2014-0095

Isa, M., Rahman, M. M. G. M. A., Sipan, I., & Hwa, T. K. (2013). Factors affecting green office building investment in Malaysia. *Procedia – Social and Behavioral Sciences, 105*, 138–148. doi:10.1016/j.sbspro.2013.11.015

Joas, F., & Flachsland, C. (2014). The (ir)relevance of transaction costs in climate policy instrument choice: An analysis of the EU and the US. *Climate Policy, 16*(1), 26–49.

Johnston, R. A., Schwartz, S. I., Wandesforde-Smith, G. A., & Caplan, M. (1989). Selling zoning: Do density bonus incentives for moderate-cost housing work. *Wash U Law Journal of Urban & Contemporary Law, 36*, 45–61.

Kats, G., Alevantis, L., Berman, A., Mills, E., & Perlman, J. (2003). *The costs and financial benefits of green buildings. A Report to California's Sustainable Building Task Force.* Report by Lawrence Berkeley National Laboratory, USA. Retrieved from http://www.usgbc.org/Docs/Archive/General/Docs1992.pdf

Kayden, J. S. (1978). *Incentive zoning in New York City: A cost-benefit analysis.* Cambridge, MA: Proquest Information & Le.

Küçükmehmetoğlu, M., & Büyükgöz, A. (2013). *Consensus building via cooperative game theory in the process of urban redevelopment.* ERSA conference papers, European Regional Science Association.

Lam, P. T., Chan, E. H., Poon, C., Chau, C., & Chun, K. (2010). Factors affecting the implementation of green specifications in construction. *Journal of Environmental Management, 91*(3), 654–661. doi:10.1016/j.jenvman.2009.09.029

Larsson, N., & Clark, J. (2000). Incremental costs within the design process for energy efficient buildings. *Building Research & Information, 28*(5–6), 413–418. doi:10.1080/096132100418573

LBNL. (2007). *Assessing transaction costs of project-based greenhouse gas emissions trading.* Berkeley, CA: Lawrence Berkeley National Laboratory.

Liu, Y., & Lau, S. (2013). *What is green policy? An anatomy of green policy through two Asian cities: Hong Kong and Singapore,* The World Sustainable Building (SB) Conference, Research Publishing, 323.

Love, P. E., Niedzweicki, M., Bullen, P. A., & Edwards, D. J. (2011). Achieving the green building council of Australia's world leadership rating in an office building in Perth. *Journal of Construction Engineering and Management, 138*(5), 652–660. doi:10.1061/(ASCE)CO.1943-7862.0000461

Matthews, R. C. (1986). The economics of institutions and the sources of growth. *The Economic Journal, 96*(384), 903–918.

McCann, L., Colby, B., Easter, K. W., Kasterine, A., & Kuperan, K. V. (2005). Transaction cost measurement for evaluating environmental policies. *Ecological Economics, 52*(4), 527–542. doi:10.1016/j.ecolecon.2004.08.002

Meacham, B. J. (2010). Accommodating innovation in building regulation: Lessons and challenges. *Building Research & Information, 38*(6), 686–698. doi:10.1080/09613218.2010.505380

Michaelowa, A., & Jotzo, F. (2005). Transaction costs, institutional rigidities and the size of the clean development mechanism. *Energy Policy, 33*(4), 511–523. doi:10.1016/j.enpol.2003.08.016

Miller, N., Spivey, J., & Florance, A. (2008). Does green pay off? *Journal of Real Estate Portfolio Management, 14*(4), 385–400.

Mundaca, T. L., Mansoz, M., Neij, L., & Timilsina, G. R. (2013). Transaction costs analysis of low-carbon technologies. *Climate Policy, 13*(4), 490–513. doi:10.1080/14693062.2013.781452

Nilsson, F. O. L. (2009). Transaction costs and agri-environmental policy measures: Are preferences influencing policy implementation? *Journal of Environmental Planning and Management, 52*(6), 757–775. doi:10.1080/09640560903083723

North, D. C. (1990a). *Institutions, institutional change and economic performance.* Cambridge, UK: Cambridge University Press.

North, D. C. (1990b). A transaction cost theory of politics. *Journal of Theoretical Politics, 2*(4), 355–367. doi:10.1177/0951692890002004001

Ocampo, J. A. (2011). *The transition to a green economy: Benefits, challenges and risks from a sustainable development perspective: Summary of background papers.* Report to Second Preparatory Meeting for United Nations Conference on Sustainable Development, Division for Sustainable Development UN-DESA, UNEP, UN Conference on Trade and Development.

Ofei-Mensah, A., & Bennett, J. (2013). Transaction costs of alternative greenhouse gas policies in the Australian transport energy sector. *Ecological Economics, 88*, 214–221. doi:10.1016/j.ecolecon.2012.12.009

Olubunmi, O. A., Xia, P. B., & Skitmore, M. (2016). Green building incentives: A review. *Renewable and Sustainable Energy Reviews, 59*, 1611–1621. doi:10.1016/j.rser.2016.01.028

Paetz, M. M. D., & Pinto-Delas, K. (2007). *From red lights to green lights: Town planning incentives for green building.* Talking and Walking Sustainability International Conference, February, 2007.

Pearce, A. R., DuBose, J. R., & Bosch, S. J. (2007). Green building policy options for the public sector. *Journal of Green Building, 2*(1), 156–174. doi:10.3992/jgb.2.1.156

Pivo, G., & McNamara, P. (2005). Responsible property investing. *International Real Estate Review, 8*, 128–143.

Qi, G. Y., Shen, L. Y., Zeng, S. X., & Jorge, O. J. (2010). The drivers for contractors' green innovation: An industry perspective. *Journal of Cleaner Production, 18*(14), 1358–1365. doi:10.1016/j.jclepro.2010.04.017

Qian, Q. K. (2010). Government measures in China for promoting Building Energy Efficiency (BEE): A comparative study with some developed countries. *The International Journal of Construction Management, 10*(4), 119–138. doi:10.1080/15623599.2010.10773157

Qian, Q. K. (2012). *Barriers to Building Energy Efficiency (BEE) promotion: A Transaction Costs (TCs) perspective* (PhD thesis). The Hong Kong Polytechnic University.

Qian, Q. K., Chan, E. H. W., & Bin Khalid, A. G. (2015b). Challenges in delivering green building projects: Unearthing the Transaction Costs (TCs). *Sustainability, 7*, 3615–3636. doi:10.3390/su7043615

Qian, Q. K., Chan, E. H. W., & Choy, L. H. T. (2013). How transaction costs affect real estate developers entering into the Building Energy Efficiency (BEE) market? *Habitat International, 37*, 138–147. Retrieved February 23, 2012, from http://www.sciencedirect.com/science/journal/aip/01973975 doi:10.1016/j.habitatint.2011.12.005

Qian, Q. K., Chan, E. H. W., Visscher, H. J., & Lehmann, S. (2015a). Modeling the Green Building (GB) investment decisions of developers and end-users with Transaction Costs (TCs) considerations. *Journal of Cleaner Production,* in Press. doi:10.1016/j.jclepro.2015.04.066

Rehm, M., & Ade, R. (2013). Construction costs comparison between 'green' and conventional office buildings. *Building Research & Information, 41*(2), 198–208. doi:10.1080/09613218.2013.769145

RIBA, Outline Plan of Work. (2007/08). Retrieved March 18, 2015, from http://www.architecture.com/Files/RIBAProfessionalServices/Practice/Archive/OutlinePlanofWork(revised).pdf

Rørstad, P. K., Vatn, A., & Kvakkestad, V. (2007). Why do transaction costs of agricultural policies vary? *Agricultural economics, 36*(1), 1–11. doi:10.1111/j.1574-0862.2007.00172.x

Simmons, R. (2015). Constraints on evidence-based policy: Insights from government practices. *Building Research & Information*, 43(4), 407–419. doi:10.1080/09613218.2015.1002355

Singh, G. (2009). *Understanding carbon credits*. New Delhi: Aditya Books.

Tang, B., & Tang, R. M. (1999). Development control, planning incentive and urban redevelopment: Evaluation of a two-tier plot ratio system in Hong Kong. *Land Use Policy*, 16(1), 33–43. doi:10.1016/S0264-8377(98)00035-0

United Nations, Department of Economic and Social Affairs, Population Division (UNDESA). (2014). World urbanization prospects: The 2014 revision, highlights (ST/ESA/SER.A/352).

United Nation Environment Programme (UNEP). (2007). *Buildings & climate change: Status, challenges and opportunities*. Retrieved from http://www.unep.fr/pc/sbc/documents/Buildings_and_climate_change.pdf

Walker, A., & Chau, K. W. (1999). The relationship between construction project management theory and transaction cost economics. *Engineering, Construction and Architectural Management*, 6(2), 166–176. doi:10.1108/eb021109

WBCSD. (2009). *Transforming the market: Energy efficiency in buildings world business council for sustainable development*. Geneva: World Business Council for Sustainable Development.

Williamson, O. E. (1981). The economics of organization: The transaction cost approach. *American Journal of Sociology*, 87, 548–577. doi:10.1086/227496

Williamson, O. E. (1985). *The economic institutions of capitalism*. New York, NY, USA: The Free Press.

World Bank. (2005). *Project document on a proposed Global Environment Facility (GEF) grant of US$ 18 million to the PRC for the heat reform and Building Energy Efficiency Project* (Report No. 20747-cn), 1–9.

Yu, S. M., & Tu, Y. (2011). *Are green buildings worth more because they cost more?* IRES Working Paper Series IRES2011-023.

Endnotes

[1]The 'utility platform' is a unique feature in Hong Kong high-rise living as there is often no outdoor or semi-outdoor space for activities such as washing or drying laundry. The utility platform is designed for residential buildings, providing space for residents to wash and dry clothes. It is often in the form of a balcony, but has different design restrictions of size, location and GFA concession. For example, the maximum area to be exempted for a utility platform is 0.75 m^2, but the total area should not be less than 1.5 m^2. For the balcony, the maximum GFA concession is 3 m^2.

[2]The Hong Kong Green Building Council (HKGBC) is responsible for GB promotion, leading the GB market transformation. BEAM Plus is the product of integrating HKBEAM and the Comprehensive Environmental Performance Assessment Scheme (CEPAS).

The new governance for low-carbon buildings: mapping, exploring, interrogating

Jeroen Van der Heijden

Acknowledging the limitations of traditional, mandatory governance instruments (building codes, planning legislation) to achieve low-carbon buildings, governments, firms and other organizations have been experimenting with alternatives. This trend has become known as the 'new governance'. This paper brings together 50 new-governance instruments to understand better this new governance for low-carbon buildings, and what may be expected from it. It finds that new-governance instruments fall short in exactly the same areas as do traditional instruments. It argues for a change in the application of new-governance instruments along three paths to improve their performance.

Introduction

Traditional building codes and planning legislation are ill-suited to accelerate the transition to low-carbon buildings. The main problems relate to the long time it takes to develop buildings codes and planning legislation; the tendency to exempt existing buildings from complying with new or mandated codes and legislation; and the difficulty of addressing user behaviour through them. In response, governments, firms and other organizations have been trialling alternative and complementary governance instruments for some decades now. Examples include the certification and classification of buildings, new forms of financing, and innovative ways of generating and disseminating information. The larger policy experiment of trialling alternative governance instruments has become known as 'new governance' (Holley, Gunningham, & Shearing, 2012; Wurzel, Zito, & Jordan, 2013).

This new governance fits a broader transition from traditional state-led direct regulatory interventions to governance approaches that allow for a broad inclusion of non-state stakeholders and the use of less-coercive regulatory instruments. Often new-governance instruments blend command-and-control-type approaches with newer voluntary mechanisms, and they are normally added as complements to existing regulatory frameworks. But what exactly does this new governance for low-carbon buildings look like around the globe? Given the popularity of new governance for low-carbon building development and transformation, it is of relevance to understand the scope of the new-governance instruments' designs, why they were introduced, how they perform and whether better outcomes may be expected of them than of traditional building codes and planning legislation.

This paper seeks to map, explore and interrogate this new governance for low-carbon buildings by drawing together insights from 50 new-governance instruments from Australia, Asia, Europe and North America. It finds that these are predominantly applied at the very top end of the construction and property sector; that their uptake is limited, particularly in the area of existing buildings; and that they have a strong focus on building technology but appear less interested in changing the behaviour of building users.

Complications of governing low-carbon buildings

The construction, maintenance and use of buildings account for 40% of global energy consumption and 35% of global carbon emissions – of which 80% relates to the operational phase for heating, cooling and ventilating buildings as well as for operating appliances. Building-related energy consumption is split roughly evenly between the residential and commercial property sectors (IPCC, 2014; US Energy Information Administration, 2013 – there is some debate about these numbers; for detailed discussions, see Crawford & Stephan, 2013; Majeau-Bettez, Strømman, & Hertwich, 2011). In short, buildings and how occupants use them contribute considerably to the process of human-induced climate change.

At the same time buildings hold a considerable mitigation potential. Technologies and design solutions are available that allow for cost-effective reductions of carbon emissions of 30–80% (IPCC, 2014; Mumovic & Santamouris, 2013). Likewise, much knowledge is available on how changes to building user behaviour can reduce energy consumption and related carbon emissions (Cabinet Office, 2011). The question is, then, how does one ensure that technology, design solutions and insights on behavioural change are taken up on a large scale and in a timely manner?

An obvious answer to that question would be: let governments introduce mandatory requirements. Governments have, after all, a long history of regulating buildings through building codes and planning legislation, and these traditional governance instruments have contributed to a relatively safe and healthy built environment – albeit these privileges are not shared equally around the world (Taylor, 2013). And indeed, governments have sought, and to some extent achieved, improved building energy efficiency and reduced building carbon intensity through buildings codes since the 1970s (IEA, 2013). Unfortunately building codes and planning legislation come with a number of constraints that hamper a rapid and large-scale transition towards low-carbon buildings. To name a few (UN, 2014; UNEP, 2007; World Bank, 2011):

- It often takes a long time to develop and implement mandatory requirements due to the many checks and balances required to ensure democratic accountability and transparency. This sometimes results in situations where mandatory requirements cannot keep up with technological developments and hamper their uptake.

- Proposals for change of mandatory regulation easily become politicized due to vested interests and sunk-costs in the construction and property sectors. Policy-makers take high risks in proposing ambitious (amendments to) mandatory requirements for low-carbon buildings, and if they do they often face a long process of lobbying and being lobbied.

- The development and implementation of mandatory requirements require substantial institutional capital – knowledgeable policy-makers, bureaucrats to process regulations, inspectors to assess building plans and construction work, and so on. Particularly in rapidly developing economies, such institutional capital is often found to be lacking. This is problematic because this is where the fastest growth of the built environment is expected and where mandatory requirements currently are lowest. That being said, the enforcement of building energy-efficiency codes is also problematic in developed economies because inspectors prioritize classic regulated areas (such as structural safety and healthiness) over the relatively novel area of resource efficiency and carbon intensity.

- New and amended mandatory requirements often exempt existing buildings from compliance – a process known as 'grandfathering'. This is problematic because today's built environment already contributes to unsustainable levels of carbon emissions. Grandfathering clauses reduce the impact of amended and new mandatory instruments, which is particularly challenging for developed economies. In these economies, the built environment transforms by at best 2% per year – implying that 70% of today's buildings will be still in use in 2050.

- Finally, mandatory requirements address objects – building parts, buildings, precincts – and not the behaviour of people using buildings. It would be unheard of for a government to require its citizens to wear an extra jumper in winter when they feel cold instead of turning up the heating. User behaviour is, however, a key aspect to reduced energy consumption and carbon emissions of the built environment, as discussed above.

A turn to new governance for low-carbon buildings

These problems are acknowledged by governments, firms and other organizations. Since the early 1990s they have been trialling alternative governance instruments, often as substitutes or compliments to traditional ones. Experimenting with alternative governance instruments is not unique to the area of low-carbon buildings. It fits a logical development in an ongoing philosophy of deregulation, government reforms and a larger shift in rethinking the role of government in governing society (DeLeon, Rivera, & Manderino, 2010; Sabel & Zeitlin, 2011). This 'new governance'[1] is particularly strong in the area of environmental and resource sustainability (Holley et al., 2012; Wurzel et al., 2013). At least three related trends distinguish this new governance from earlier approaches to governance. First, a shift away from sole-government authority in governing environmental problems towards the involvement of public and private sector stakeholders. Second, an interest in governance instruments that encourage self-organization, market solutions or both as substitutes for or complements to mandatory command-and-control-style instruments. Third, a shift towards instruments that reward voluntary compliance as opposed to enforcing mandated behaviour (for a comprehensive review of the new governance literature, see Van der Heijden, 2013).

This paper seeks to understand better this shift towards new governance for low-carbon building development and transformation. It brings together insights from 50 real-world new-governance instruments from around the globe. The author has studied these as part of a larger multi-year project that seeks to understand better new forms of governance for urban sustainability (Van der Heijden, 2014). Whilst the number of instruments examined here is substantial, by no means is it claimed that it is a perfectly representative picture of all new governance activity for low-carbon buildings around the world. The set is comprehensive enough to provide a window on the larger policy experiment of trialling these new-governance instruments that allows addressing the questions introduced above (for a comparable research design to study new-governance instruments, see Hoffmann, 2011).

Instruments were initially identified through an extensive Internet search using keywords such as 'sustainable development AND [country]', 'sustainable building AND [country]', 'green building AND [country]', 'sustainable construction AND [country]' and 'green construction AND [country]'. In addition, social media was used (predominantly, sustainable and 'green' building groups on LinkedIn) along with the author's network of policy-makers, administrators, architects, engineers, constructors, developers, investors etc. in Australia, India, Malaysia, the Netherlands, Singapore and the United States to gain additional information about potential cases identified in this Internet search. This network was further explored for additional potential cases to study. The 50 instruments discussed here were selected from the larger pool of instruments uncovered by the author based on two selection criteria. The first relates to case variance and coverage. This article seeks to present the richness of new-governance instrument designs for low-carbon buildings. As such, cases are selected to represent the breadth of instrument design. At the same time, some general patterns are highlighted that emerge from this research in terms of instrument design. This relates to the four general types of instruments that are discussed in what follows. Instruments were selected from the larger pool to ensure that every type is illustrated with sufficient examples. The second selection criterion relates to the availability and quality of information about the cases studied. This follows Bent Flyvbjerg's (Flyvbjerg, 2015) notion of information-oriented case selection, which implies that cases are selected to maximize the utility of information from a relatively small numbers of cases. The pool of 50 instruments presents narratives of individual examples and types, as well as a larger narrative of the opportunities and constraints of new-governance instruments for low-carbon buildings.

Relevant data for analysing the instruments were obtained from websites, existing reports and other documented sources. Novel data were obtained through a series of in-depth face-to-face elite interviews undertaken between 2012 and 2014 (with a number of follow-up interviews in 2015). These interviews aimed to fill in gaps in the data from other sources, to resolve conflicts in data from other sources, and to gain additional insight into the instruments under scrutiny. Interviewees were traced through additional Internet searches and through social-network websites, again particularly LinkedIn. This resulted in a pool of over 200 interviewees from various backgrounds, including policy-makers, administrators, architects, engineers, constructors, developers and investors. The data were processed by means of a systematic coding scheme and qualitative data analysis software (Atlas.ti). By using this approach the data were systematically explored to gain an understanding of the performance of the instruments studied. That said, the insights provided here were predominantly built on data obtained from websites, existing reports and other documented sources (for a more extensive discussion of the methodology underlying this research project, see Van der Heijden, 2014, pp. 165–175).

There is not enough space here to discuss each of these 50 instruments in depth. Therefore, the dominant types of instruments are introduced and examples provided for illustrative purposes. The Appendix in the supplemental data online provides a brief description of each instrument (see Table A1) and a summary of the main characteristics of the instruments (see Table B1); for each instrument a hyperlink is provided for interested readers to follow up on (see Table A1).[2] It also goes without saying that the boundaries of the different instrument types overlap. The types that are introduced in what follows are ideal types. Real-world instruments might have characteristics of more than one ideal type.

Certification and classification instruments

The dominant type of new-governance instruments for low-carbon buildings is certification and classification (Cole & Valdebenito, 2013; Fowler & Rauch, 2006). In general, such instruments allow for the assessment of buildings against a number of criteria. If these are met, then a certificate is issued to indicate compliance. Its application for low-carbon buildings comes with a specific twist: classification (Pérez-Lombard, Ortiz, González, & Maestre, 2009). Classification allows for an indication of the relative performance of buildings within the same certification instrument. Three forms of classification are identified (Pérez-Lombard et al., 2009):

- *Benchmarking*
  Indicates that a building, building part or even building users meet the rules of that instrument. An example is Eco-Office in Singapore; an instrument that assesses office users' energy, water and paper consumption against a baseline and issues a certificate if that baseline is met. The instrument does not distinguish based on different classes of certification.

- *Rating*
  Uses the relative performance of a building within the set of buildings certified to classify the building within that set. Identifiers such as stars or colours are often used to indicate this relative performance – the higher its relative performance compared with other certified buildings, the higher the number of stars awarded, indicating the specific class of certification. An example is the National Australian Built Environment Rating System (NABERS), which assesses the energy performance or water performance of buildings, and issues certificates in six classes using a star rating to highlight differences among certified buildings. If a building is certified for energy and water performance, two different certificates are issued.

- *Labelling*
  Builds on a holistic approach to classification. It does not certify energy efficiency or water consumption or carbon intensity individually, but seeks to classify buildings based on overall performance. Well-known instruments such as the Building Research Establishment Environmental Assessment Method (BREEAM) and the Leadership in Energy and Environmental Design (LEED) award credits for different sustainability credentials, and the total number of credits awarded is the basis for the class of certification.

Differences in classification might raise information barriers, rather than reduce them. Various studies highlight that it is next to impossible to compare the performance of distinct certification and classification instruments simply because they are built on a different form of classification, or because the rules underlying the classification system are so diverse that a similar building could achieve a low class of certification in one instrument, but a high class of certification in another (Roderick, McEwan, Wheatley, & Alonso, 2009; Van der Heijden, 2015).

To complicate things further, different forms of certification exist: 'as designed', 'as built' and 'in operation'. The first form certifies the expected (*i.e.* hypothecated) performance of a building design; the second form certifies the performance of a building built in compliance with that design; and the final form certifies the achieved performance after a specified period of use – it is often subject to periodical renewal. The first two forms are dominant; the latter form was developed in response to critiques of these two: buildings are often not built in compliance with their (certified) design; during construction many flaws might be made that instrument administrators or their inspectors do not notice; or operator and user behaviour may undo the low-carbon credentials of a building – all of which could result in a certified design or completed building not meeting its expected performance. Such problems are often reported in the literature, as are problems of developers and property owners gaming the instruments by seeking easy but not necessarily low-carbon solutions to achieve high classes of certification (Scofield, 2013).

Certification and classification instruments have, thus far, achieved most promising results in the area of high-profile new commercial building development (such as offices in central business districts of major cities; Van der Heijden, 2015). It is here where developers and property owners expect (and get) high returns on their investments. These instruments have thus far been less successful in changing the market for residential buildings and existing buildings – but they also have not had a transformative impact in the area of less prestigious commercial building development (Yudelson & Meyer, 2013). Moreover, when looked at in relative numbers the overall performance of this type of instruments is not too promising: even one of the best-performing instruments in the world, LEED in the United States, has over the course of 15 years certified fewer than 4% of all new commercial property developed in that period.[3]

Information generation and dissemination instruments

A second type of instrument seeks to generate and disseminate knowledge on how to construct and retrofit low-carbon buildings, and how building users' behaviour can be modified to achieve reduced energy consumption. A typical example is Green Lights in the United States, implemented by the US Environmental Protection Agency (USEPA) in the early 1990s (USEPA, 1994). Through Green Lights the USEPA sought to overcome initial resistance to and unfamiliarity with energy-efficient lightning. It made visible to building users the ease of reducing energy consumption and supported them in generating knowledge relevant for running their business: cost savings. Green Lights participants committed to installing energy-efficient lighting in 90% of their facilities – but only where this was profitable to do so. In return, the USEPA provided participants with tools (software predominantly) to carry out assessments and keep track of energy savings, helped them connect with lighting retrofitting services, and pointed out potential funding opportunities. Green Lights has witnessed a wide uptake throughout the United States, making it easier for participants to reduce operation costs and to brand themselves as environmentally aware (Moon, Bae, & Jeong, 2014; Videras & Alberini, 2000). It is now incorporated in Energy Star Buildings, a certification and classification instrument.

Besides supplying information directly from instrument-administrators to instrument-participants some instruments challenge participants to generate information on how to develop, retrofit and use low-carbon buildings and share this information with instrument-administrators and other participants (Gollagher & Hartz-Karp, 2013). The Better Building Partnership in Sydney's central business district is a clear example. It brings together Sydney's 14 major property owners and the city government and aims for significant building-related carbon emission reductions through, among others, building energy retrofits. The combined building stock of these property owners is responsible for 50% of all carbon emissions of the business district. Under the partnership the property owners and city government seek to reduce carbon emissions by 70% by 2030. The partnership was launched in 2011, and by 2015 it was already halfway towards its envisioned target (Better Buildings Partnership, 2015). By working closely together and sharing knowledge on how they have upgraded their individual property, all participating property owners gain from each other's experiences.

Information generation and dissemination instruments appear most promising in exactly these two situations: broad instruments that set low and relatively easy requirements and provide considerable (financial) gains for participants, and elite instruments that set challenging requirements and provide considerable collective gains for participants (Van der Heijden, 2016). In other situations less successful outcomes are reported. The United States-based Better Buildings Challenge, for example, is an instrument that seeks to incentivise participants to reduce the energy consumption of their offices by 20% over a 10-year period and to share their experiences. In return they are supported by the US Department of Energy to acquire the funds needed to retrofit their buildings. This instrument was also launched in 2011, and by 2015 some 32,000 buildings were participating. Whilst this number appears impressive, it reflects only 0.5% of the close to 6 million commercial buildings in the United States.[4]

Financing instruments

A third type of new-governance instruments for low-carbon buildings are particularly concerned with financing. Property developers and property owners often face difficulties in obtaining funds for the development or retrofitting of low-carbon buildings. Banks and other fund providers are concerned that the additional costs that (may) come with developing low-carbon buildings will not be represented in these buildings' future market value, and fear that lenders will not be able to pay back loans provided. The idea that future financial gains of low-carbon buildings – reduced operation costs, among others – improve these owners' ability to pay back loans does not fit current business models based on cash flows from production – and not reduced consumption – and growth (World Bank, 2011). Likewise, building owners might be concerned they will not see a return on their investments in low-carbon buildings because they do not own a property long enough, or they might not value long-term gains over short-term costs – known as hyperbolic discounting (Ameli & Brandt, 2015).

A variety of instruments addresses these issues. Revolving loan funds are one of these. Revolving loan funds for low-carbon buildings often consist of money lent to property developers and property owners to fund low-carbon building solutions. The assumption is that these solutions result in reduced operation costs of buildings, allowing owners to pay back the loan. Once the fund has recouped the money it can lend it again to others (Boyd, 2013). The Billion Dollar Green Challenge is one of the most ambitious examples of revolving loan funds. It challenges educational organizations, predominantly universities and colleges, in the United States to invest a total of US$1 billion in self-managed revolving funds to finance energy-efficiency upgrades of their buildings. By participating in the challenge these organizations are supported in managing their funds, and share information on how to carry out energy-efficiency upgrades (Green, 2013). The instrument was launched in 2011, and by 2015 it has attracted 52 participating organizations, who together have committed US$114 million to the challenge – again not outstanding performance, keeping in mind there are about 5300 universities and colleges in the United States.[5]

Other financing instruments are built on climate bonds or related forms of tripartite financing. The instrument 1200 Buildings in Melbourne, for example, builds on tripartite financing. As of 2009, it allows the city government to act as a 'middle person' between finance providers and commercial property owners who seek funds for building retrofits. The city government lends funds and supplies these to building owners – who commit to reduce their buildings' carbon emissions by at least 38% – and recoups the loan through a special property tax. In doing so the city government takes away the risk experienced by the finance provider. A comparable example is Property Assessed Clean Energy (PACE) launched in 2008 in the United States. The instrument allows (local) governments to issue bonds to fund building retrofits and supply funds to property owners. Funds are, again, recouped through a property tax. The instrument initially sought to finance retrofits of commercial and residential buildings, but the subprime mortgage crisis has dealt a major blow. US mortgage authorities no longer finance mortgages for upgrades of residential buildings under the instrument. Neither instrument has achieved impressive results yet: by 2015, fewer than 50 buildings were participating in 1200 Buildings, and PACE has resulted in some 350 retrofitted commercial buildings throughout the United States.

It may very well be that 'just' supporting property owners in finding financing is not the answer to the more fundamental question why property owners do not want to retrofit their existing buildings or demand low-carbon new buildings in the first place. Another example of a financing instrument is illustrative here. In 1995 the US Federal Housing Administration implemented the Energy Efficient Mortgage Program. The programme allows (future) homeowners to borrow additional funds to improve the energy-efficiency of their (future) home. Homeowners are, however, hardly interested in the programme: in its best year thus far, 2011, of close to 15 million mortgages issued throughout the United States a mere 1065 – about one in 15,000 – was issued under the programme (Federal Housing Administration, 2014). Homeowners are found reluctant to undertake energy-efficiency retrofits of their homes, and those who are interested often have funds for retrofits or consider the paperwork and other related administrative efforts too much hassle for the relatively small (additional) mortgage provided through the programme (Kolstad, 2014).

Accelerators and bridging instruments

A fourth and final type of instruments seeks to accelerate the uptake of the earlier ones discussed, and seeks to bridge different instruments (traditional and new ones) aiming for synergies that make the whole of

governance instruments for low-carbon buildings larger than the sum of its parts. Examples of accelerators are a range of incentives in place that seek to speed up the application of certification and classification instruments such as BREEAM and LEED. Some counties and cities in North Carolina give density bonuses, for example, to builders who comply with LEED criteria, and other states have in place tax incentives for construction work that receives LEED certification. Likewise, the Dutch government has in place Sustainable Public Procurement criteria that require that future government buildings or office space needs to be BREEAM certified or meet standards comparable with BREEAM certification. Other examples are the City of New York's 1000 Superintendents and the Australian Supply Chain Sustainability. Both instruments seek to (re)educate specific actors in the construction and property sectors about the advantages of low-carbon buildings. 1000 Superintendents builds on the assumption that through interactions with property owners and tenants, superintendents might be able to change their mindset about low-carbon buildings and ways to reduce building energy consumption.

To give an example of a bridging instrument: the participants of the Better Building Partnership in Sydney are experimenting with green leases as a bridge between property owners participating in new-governance instruments and tenants participating in other instruments. A green lease is, like any lease, an agreement between a property owner (landlord) and a tenant, but it includes specific clauses about both parties' responsibilities for low-carbon and other (environmental) sustainability credentials of a building. What participants of the partnership realized is that property owners often need the commitment of their tenants to use a low-carbon building in a specific manner to maintain, for instance, a high level of 'in operation' certification under a certification and classification instrument. At the same time tenants might need their landlord to make changes to their building when they participate in an instrument such as Eco-Office in Singapore – after all, there is only so much they can do through changed user behaviour in terms of energy reduction. Without commitment from the other party, landlords or tenants might refrain from participating in these instruments. By promoting green leases throughout Australia, the participants of the partnership hope these problems can be overcome (Blundell, 2014).

Four new-governance trends

From exploring and mapping new-governance instruments for low-carbon buildings, it has become clear there is no shortage of these instruments. They come in a wide range of designs and are implemented in a variety of settings (see further in the supplemental data online). At first glance they appear hopeful complements to traditional, mandatory governance instruments such as building codes and planning legislation because they are often tailored to a specific problem, move beyond the one-size-fits-all approach of traditional instruments, and include the relevant actors to address that problem and utilize their tacit knowledge. That being said, the 50 instruments studied point to four trends that raise questions about the overall ability of new-governance instruments to accelerate a rapid and large transition to low-carbon buildings (see also Table B1 in the supplemental data online):

- Whilst the number of these new-governance instruments is vast, their overall impact is limited. Even the most widely applied instruments, certification and classification instruments such as LEED and BREEAM have achieved a marginal uptake in their potential market at best. This holds for the vast majority of instruments studied. The majority of instruments has not yet been able to move beyond the absolute leaders in the construction and property sectors.

- The majority of instruments studied have a sole focus on or dominant application in the high end of the commercial property market: office buildings in central business districts of major cities and government-owned and -leased property. This holds particularly for certification and classification instruments. Whilst these instruments allow for certification of other building types, including residential buildings, they face difficulties penetrating these other markets (Cole & Valdebenito, 2013). There appears a clear logic to why new-governance instruments are dominant in the high end of the commercial property market: these are the buildings commissioned and leased by relatively large firms to whom leadership matters, who have in place social corporate responsibility policies, and who have to justify their behaviour to internal and external stakeholders. They are more likely to demand low-carbon buildings and are more willing to pay a premium for these than smaller firms and households that do not face such pressures (Dixon, Ennis-Reynolds, Roberts, & Sims, 2009).

- Whilst the instruments studied focus evenly on new and existing buildings, the uptake of instruments with a focus on existing buildings is worse than for those with a focus on new buildings. Two issues stand out. First, even when a landlord can acquire funds to retrofit a building, she might be held back in doing so because of the typical split-incentive problem – the gains of the retrofit come to the tenant, but the landlord has to bear the costs (Hakkinen & Belloni, 2011). Green leases could help landlords overcome this problem, but financing instruments often have a sole focus on financing retrofits – 1200 Buildings in Melbourne and PACE in the United States are typical examples. Second, particularly for small and medium-sized firms and households the process of going through a retrofit might be too much trouble to make up for the longer-term gains. This 'hassle factor' is not addressed by the instruments studied (cf. Cabinet Office, 2011).

- A final trend that stands out is that the majority of instruments studied seek to reduce the carbon intensity and energy consumption of buildings through technology. Only a few seek improvements through behavioural change of building users. The research did not suggest a clear explanation for this trend. It could be that a larger paradigm of ecological modernization, popular in many of the countries that provide contexts for these instruments, skews governments, firms and other organizations towards thinking in terms of technological solutions instead of behavioural ones (Hayden, 2014). Another explanation is that most instruments focus on 'per square metre' performance of a building, and not on a 'per capita' basis (Janda, 2011; Stephan & Crawford, 2014).[6] A less prosaic explanation would be that firms producing technological solutions drive the new governance agenda because they have a strong interest in seeing this type of instrument implemented. Future research may wish to explore this area further.

Conclusions and discussion: beyond the limitations of new governance for low-carbon buildings

It goes without saying that the findings presented should be considered in the light of the approach taken to the research. The 50 instruments studied built on a stratified sample. It is likely that a broader variety of instruments may be found than the four ideal types discussed here. The paper is further biased towards discussing instruments from Australia and the United States – roughly 60% of instruments discussed are from these two countries (see Table A1 in the supplemental data online). Studies from other countries may result in different findings. Also, this article has largely excluded the impact of instrument contexts from the analysis, and some of the generalizations presented may not hold in specific instrument contexts. This is, ultimately, an issue for further research.

That having been said, the new governance for low-carbon buildings – illustrated by the sample of 50 instruments – appears most promising in the high end of new commercial property development, and has limited impact in other areas of the construction and property sectors. It has achieved marginal results in the areas of residential buildings, existing buildings and building user behaviour. This is problematic because these are the exact areas where traditional, mandatory governance instruments such as building codes and planning legislation also fail to deliver results.

Does this imply the new governance for low-carbon buildings should be considered a failed policy experiment? I think not. It has provided leaders in the construction and property sectors a context and means to showcase the possibilities in terms of low-carbon buildings. A wealth of best practices is now available that evidence the possibilities to design, construct and use buildings in ways that result in considerable lower carbon emissions than conventional practice. These buildings and the media attention generated by new-governance instruments help to change norms in the construction and property sectors and help to make low-carbon buildings less alien (Yudelson & Meyer, 2013).

A shortcoming of new governance for low-carbon buildings, as a larger policy experiment, is that the focus has predominantly been on generating and showcasing leadership in the construction and property sectors. Time and again the mission statements of the instruments stress the need of illustrating that low-carbon buildings that move well beyond conventional practice are possible – whether at net-cost benefit, with existing technology, within existing legal settings, through changed behaviour, or otherwise (see the various links to webpages in the supplemental data online). What the instruments have in general failed to do, however, is think beyond attracting leaders and explore how they can move from leaders to other players in the construction and property sectors. At the end of the day it is relevant that the masses make a change, and not the leaders only.

How then can new-governance instruments be used to accelerate a large-scale transition to low-carbon buildings? Three complementary paths are posited. The first is to rethink their purpose. The current generation of instruments is preoccupied with generating and showcasing leadership, a second generation of instruments might wish to focus on taking the next step: moving from leaders to other players. Future research may seek to understand better why other players are less enthusiastic about the new-governance instruments. Do they face different barriers than leaders? Are they less interested than leaders? Do they need different incentives than leaders? Understanding and addressing these questions should be the ambition of a next generation of new-governance instruments.

The second path is to rethink these instruments' position in building regulatory frameworks at local, national and international levels. Most of the current instruments are voluntary in nature – after all, they built on the notion of incentivising leadership through positive rewards and not by enforcing leadership through penalties. A future generation of instruments might wish to move more to the mandatory space, or at least become less voluntary. Future research and future new-governance instruments may wish to build more on the state of the art in the current governance literature. For example, what would happen if these instruments move from voluntary opting in to voluntary opting out (*cf*. Kosters & Van der Heijden, 2015)? This may improve the uptake of instruments such as the Energy Efficiency Mortgages in the United States. What would happen if participation becomes mandatory, but the level of performance remains voluntary? This resembles the design of the certification and classification instruments NABERS in Australia and Green Mark in Singapore, both showing considerable uptake and performance when compared with related instruments such as LEED in the United States and Green Star in Australia.[7]

The third path is to think more carefully about the interaction between new-governance instruments and existing policy mixes. Where can they contribute or fill in gaps? Can positive synergies between new-governance instruments and existing regulation be designed, and, if so, how? Would it even be possible to introduce new-governance instruments with the aim of ultimately replacing existing regulation and legislation? A challenging design would be to think of new-governance instruments as part of rolling rule regimes (*cf*. Sabel, Fung, Karkkainen, Cohen, & Rogers, 2000). Under such a regime, today's leading practice would be the future's mandatory bottom line. This still rewards leaders as they have the opportunity to set the future bottom line and have the experience of achieving it, but it would actively push others to follow suit within the time frame set.

Acknowledgements

The article builds on a research project that is discussed in more detail in Van der Heijden (2014). The current article presents the main findings from this larger project, and expands on them by presenting a number of ideas that may help policy-makers to move beyond the limitations of new governance for low-carbon buildings. Many thanks to three anonymous reviewers for suggested changes to an earlier version of the manuscript. Special thanks to Shane Chalmers for proofreading an earlier version of this article.

Disclosure statement

No potential conflict of interest was reported by the author.

Supplemental data

Supplemental data for this article can be accessed at 10.1080/09613218.2016.1159394.

References

Ameli, N., & Brandt, N. (2015). *What impedes household investment in energy efficiency and renewable energy?* Paris: OECD Publishing.

Better Buildings Partnership. (2015). *Better buildings partnership. Annual Report 2013–2014.* Sydney: City of Sydney.

Blundell, L. (2014). *The tenants & landlords guide to Happiness.* Sydney: The Fifth Estate/Better Buildings Partnership.

Boyd, S. (2013). Financing and managing energy projects through revolving loan funds. *Sustainability: The Journal of Record, 6*(6), 345–352. doi:10.1089/SUS.2013.9826

Cabinet Office. (2011). *Behaviour change and energy use.* London: Cabinet Office.

Cole, R., & Valdebenito, M. J. (2013). The importation of building environmental certification systems: International usages of BREEAM and LEED. *Building Research & Information, 41*(6), 662–676. doi:10.1080/09613218.2013.802115

Crawford, R., & Stephan, A. (2013). *The significance of embodied energy in certified passive houses.* ICCBM 2013: International Conference on Construction and Building Materials, Copenhagen, 13–14 June, 473–479.

DeLeon, P., Rivera, J., & Manderino, L. (2010). Voluntary environmental programs: An introduction. In P. DeLeon & J. Rivera (Eds.), *Voluntary environmental programs* (pp. 1–10). Plymouth: Lexington Books.

Dixon, T., Ennis-Reynolds, G., Roberts, C., & Sims, S. (2009). Is there a demand for sustainable offices? An analysis of UK business occupier moves (2006–2008). *Journal of Property Research, 26*(1), 61–85. doi:10.1080/09599910903290052

Federal Housing Administration. (2014). *Annual management report: Fiscal year 2014.* Washington, DC: US Department of Housing and Urban Development.

Flyvbjerg, B. (2015). *Making social science matter* (17th ed.). Cambridge, Cambridge University Press.

Fowler, K. M., & Rauch, E. M. (2006). Sustainable Building Rating Systems. Retrieved from http://www.pnl.gov/main/publications/external/technical_reports/PNNL-15858.pdf

Gollagher, M., & Hartz-Karp, J. (2013). The role of deliberative collaborative governance in achieving sustainable cities. *Sustainability, 5*(6), 2343–2366. doi:10.3390/su5062343

Green Billion. (2013). *The billion green dollar challenge.* Retrieved from http://greenbillion.org/

Hakkinen, T., & Belloni, K. (2011). Barriers and drivers for sustainable building. *Building Research & Information, 39*(3), 239–255. doi:10.1080/09613218.2011.561948

Hayden, A. (2014). *When green growth is not enough: Climate change, ecological modernization, and sufficiency.* Quebec: McGill-Queen's University Press.

Hoffmann, M. (2011). *Climate governance at the crossroads.* Oxford: Oxford University Press.

Holley, C., Gunningham, N., & Shearing, C. (2012). *The new environmental governance.* London: Routledge.

IEA. (2013). *Modernising building energy codes.* Paris: United Nations Development Programme.

IPCC. (2014). *Climate change 2014: Impacts, adaptation, and vulnerability.* Cambridge: Cambridge University Press.

Janda, K. (2011) Buildings don't use energy: People do. *Architectural Science Review, 54*(1), 15–22. doi:10.3763/asre.2009.0050

Kolstad, L. (2014). *Designing a mortgage process for energy efficiency.* Washington, DC: American Council for an Energy-Efficient Economy.

Kosters, M., & Van der Heijden, J. (2015). From mechanism to virtue: Evaluating Nudge theory. *Evaluation, 21*(3), 276–291. doi:10.1177/1356389015590218

Majeau-Bettez, G., Strømman, A. & Hertwich, E. (2011) Evaluation of process- and input–output-based life cycle inventory data with regard to truncation and aggregation issues. *Environmental Science & Technology, 45*(23), 10170–10177. doi:10.1021/es201308x

Moon, S.-G., Bae, S., & Jeong, M.-G. (2014). Corporate sustainability and economic performance: An empirical analysis of a voluntary environmental program in the USA. *Business Strategy and the Environment, 23*(8), 534–546. doi:10.1002/bse.1800

Mumovic, D., & Santamouris, M. (2013). *A handbook of sustainable building design and engineering.* Abingdon: Routledge.

Pérez-Lombard, L., Ortiz, J., González, R., & Maestre, I. R. (2009). A review of benchmarking, rating and labelling concepts within the framework of building energy certification schemes. *Energy and Buildings, 41*(3), 272–278. doi:10.1016/j.enbuild.2008.10.004

Roderick, Y., McEwan, D., Wheatley, C., & Alonso, C. (2009). *A comparative study of building energy performance assessment between LEED, BREEAM and Green Star schemes.* Paper presented at the Building Simulation 2009, Glasgow.

Sabel, C., Fung, A., Karkkainen, B., Cohen, J., & Rogers, J. (2000). *Beyond backyard environmentalism.* Boston: Beacon Press.

Sabel, C., & Zeitlin, J. (2011). Experimentalist governance. In D. Levi-Faur (Ed.), *The Oxford handbook of governance* (pp. 169–185). Oxford: Oxford University Press.

Scofield, J. (2013). Efficacy of LEED-certification in reducing energy consumption and greenhouse gas emissions for large New York City office buildings. *Energy and Buildings, 67,* 517–524. doi:10.1016/j.enbuild.2013.08.032

Stephan, A. & Crawford, R. (2014) A comprehensive life cycle water analysis framework for residential buildings. *Building Research & Information, 42*(6), 685–695.

Taylor, P. (2013). *Extraordinary cities: Millennia of moral syndromes, world-systems and city/state relations.* Cheltenham: Edward Elgar.

UN. (2014). *World urbanization prospects: The 2014 revision.* New York: United Nations.

UNEP. (2007). *Buildings and climate change. Status, challenges and opportunities.* Paris: United Nation Environment Programme.

USEPA. (1994). *EPA's financial management status report & five-year plan.* Washington, DC: US Environmental Protection Agency.

US Energy Information Administration. (2013). *International energy outlook.* Washington, DC: US Energy Information Administration.

Van der Heijden, J. (2013). Looking forward and sideways: Trajectories of new governance theory. Amsterdam Law School Research Paper No. 2013-04, *General Subseries Research Paper No. 2013-01.* Retrieved from http://papers.ssrn.com/sol3/papers.cfm?abstract_id = 2204524

Van der Heijden, J. (2014). *Governance for urban sustainability and resilience: Responding to climate change and the relevance of the built environment.* Cheltenham: Edward Elgar.

Van der Heijden, J. (2015). On the potential of voluntary environmental programmes for the built environment: A critical analysis of LEED. *Journal of Housing and the Built Environment, 30*(4), 553–567. doi:10.1007/s10901-014-9428-z

Van der Heijden, J. (2016). Experimental governance for low-carbon buildings and cities: Value and limits of local actions networks. *Cities.* Early view available online. doi:10.1016/j.cities.2015.12.008

Videras, J., & Alberini, A. (2000). The appeal of voluntary environmental programs: Which firms participate and why? *Contemporary Economic Policy, 18*(4), 449–460. doi:10.1111/j.1465-7287.2000.tb00041.x

World Bank. (2011). *Climate change and the world bank group: The challenge of low-carbon development*. Washington, DC: World Bank.

Wurzel, R., Zito, A., & Jordan, A. (2013). *Environmental governance in Europe*. Cheltenham: Edward Elgar.

Yudelson, J., & Meyer, U. (2013). *The world's greenest buildings*. Abingdon: Routledge.

Endnotes

[1]Using the term 'new' when pointing out an empirical phenomenon is always risky. After a while it inevitably loses it 'newness'. This is also the case for new governance and new-governance instruments: they have been trialled for more than two decades and some of the new-governance instruments studied – certification and classification, for example – have become fairly normal in the governing of low-carbon buildings.

[2]See the supplemental data online.

[3]Data are from www.usgbc.org and www.census.gov (accessed on 7 July 2015).

[4]Data are from www.census.gov (accessed on 7 July 2015).

[5]Data are from www.washingtonpost.com/news/grade-point/wp/2015/07/20/how-many-colleges-and-universities-do-we-really-need/ (accessed on 19 December 2015).

[6]The author thanks an anonymous reviewer for this suggestion.

[7]Data are from www.nabers.gov.au and www.greenmark.sg (both accessed on 14 July 2015).

Reducing CO$_2$ emissions from residential energy use

Paul Drummond and Paul Ekins

To achieve European Union (EU) greenhouse gas emissions of 80–95% below 1990 levels by 2050, CO$_2$ emissions from residential energy consumption must be substantially reduced. Recognition of this has led to the introduction of a range of policy instruments at both EU and member state level. These policies are examined for the EU and the UK, first by grouping them into three 'pillars of policy' – standards and engagement, markets and pricing, and strategic investment (each of which focus on different 'domains of change' embodying different economic processes) – and then by assessing the strengths and weaknesses of each pillar in terms of instrument coverage and effectiveness. Strengths and weaknesses common to both UK and EU policy landscapes are found, including a comprehensive but broadly ineffective standards and engagement pillar of policy, and an ineffective markets and pricing landscape (including effective subsidization of energy consumption in the UK, permitted by the EU), with poor coverage. The strategic investment landscape is found (until recently) to be substantially stronger in the UK compared with EU instruments and requirements. Priority reform actions are also proposed to address the weaknesses identified. The paper also offers discussion of recent policy developments in the UK.

Introduction

This paper is concerned with CO$_2$ emissions from energy use by the European Union (EU) and UK residential sector.[1] This includes both regulated energy, defined as energy use by a building, including (space) heating, ventilation and air-conditioning (HVAC), domestic hot water and lighting; and non-regulated (household) energy, defined as energy use in a building particularly by appliances and equipment bought by the occupier. Direct CO$_2$ emissions alone from the residential sector accounted for 11% of total CO$_2$ emissions from the EU-28 in 2012 (EEA, 2015a), with upstream emissions associated with residential electricity consumption adding significantly to this. If the stated goal of a reduction in EU greenhouse gas (GHG) emissions of 80–95% below 1990 levels by 2050 is to be achieved (EC, 2011), the residential sector must therefore decarbonize very substantially over the coming decades.

Figure 1 illustrates historic residential CO$_2$ emissions from the EU-28, and projected requirements to 2050 based on the European Commission's Energy Roadmap 2050 ('Diversified Supply Technologies' Scenario), which seeks to map the contribution required by each sector if the overarching GHG targets, described above, are to be achieved (EC, 2011). This scenario envisages an 85% reduction in direct residential CO$_2$ emissions from 1990 levels by 2050. Between 1990 and 2010, such emissions reduced by an average of 1.8% annually. To achieve the 2050 target, this value must increase to 4.4% from 2010 onwards.[2] To compound this already substantial challenge, the number of households is projected to increase by around 25% over this time (IEA, 2012).

However, the EEA (2015b) projects that direct CO$_2$ emissions from buildings across the EU will only reduce by an annual average of around 1.3% between 2010 and 2030 under existing measures, and around 1.5% with the addition of 'planned measures'. Whilst this projection pertains to all buildings, as the residential sector currently accounts for around 70% of direct emissions from all buildings (EEA, 2015a), it is reasonable to conclude that the rate for the

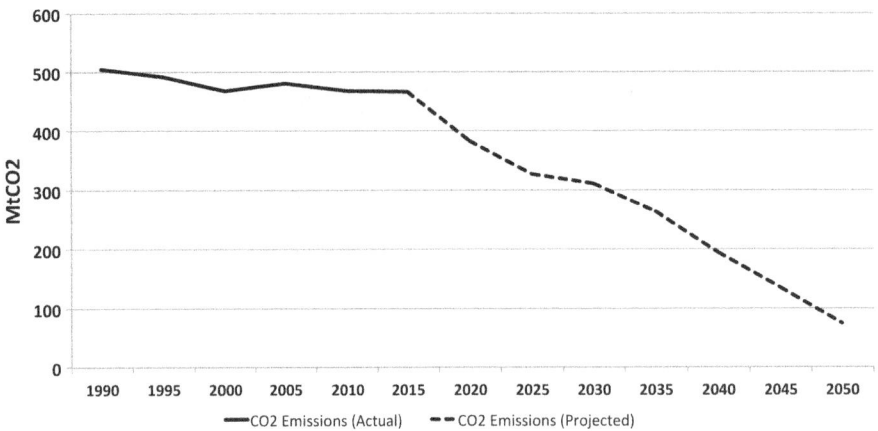

Figure 1 Actual and projected CO_2 emissions from the residential sector 1990–2050 for the EU-28

residential sector only would not deviate substantially from this aggregate projection. In addition, EC (2011), along with much of the wider literature, foresee such decarbonization to be achieved in large part through electrification of key energy services (particularly space heating). As such, it is important that CO_2 emissions from electricity generation also decrease significantly over time (over 95% by 2050, from 1990 levels).

It is clear that an effective, well-coordinated and strengthened policy instrument mix must be in place, both at the EU level and within member states, to drive such an ambitious transformation. The paper is structured as follows. It first presents a conceptual framework regarding the three 'domains of change' and related 'pillars of policy', developed by Grubb, Hourcade, and Neuhoff (2014), that are required to deliver a low-carbon transition. The method the assessment undertakes is then presented, followed by a mapping of the existing climate and energy policy instrument mix related to the residential building sector, first at the EU level and then in a key member state (the UK), onto this conceptual framework. Using the method outlined, key strengths and weaknesses will be identified in each instrument mix against this framing, and suggestions made for improvement to instruments and specific actions. The final section provides conclusions for policy development and instrument efficacy to enable ambitious CO_2 reduction targets to be realized.

Three pillars of policy

The need for a combination of policy instruments in order to address the multiple market failures that lead to the excessive generation of environmental pollutants has long been recognized in the literature (*e.g.* Lipsey & Lancaster, 1956). A number of different environmental policy instrument typologies have also

been developed. For example, Jordan, Wurzel, and Zito (2003) group instruments under four generic headings: market/incentive-based (also called economic) instruments; classic regulation instruments; voluntary (also called negotiated) agreements; and information/education-based instruments. The policy instrument framework of the OECD (2008) consists of direct environmental regulation, environmentally related taxes, tradable permits, public financial support for environmental goods and services, instruments to promote technological development, and information-based and voluntary approaches. Wurzel, Zito, and Jordan (2013, p. 26) provide a recent overview of policy instrument typologies from which it is clear that no single typology can be considered 'correct'; the choice of which to use should depend on the purpose in hand.

Regarding climate policy specifically, the landmark Stern Review on the Economics of Climate Change considered that a policy framework for CO_2 abatement should have three elements: carbon pricing, technology policy and the removal of barriers to behaviour change (Stern, 2006). For the purposes of this paper, this is an attractive typology because each kind of instrument reflects a different kind of barrier to the efficient market implementation of energy-efficiency and low-carbon measures: characteristics of human behaviour that do not accord with economic rationality, negative externalities from energy use that cause energy-efficiency and low-carbon technologies and behaviours to be undervalued, and positive externalities from innovation that prevent new energy-efficient and low-carbon technologies from being developed. Recently, Grubb et al. (2014) have developed the three-fold Stern categorization into a new framework to identify and analyse the policy elements that are likely to be required to achieve a low-carbon transition successfully. This framework, adapted for use in this paper, is illustrated in Figure 2.

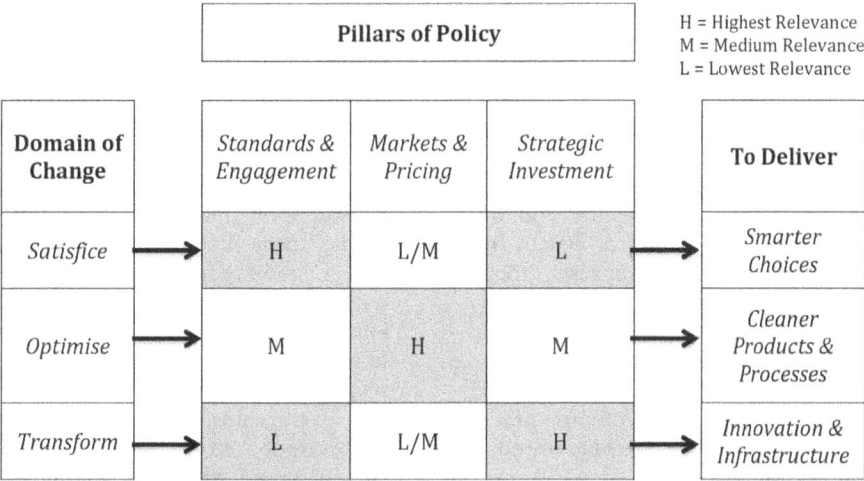

Figure 2 Three 'domains of change' and 'pillars of policy'
Source: Adapted from Grubb et al. (2014)

Figure 2 shows three 'domains of change' and corresponding 'pillars of policy' for tackling CO_2 (and other GHG) emissions. Each domain of change reflects a distinct sphere of economic decision-making and development. The first, 'satisficing', draws on the insights from behavioural and organizational economics. Evidence from these fields demonstrates that individuals (and organizations) are not always 'rational optimizers'; they do not always respond to incentives to maximize their economic welfare over time. Four broad insights that produce this conclusion may be distilled. The first is that individuals are subject to 'bounded rationality', where decisions are constrained by cognitive processes and available information (which produces 'satisficing' behaviour, whereby individuals adopt behaviour that leads to outcomes that meet a threshold of satisfaction, rather than the economically optimal outcome) (Simon, 1956). The second insight is that individuals tend to have hyperbolic discount rates, meaning that individuals do not discount the value of future costs and benefits at a consistent rate of over time. Such a discount rate is 'present biased', meaning that the present value of long-term costs and benefits is much lower than the discount rate commonly used in economic models, which discounts future costs and benefits at a constant rate over time (Laibson & Laibson, 1997). The third insight is that individuals tend to practise 'mental (or psychological) accounting', evaluating differently the utility received from different means of expenditure (*e.g.* via cash or credit card) (Morewedge, Holtzman, & Eplley, 2007). The fourth insight is that under uncertainty, individuals tend to rely on decision-making on heuristics, including 'rules of thumb', 'anchoring' (cognitive bias in future judgements based on the first piece of information provided), and the 'availability' of information in terms of the ease with which it is recalled (Kahneman, Slovic, & Tversky, 1982).

The economically suboptimal outcomes that result from these behavioural influences on decision-making may be addressed by policies that seek to produce 'smarter choices', either as default (*e.g.* through minimum standards), or through encouraging positive active choices (*e.g.* through engagement, and the provision of information). Such instruments form the standards and engagement pillar of policy in Grubb et al.'s (2014) framework.

The second domain, 'optimizing', draws upon the theories and assumptions of neoclassical and welfare economics. These schools of thought hold that individuals and organizations seek to maximize their welfare (utility) over time in response to economic incentives, with rational foresight and stable preferences and technological options. Where markets are functioning perfectly, this results in an optimal 'general equilibrium' state for economic systems, in which private and social welfare is aligned and maximized from the most efficient use of a fixed set of resources. However, it is acknowledged that private and social incentives and welfare may be misaligned, or resources used inefficiently, due to the presence of market (and institutional) failures. A key example is (positive and negative) market externalities, where the socialized economic cost (or benefit) of an action is not internalized (*i.e.* priced in) to the private cost (or benefit). In the context of this paper, the predominant example is the market externality of CO_2 emissions. The economically suboptimal outcomes that result from such situations may be addressed by policies that alter the economic calculus of actors, particularly pricing instruments (*e.g.* carbon pricing), to make 'cleaner products and processes' the economically rational choice. Instruments that alter the structure of the market are also important. Together, these form the markets and pricing pillar of policy (Grubb et al., 2014).

The third and final domain, 'transformation', draws upon the insights of evolutionary and institutional economics. The studies under these broad fields (and particularly the former) emphasize how technological and economic systems evolve over time as a result of a dynamic interaction between historic developments, the direction of innovation, inherited infrastructure, and the nature of societal norms, values and institutions (Grubb et al., 2014; Norgaard, 2010). Institutional economics, as its name suggests, focuses specifically on the role of institutions:

> including the state, political parties, courts, unions, firms, churches, and the like [. . . with their] rules, regulations, customs, common practices and laws that regulate the actions of individuals and concerns.
>
> (Rutherford, 1983, p. 723)

in setting the 'rules of the game', including long-term direction and expectations. Economic and technological systems exhibit 'path dependency' and 'lock-in' arising from earlier decision-making, particularly in terms of the nature and availability of infrastructure, which to some extent at least determine future possibilities (Grubb et al., 2014, p. 312). Private investment is likely to be constrained according to this dependency and lock-in. Without change at the institutional level, particularly through interventions by the public sector, with its unique role in seeking to maximize societal welfare through policy, law and the use of public funds, paradigmatic transitions (such as decarbonization) are difficult or impossible to achieve. A 'strategic investment' pillar of policy, which looks beyond short-term returns to invest in ways that support the evolution of more efficient and low-carbon technologies and systems (such as renewable energy), to produce appropriate 'innovation and infrastructure', is therefore required.

Each of the three domains and policy pillars, whilst presented as conceptually distinct, interact through numerous channels. As Figure 2 illustrates, whilst the impact is strongest in one, each pillar of policy has at least some influence on all three domains of change. Because household energy use exhibits characteristics of all domains, for an effective approach to decarbonization policies from each of the three pillars need to be applied in a *policy mix*, which is an increasingly advocated approach to environmental policy more generally (OECD, 2007). Grubb et al. (2014) consider all three domains, and by extension all three pillars of policy, are of largely equal importance in delivering a low-carbon energy system and economy.

Method

Under the following two sections, Tables 1 and 2 assign the key EU and UK-level instruments in the climate and energy policy landscape, as related to the residential building sector, to each of the three pillars of policy in Grubb et al.'s (2014) categorization (standards and engagement, markets and pricing, and strategic investment). The instruments selected are all those that explicitly seek to impact residential energy consumption, CO_2 emissions from residential energy consumption (including direct emissions from households, those from upstream electricity generation, and emission reduction through households' own generation from renewables), and those that directly enable the control of energy use and CO_2 emissions (*e.g.* requirements for the installation of smart meters). Requirements for the use of usually unspecified overarching strategies and plans, for example, are excluded from the analysis, as are instruments that support technological development but not deployment (*e.g.* funding for basic research). Tables 1 and 2 also act as a 'list of abbreviations' for the text that follows. Whilst Table 2 also summarizes the results of evaluations of the instruments listed for the UK, such information is not provided in Table 1 for the EU as comprehensive *ex-post* evaluations are not commonly available (and actual implementation for many instruments often varies substantially across member states).

Each section below proceeds to highlight the key strengths and weaknesses of the associated policy mix, based on an assessment of instrument coverage within each pillar of policy across sources of energy consumption and CO_2 emissions, and the effectiveness of the instruments in each pillar in achieving their objectives (either stated targets, or in terms of their stated purpose or theoretical justification). Such an assessment is made based on a review of the scientific literature and official evaluation reports, and where such literature is not available, by recourse to theoretical considerations. In the absence of full evaluations, these assessments inevitably have a subjective element in their characterization of 'strengths' and 'weaknesses'. Each section then concludes with proposals for suggestions for priority reforms that may be introduced to correct the key weaknesses identified.

Policy landscape – European Union
Key strengths
The focus of the existing instrument mix at the EU level is on the standards and engagement pillar of policy. Almost all such instruments focus on encouraging energy efficiency, directly or indirectly, rather than the reduction of CO_2 emissions per se. The regulatory push/pull effect (Rennings, 2000) is repeatedly employed, with minimum energy performance standards (MEPS) for new buildings and energy-efficiency obligation schemes (EEOS) for existing buildings providing a technology push towards higher efficiency in

Table 1 Policy landscapes across the 'three pillars' – European Union

Type and name of intervention	Description of intervention
Standards and engagement	
Energy Performance of Buildings Directive (EPBD) (2010/31/EC)	• Member states must set cost-effective minimum energy performance standards (MEPS) (*Article 4*) • Any building sold or rented to a new tenant must be issues with an energy performance certificate (EPC) illustrating energy performance data, reference values (such as minimum performance requirements), recommendations for cost-effective improvement options and information on where to find further information (*Article 11*)
Energy Labelling Directive (ELD) (2010/30/EU)	• Energy-using products and energy-related products (those which do not consume but impact the consumption of energy) are subject to energy-efficiency labelling. Products must be subject to a delegated act before they are covered by this directive
Ecodesign Directive (2009/125/EC)	• Energy-using products and energy-related products (described above) are subject to minimum environmental performance standards (usually related to in-use energy consumption). Products must be subject to an implementing measure before they are subject to this directive
Energy Efficiency Directive (EED) (2012/27/EU)	• Member states must introduce energy-efficiency obligation schemes (EEOS) for energy distributors and supplies to achieve annual savings of 1.5% total average energy sales by volume over 2009–12, each year over the period 2014–20 (*Article 7*) • 'Smart' meters must be provided when a unit is replaced or a new connection is made, when technically feasible and cost-effective (*Article 9*) • If consumers do not have smart meters, mandatory billing information must contain (1) data based on actual consumption and information, such as daily consumption profiles, be made available free of charge (*Article 10*), (2) comparisons with average or benchmarked consumption, and (3) details on where to receive information on energy efficiency • Member states must implement programmes to raise awareness of the benefits and availability of energy audits amongst households (*Article 8*) • Other consumer information and empowerment programmes to promote energy efficiency amongst small energy consumers (including domestic consumers). This may include fiscal incentives and grants • Qualification, accreditation and certification schemes must be provided to providers of energy services, audits and installers of energy-efficiency elements, in member states where technical competence, objectivity and reliability are insufficient (*Article 16*) • Information and training on energy-efficiency mechanisms and financial and legal frameworks must also be provided to all relevant market actors (including consumers, builders, architects, auditors, etc.) (*Article 17*)
Third Energy Package – Electricity (D2009/72/EC)	• 80% of electricity consumers (across all sectors) must be fitted with smart meters by 2020 in each member state, where cost-effectiveness is assessed positively (*Annex I*)
Renewable Energy Directive (RED) (2009/28/EC)	• The presence of renewable energy support schemes must be publicized through awareness-raising and training, including to consumers, builders and architects (*Article 14*) • Qualification and certification schemes must be made available for installers of small-scale biomass boilers and stoves, solar photovoltaic and solar thermal systems, shallow geothermal systems and heat pumps (*Article 14*)
Markets and pricing	
EU ETS (2009/29/EC)	• The European Union Emissions Trading System (ETS) is a cap-and-trade instrument applicable to the power and heavy industry sectors. As a consequence, residential electricity consumption is subject to an upstream carbon price

(Table continued)

Table 1 Continued

Type and name of intervention	Description of intervention
Third Energy Package – Electricity (D2009/72/EC)	• Electricity generators and suppliers must be 'unbundled' to produce a competitive electricity market price (*Article 31*) • Electricity wholesale and retail prices must be the product of market competition in all member states (*several Articles*)
Strategic Investment	
Energy Performance of Buildings Directive (EPBD) (2010/31/EC)	• All new buildings must be classified as 'nearly zero-energy buildings' (NZEBs) by 2020, with any remaining energy requirements substantially satisfied by renewables (*Article 9*)
Renewable Energy Directive (RED) (2009/28/EC)	• Renewable energy support schemes must be introduced to meet member state renewable energy targets for 2020.

regulated energy consumption (through more efficient building envelopes and heating systems, for example), whilst the presence of energy performance certificates (EPCs) and display energy certificates (DECs), by altering consumer preferences through increased awareness, engenders a market pull. The effectiveness of DECs in the UK could be strengthened by mandating their use in private as well as public buildings (Cohen & Bordass, 2015). The Ecodesign and Energy Labelling Directives employ the same dynamic for energy-using and energy-related products (and thus, largely, non-regulated energy). Whilst these instruments seek to produce technological change, other instruments, such as the requirement for the installation of smart meters and the use of nudging information (Thaler & Sunstein, 2008) on energy bills (through the provision of benchmarked consumption information), again employ push/pull dynamics to encourage reduced energy consumption through behavioural change. A further collection of instruments seeks to provide confidence in those supplying energy-efficiency and renewable technologies (and encourage high standards) through certification and accreditation mechanisms.

A comprehensive set of standards and engagement instruments to encourage improved energy efficiency is the key strength of the existing instrument mix. A substantial energy-efficiency gap, a term first coined by Hirst and Brown (1990), but more recently defined by Allocott and Greenstone (2012, p. 4) as the 'wedge between the cost-minimising level of energy efficiency and the level actually realized', exists in the EU's residential sector. Wesselink, Harmsen, and Eichhammer (2010) estimate the presence of around 80 Mtoe of cost-negative (final) energy-efficiency measures available from the use of more efficient appliances, building envelope and

system efficiency measures (including heating systems) in the EU residential sector by 2020. This equals around 27% of residential final energy consumption across the EU in 2010 (EC, 2011), the reduction of which would reduce both direct and indirect CO_2 emissions (the extent to which depends on the specific measures introduced, the equipment replaced and the fuel saved). The presence of information failures, split incentives (particularly the 'landlord–tenant' situation; Janda, Bright, Patrick, Wilkinson, & Dixon, 2016) and satisficing behaviour (discussed above) are all key explanatory factors as to why such unexploited savings opportunities exist (Gillingham & Palmer, 2013; Grubb et al., 2014; Hirst & Brown, 1990; Jaffe & Stavins, 1994). As illustrated in Figure 2, instruments under the standards and engagement pillar of policy are those most appropriate to reduce the impact of such issues. The cost-negative nature of many of the actions these instruments induce or require makes them politically attractive, as evidenced by their strong presence in the instrument mix, but actually realizing the identified energy savings can be problematic in practice, as described below.

Key weaknesses
Notwithstanding their theoretical appropriateness, the practical effectiveness of many of the instruments under the standards and engagement pillar is in many cases unclear. This is the first key weakness of the existing instrument mix. There are two principal reasons for this. The first is low quality or non-implementation by member states, and poor monitoring and enforcement. For example, MEPS for new buildings required under the Energy Performance of Buildings Directive (EPBD) are often poorly enforced, with compliance often found to be relatively low across member states (Pan & Garmston, 2012). At present, 17 member

Table 2 Policy landscapes across the 'three pillars' – UK

Type and name of intervention	Description of intervention	Effectiveness evaluation energy/carbon saved; cost savings; renewable power
Standards and engagement		
Energy Efficiency Standards of Performance (EESoP)	EESoP ran in three phases from 1994 to 2002, and led to the principle of energy suppliers being obligated to help their customers save energy through energy-efficiency measures being incorporated in UK law through the Utilities Act (2000). The main measures delivered were insulation, lighting, heating and appliances	The National Audit Office (NAO) in 1998 calculated that the overall net financial benefit of EESoP1 (1994–98) was £250 million (Ofgem and Energy Saving Trust, 2003)
Energy Efficiency Commitment (EEC)	EEC, a supplier obligation that built on the experience of EESoP was implemented in two phases. EEC1 ran from 2002 to 2005, EEC2 from 2005 to 2008, with 'lifetime' energy savings targets of 62 and 130 TWh, and projected costs of £500 m and £1.2 billion, respectively (Rosenow, 2012)	Estimated energy savings from EEC1 were substantially in excess of its targets, and were delivered for electricity and gas at a cost of 1.3 and 0.5 p/kWh respectively, far less than the retail price (Lees, 2006, p. 6). The targets of EEC2 were also substantially exceeded, with EEC1 and EEC2 together estimated to be saving by 2010 6.1 and 7.4 TWh of electricity and natural gas, with the EEC2 cost being 2.1 and 0.6 p/kWh, respectively – less than a quarter of the average price to consumers of those fuels in EEC2. The lifetime CO_2 savings from meeting the EEC2 target were estimated as 59 $mtCO_2$, yielding a net present value of £53/tCO_2 saved (Lees, 2008, p. 6)
Carbon Emissions Reduction Target (CERT) and Community Energy Saving Programme (CESP)	The target for CERT, introduced in 2008, was lifetime carbon savings of 293 $mtCO_2$, estimated to amount to 104 TWh of energy. The estimated cost over 2008–12 was £5.5 billion (Ofgem, 2015a). This implies a cost of carbon saving of £18.8/tCO_2. The lifetime CO_2 savings target for CESP was set at 19.25 $mtCO_2$, at an estimated cost of £332 million (Martin, de Preux, & Wagner, 2011)	CERT carbon savings were estimated as 296.9 $mtCO_2$, 101% of the target. For CESP the estimated delivery against the target was 16.31 $mtCO_2$, or 84.7% of the target. CERT savings were delivered at an estimated cost of £3.65 billion, well below the *ex ante* estimation. The cost of carbon savings was estimated at £13.79/tCO_2, also well below the *ex ante* estimation. In contrast, the cost of CESP was estimated at £702 million, more than twice the *ex ante* estimation, while carbon savings of 20.2 $mtCO_2$ just exceeded the target, implying a cost of carbon saving of £34.8/tCO_2 (Ipsos Mori et al., 2014)
Energy Companies Obligation (ECO, ECO2)	Following on from CERT, ECO, running from 2013 to 2015, required obligated suppliers to achieve carbon and cost savings in respect of three distinct obligations: the carbon emissions reduction obligation (CERO), which promotes the installation of solid-wall and hard-to-treat cavity wall insulation; the carbon-saving community obligation (CSCO), which promotes the installation of insulating measures and connections to district heating systems in areas of low income and rural areas; and the home heating cost-reduction obligation (HHCRO, which promotes the installation of measures, including the repair and replacement of boilers, to homes in receipt of certain benefits, to reduce the overall cost of space heating). ECO2, with the same obligations, was introduced to run from 2015 to 2017. Targets under the different obligations for ECO and ECO2 were, respectively: (1) CERO 20.9 (later reduced to 14.0) and 12.4 $mtCO_2$; and (2) CSCO 6.8 and 6.0 $mtCO_2$; HHCRO £4.2 and £3.7 billion. (Ofgem, 2012, 2015a)	The monthly ECO Compliance Update for April 2015 produced by Ofgem showed that the targets to the end of March 2015 had been over-achieved, with the exception of the HHCRO rural sub-obligation, which had achieved a 99% compliance, but with many measures in the pipeline (Ofgem, 2015a)

(Table continued)

Table 2 Continued

Type and name of intervention	Description of intervention	Effectiveness evaluation energy/carbon saved; cost savings; renewable power
Green Deal	The Green Deal policy was enabled through the Energy Act 2011. Its main novel provisions were the ability of a householder to take out a loan for energy-efficiency measures which was repaid through the energy bill of that property, even if the householder moved, when the repayments would continue for the new householder in the property. The measures needed to be those recommended and installed by individuals and companies with appropriate certification, and only those measures would be recommended that were calculated to pay back their cost within their lifetime – the 'Golden Rule' (DECC, 2010)	The Green Deal achieved far less take up than had been hoped for – only 15 000 by 2015. In July 2015 the new Conservative government in the UK announced that it would not continue funding the Green Deal through the Green Deal Finance Company that had been set up for this purpose. While in principle Green Deals could continue to be financed privately, the announcement was widely interpreted as amounting to the end of the policy
Markets and pricing		
Climate Change Levy (CCL)	The CCL was introduced in 2001 as a combined carbon–energy tax on business. Energy-intensive firms were largely exempted provided they signed up to climate change agreements (CCAs). The exemption for renewables was ended in 2015, making the CCL almost wholly an energy tax	The UK NAO evaluation in 2007 agreed with 'the most recent estimate of annual savings of 3.5 MtC in 2010' from the CCL, and 1.9 MtC from the CCAs (NAO, 2007). A different 2011 evaluation found that the CCL 'caused plants paying the full rate to reduce CO_2 emissions by between 9.6% and 22.6% compared to plants [covered by the CCAs] that paid the reduced rate' (Martin et al., 2011, p. 28). No evidence was found that the CCL had adversely affected the competitiveness of UK manufacturing
Carbon Price Floor (CPF)/Carbon Support Price (CSP)	The CPF was introduced in 2011 and was intended to provide a minimum price for carbon in the power sector, rising from £16/tCO_2 in 2013. It consists of the EU ETS allowance price, plus the CSP, which is calculated two years in advance and levied on the fossil fuel inputs to power generation, to deliver the projected minimum price, originally intended to be £30/tCO_2 in 2020, and £70/tCO_2 in 2030. Because of the weakness of the EU ETS price, the CPS was frozen at £18/tCO_2 in 2014	No evaluation is yet available, but the government estimated in its impact assessment that the CSP would cumulatively reduce emissions by 261 mtCO_2 over 2013–30 (HM Treasury & HM Revenue and Customs, 2010, p. 13)
Strategic investment		
Non-Fossil Fuel Obligation (NFFO)	NFFO was introduced in England and Wales through the 1989 Electricity Act (with similar legislation being introduced in Scotland and Northern Ireland). The act allowed the government to require public electricity suppliers (PES) to purchase a certain amount of electricity produced from non-fossil fuels, and levied a charge on consumers, the fossil fuel levy (FFL), to compensate the PES for the NFFO. Most of the revenues from the FFL went to the subsidize the nuclear industry, but NFFO itself supported renewables through five orders, the first in 1990, the fifth in 1998, paving the way for the introduction of the more ambitious renewables obligation in 2002	The total NFFO renewable generation target was 1500 MW, but by 2003 only 1098 MW had been built (Tovey, n.d), with many contracted projects not proceeding to construction. The NFFO installations generated around 3% of 2003 total UK electricity (around 380 TWh)

Policy	Description	
Renewables Obligation (RO)	The RO came into force in 2002 in England, Wales and Scotland, and 2005 in Northern Ireland. It places an obligation on the main electricity suppliers to source an increasing proportion of their generation from renewable sources. The obligation is met by presenting RO certificates (ROCs) at the end of year, or paying a 'buy out price' per MWh of shortfall in renewables supply. The RO is due to terminate in 2017, to be replaced for the Contracts for Difference (CfD)	Renewable electricity generation increased from less than 3% of total generation in 2000 to around 19% of generation in 2014 (DECC, 2015a). Pursuant to its manifesto commitment, the new Conservative government in the UK announced in July 2015 that it was closing the RO for new onshore wind in 2016. It also announced that in order to reduce subsidy costs, it was limiting the eligibility of biomass plants that were converting from fossil fuels
Contracts for Difference (CfD, a premium Feed-in-Tariffs (FiT))	CfD were introduced as part of the electricity market reform in the UK Energy Act 2013. They comprise a payment to generators of the difference between the average wholesale price or electricity and a (higher) technology-specific 'strike price', to apply to low-carbon technologies (renewables, nuclear and carbon capture and storage). CfDs will replace the RO from April 2017	The first CfD auctions, in February 2015, awarded contracts to 27 projects (onshore and offshore wind, advanced conversion, solar photovoltaics (PV) and energy from waste with combined heat and power (CHP)), with a total capacity of 2.1 GW and an estimated subsidy of £315 million/ annum by 2020/21. The strike prices were substantially below those anticipated in advance (e.g. £115–120/MWh for offshore wind, as against £140/MWh)
Feed-in-Tariffs (FiT)	While the RO and CfDs are intended to support larger-scale renewables, FiTs, first introduced in the UK in 2010; they apply to smaller projects for solar PV, wind, hydro, domestic CHP and anaerobic digestion	By the end of 2014 FiTs had led to the accreditation of 3.25 GW of renewables capacity, 83% of which was solar PV (DECC, 2015a). The tariff rate has been repeatedly reduced (e.g. for solar PV < 4 kW from over 40 p/kWh in 2010 to around 13 p/kWh in October 2015) (Ofgem, 2015b)
Renewable Heat Incentive (RHI)	The RHI applies to both non-domestic and domestic installations of eligible renewable heat technologies (including solid biomass, heat pumps, geothermal, solar thermal. biogas combustion and biomethane injection), with the former introduced in 2011 and the latter in 2014	By mid-2015 1 GW of renewable heat had been installed under the non-domestic scheme, nearly 99% of which was solid biomass (Ofgem, 2014b). In the first six months of its operation (by October 2014), there were 10 000 domestic RHI accreditations (Ofgem, 2014a)
Zero Carbon Homes (ZCH)	The ZCH policy, introduced in 2006, required all new homes from 2016 to avoid or mitigate all regulated emissions using a combination of on-site energy-efficiency measures (such as insulation and low-energy heating systems), on-site zero-carbon technologies (such as solar panels), and off-site measures to deal with any remaining emissions	In July 2015 the new Conservative government in the UK announced that it did not intend to proceed with the 'zero-carbon' carbon offsetting scheme, or the proposed 2016 increase in on-site energy-efficiency standards, but would keep energy-efficiency standards under review. It is not clear what the implications of this are for the NZEB requirements for the European EPBD

states have, or plan to implement, an EEOS (often in combination with other instruments) to satisfy the requirements of Article 7 of the Energy Efficiency Directive (EED).[3] However, of these instruments, eight have major credibility issues, six have minor credibility issues and only two have no issues.[4] Additionally, most monitoring, verification, control and compliance regimes have been judged to be inadequate (Rosenow et al., 2015, 2016). The second factor is that of initial instrument design. For example, whilst all member states have a functioning EPC scheme (as also required under the EPBD), their design and structure differ somewhat (often resulting from a slow and partial implementation of the EPBD), producing different levels of clarity and effectiveness (Economidou et al., 2011).

Whilst labels required under the Energy Labelling Directive (ELD) are of a harmonized design across the EU, it is unlikely that this instrument had any significant impact on overall market sizes, structure or product choices amongst consumers for the products covered (Ecofys, 2014b). Various factors have been identified as contributing to this. First, the introduction of A+ to A+++ labels appears to have produced confusion and a feeling of diminishing returns amongst consumers, reducing its efficacy compared with the simpler 'A–G' scale used in the first incarnation of the ELD in 1992 (Heinzle & Wüstenhagen, 2012; London Economics & Ipsos, 2014). Secondly, around 90% of appliances covered by implementing measures fall into the 'A' category (Heinzle & Wüstenhagen, 2012), reducing the ability for consumers to differentiate between products. Thirdly, Waide and Watson (2013) found that although energy efficiency is commonly an important factor in purchase decisions for products with energy labels, factors such as capital cost often hold higher importance.

This links to the second key weakness of the existing instrument mix. Even if instruments under the standards and engagement pillar are well designed, implemented and enforced, they may generally only be as effective as economic incentives permit them to be (even required MEPS are subject to cost-efficiency clauses[5]). This requires the correct incentives to be delivered, primarily by instruments under the markets and pricing pillar of policy.

As highlighted in Table 1, carbon pricing is only applied to electricity at the EU level via the EU Emissions Trading Scheme (ETS). Other (non-renewable) fuels used in households (such as gas, coal and oil) account for over 60% of energy consumption (and all direct net CO_2 emissions) from the residential sector,[6] mainly for space and water heating. These fuels may be exempt from any form of taxation under provisions of the Energy Taxation Directive

(2003/96/EC) (except value added tax (VAT), although this may be set at a reduced rate of 5%). As a result, markets and pricing instruments are largely absent at the EU level, with (CO_2-intensive) energy consumption able to be effectively subsidized through a reduced-rate VAT (permitted by the VAT Directive), and a failure to internalize the market externality of CO_2 emissions. As such, the incentives for private and socially optimal levels for the installation of energy-efficiency measures, reduced demand through behaviour change, and the installation of low-carbon heating and electricity generating technologies are misaligned. Additionally, despite legal obligations to the contrary, many member states maintain regulated electricity prices (EC, 2014), preventing the full communication of the (albeit currently low) carbon price generated by the EU ETS.

Instruments present under the strategic investment pillar of policy in Table 1 do relatively little to alter this picture. This is the third key weakness of the existing instrument mix. The near-zero energy buildings (NZEB) requirement for new buildings by 2020 is an extension of the existing MEPS requirement, but is included under the strategic investment pillar due to its apparently ambitious nature and the use of currently niche or immature technologies that would likely be required to achieve it. The EPBD allows member states to set the definition of 'nearly zero', with many setting such definitions at 45–50 kWh/m²/year for residential properties (with some member states yet to set a definition at all; BPIE, 2015). Additionally, few member states require any proportion of the remaining energy consumption to be satisfied by renewables, as required by the EPBD (Ecofys, 2014a). In 2012, the average energy intensity of the existing EU residential stock was 185 kWh/m² (Gynther, Lapillonne, & Pollier, 2015). As such, it appears that in practice NZEB requirements are not as ambitious as the name may at first suggest (indeed, Ecofys, 2012, suggests that at 2010 energy and technology prices a cost-effective level of energy consumption is 15 kWh/m² or lower in four of five climatic zones across the EU, with < 25 kWh/m² recommended for the coldest climatic zone, which includes Stockholm, Helsinki and Riga). This is particularly the case if enforcement issues associated with existing MEPS continue.

Twenty-seven of 28 EU member states provide financial support for renewable electricity for residential or community-level installations, for a range of technologies (commonly via feed-in tariffs, or increasingly feed-in premiums).[7] Twenty-three of 28 also provide some form of support mechanism for renewable heating sources and technologies. However, many of these renewable heating support schemes either do not apply to household or community-level installations (or only indirectly, such as support for the production of biogas in agriculture to feed into the gas grid), or support some technologies but not others (*e.g.* biomass but not solar thermal), or are relatively weak, for example, only offering loans or exemptions from other instruments (*e.g.* CO_2 taxation), rather than direct financial subsidy.

As such, the focus of the strategic investment pillar of policy at EU level is largely limited in practice to the energy efficiency of new buildings (although with relatively low practical ambition), and the deployment of renewable electricity generation. The requirement for an EEOS under the EED is considered a standards and engagement instrument rather than strategic investment due to the relatively low level of energy savings targeted, the flexibility in defining the target at member state level and the presence of alternative compliance mechanisms.

In summary, the existing EU policy mix places a substantial focus on encouraging short-term uptake of energy-efficiency measures that are already cost efficient at prevailing (albeit effectively subsidized in many member states) energy prices. However, various design and implementation issues hamper this objective. Additionally, whilst important, considerably more must be done to move beyond this goal if emissions targets are to be met. Most significantly, medium- and long-term requirements and incentives to reduce meaningfully demand for and decarbonize space and water heating in existing properties are either absent or insubstantial. This is by far the largest source of energy demand and CO_2 emissions from the residential building sector, and without attention in the short-term to prevent continued high-energy and high-carbon lock-in, achieving the CO_2 abatement trajectory illustrated in Figure 1 may become impossible in practice.

Suggested priority reforms
To provide both long- and short-term requirements and incentives for reducing energy consumption and CO_2 emissions from the residential sector, the existing EU policy mix must be strengthened and rebalanced across all three pillars of policy. The key weaknesses of the EU-level policy mix highlighted above suggest several priority actions that may begin to achieve this:

- *Ensure adequate enforcement and encourage high-quality implementation* of existing and future directives and provisions (within constraints of the subsidiarity principle). This is a key weakness in the existing mix, under the standards and engagement pillar in particular. Without enforcement, little benefit derives from extending or strengthening existing requirements. High-quality implementation may be encouraged both by specifying a small number of acceptable compliance

mechanisms and by the dissemination of best-practice approaches

- *Redesign key 'engagement' instruments* to increase their potential effectiveness. For example, the ELD may be revised to re-specify the A–G ratings and ensure an even distribution of products across categories. In July 2015, the European Commission released a proposal to do just this (EC, 2015). Standardization or dissemination of best practice for other such instruments, such as EPCs, would also likely be beneficial.

- The role of the markets and pricing pillar of policy should be extended by *introducing carbon pricing to heating fuels and reducing market distortions*. This would broaden the use of carbon pricing beyond electricity and reduce the implicit subsidization of fossil fuels and energy consumption. The former may be achieved through various means, such as a revision of the Energy Taxation Directive, or the expansion of the EU ETS (applied upstream). The latter would include ensuring regulated electricity prices are removed (as per current requirements), and the removal of the ability for member states to impose a reduced rate of VAT.

- The role of the strategic investment pillar of policy should also be substantially expanded beyond (relatively low ambition) standards for new buildings (which are likely to account for a minority of the total residential building stock by 2050[8]). Three key actions may be taken to achieve this. The first is to *reform NZEB requirements from 'nearly-zero energy' to 'net-zero energy'*. This would increase ambition against current requirements, and may reduce flexibility surrounding the regulatory limits that may be set by member states (*i.e.* interpretations of the meaning of 'nearly zero', which currently allow for requirements that are significantly less ambitious than the literature suggests would be most cost effective).

- The second action to expand the role of the strategic investment pillar of policy is to *extend EEOS requirements beyond 2020*, and introduce mechanisms to encourage deep retrofits (*e.g.*, through a mandatory proportion of total compliance requirements, or through a compliance 'uplift factor'). An explicit focus *on residential buildings* and a restriction on the ability to use alternative compliance mechanisms, as discussed above, may also prove beneficial.

- The third action to expand the role of the strategic investment pillar of policy is to *require or more directly encourage support mechanisms for residential renewable heating*. Alongside expanded EEOS requirements, this encourages the installation of renewable heating technologies in existing buildings, individually or as district heating, which, as discussed above, is a significant shortcoming of the existing policy mix, particularly since heating accounts for the vast majority of the non-electric energy consumption in residential properties.

The likely impact of such actions depends substantially on the specific design such instruments take, and the specific implementation in individual member states, and as such should be assessed in future work. Regardless, it is likely that such actions alone would not be sufficient to achieve the rates of decarbonization identified above in the first section. Beyond the policy instrument landscape as defined here, appropriate long-term planning and governance infrastructures will also be required. However, such aspects are beyond the scope of this paper.

Policy landscape: UK

This section describes the experience of the UK since 1997[9] in its implementation of measures to improve the energy efficiency and reduce (direct and indirect) CO_2 emissions from the residential sector. Many of the instruments introduced were in response to EU legislation, described in the previous section, and therefore illustrate how one member state sought to transpose such legislation into national policy. Other instruments precede such requirements (*e.g.*, the UK energy-efficiency obligations), and may even have led to, the EU legislative provisions described above. The UK is taken as an example because of its strong national legislation on climate change with ambitious emission reduction targets. Other member states will, of course, have responded differently to the EU legislation described above.

Table 2 lists the various instruments introduced in the UK to promote the decarbonization and more efficient use of residential energy, with two pre-1997 instruments – energy efficiency standards of performance (EESOP) and the non-fossil fuel obligation (NFFO) – listed as important precursors of subsequent instruments.

Key strengths

Figure 3 shows that UK residential energy use has fallen consistently since 2005, despite periods of strong economic growth, showing a clear departure from the energy–economy coupling of the 1990s. By 2013, although total residential energy use was 9% above that in 1990, energy use per household was 8% lower, and 35% lower per unit of disposable income.[10] Figure 3 also presents the evolution of CO_2 emissions arising from this energy use. Two-thirds of residential energy consumption in the UK is fossil fuel (primarily natural gas) combustion for space and water heating.[11] Therefore, both direct and total CO_2 emissions track energy-use trends closely. The divergence in total emissions is due to the decarbonization of UK electricity from around 740 gCO_2/kWh in 1990 to 459 gCO_2/kWh in 2013.[12] Despite the decrease over time, total direct CO_2 emissions from the UK's residential sector was second only to Germany in 2012, and the fourth highest in proportion to total domestic CO_2 emissions.[13] The UK must therefore continue to drive substantial reductions over time, driven by an appropriate policy framework.

There is as yet no convincing evidence to attribute specific CO_2 emission reductions to specific policy instruments and other (non-policy) influences, but it is likely that both aspects have contributed to the trends illustrated in Figure 3. However, in the 1990s the major cause of the reduction in total residential

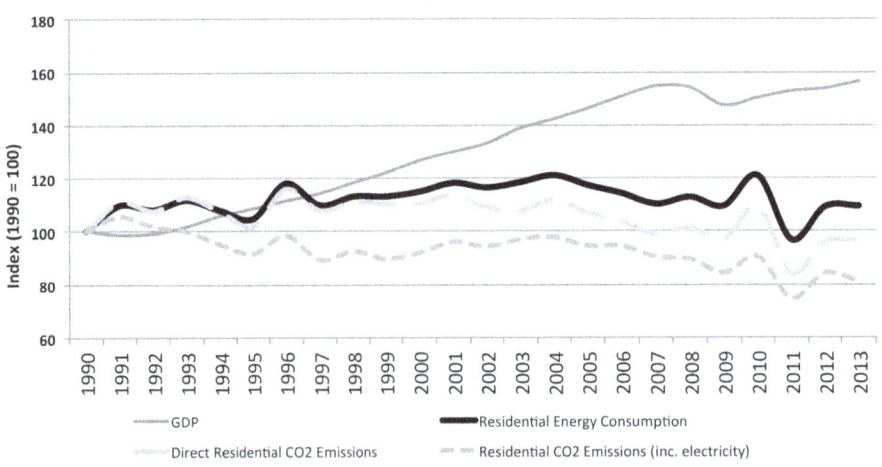

Figure 3 Indices of gross domestic product (GDP), household energy use and CO_2 emissions, 1990–2013 (1990 = 100)
Data sources: UK Office of National Statistics (ONS) and Energy Consumption in the UK (domestic data tables)

CO_2 emissions was undoubtedly the substitution of gas for coal in power generation (driven by market forces in the newly privatized sector, rather than policy drivers), producing the reduction in CO_2 intensity described above. A reverse in this trend and the 2009–11 recession contributed substantially to the volatile trend experienced in 2008–12. However, it is clear that a comprehensive strategic investment pillar of policy had an effect. This is the first key strength of the UK's policy mix.

The Renewables Obligation (RO), and, most recently, a premium Feed-in-Tariff (FiT) called Contracts for Difference (CfD), are the UK's main strategic investment instruments for incentivizing the deployment of large-scale low-carbon power generation, with a fixed-rate FiT encouraging mainly smaller-scale low-carbon power generation. As noted in Table 2, the RO required an increasing proportion of power generation to come from renewables, with penalties paid for a shortfall, while CfD operates through a guaranteed 'strike price' to come from the wholesale power market supplemented by a levy on consumers. The FiT provides both a generation tariff (differentiated by technology and installation size) and an export tariff (equalized across all installations). Figure 4 shows the evolution of the UK electricity mix from 2005 to 2014, and the uptake by households of the FiTs for solar photovoltaics (PVs) since their introduction in 2010. It can be seen that both the RO (for large-scale renewables) and FiTs have been highly successful in incentivizing the deployment of renewable electricity capacity (such data are as yet unavailable for the CfDs), which in turn will have contributed to the reduction in CO_2 intensity of power generation (and subsequent residential CO_2 emissions from associated electricity consumption) shown in Figure 3.

These instruments are in place to meet the requirements of the EU Renewable Energy Directive (and the preceding 2001 Renewable Electricity Directive), as described in the previous section (although, the UK's NFFO, which also sought to deploy renewable electricity capacity, preceded these obligations by over a decade).[14] By encouraging the deployment of renewable heating technologies in a manner similar to the FiT for electricity generation (generation tariff), the (domestic) renewable heat incentive (RHI) was the first of its kind in the world when introduced in 2011, and also seeks to contribute to the UK's obligations under the Renewable Energy Directive. The RHI remains a unique instance in the EU of incentivizing the deployment of renewable heating in such a direct manner.

The remaining instrument in the strategic investment pillar of policy is the zero-carbon homes (ZCH) requirement, through which new residential properties from 2016 needed to achieve zero net CO_2 emissions through a combination of an increase in regulated energy-efficiency requirements, on-site renewable electricity generation (*e.g.*, through rooftop solar PV) and off-site 'allowable solutions' to offset remaining CO_2 emissions. The instrument was intended to align with the EU's NZEB requirements (described above). However, it was cancelled by the UK government in 2015 (a year before it was due to come into force). It is now unclear how the UK intends to comply with NZEB requirements (although MEPS remain as part of the UK's building regulations (Part L), satisfying current EPBD requirements). Also, as with the EU-level policy mix, instruments to tackle energy consumption in existing buildings in the strategic investment pillar of policy are lacking.

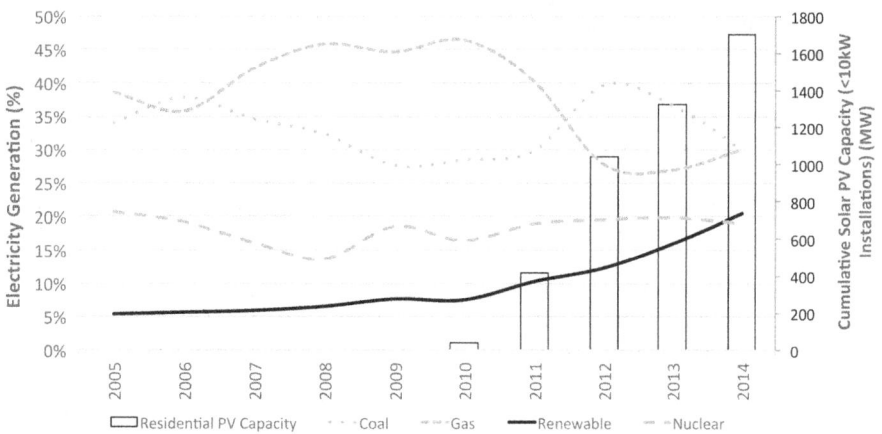

Figure 4 UK electricity generation sources and residential solar photovoltaic capacity
Data sources: Digest of United Kingdom Energy Statistics (DUKES) and Department of Energy and Climate Change (DECC) statistics

The second key strength of the UK's policy mix, as at the EU level (and deriving directly from it), is a comprehensive standards and engagement pillar of policy. This includes both instruments that seek to achieve the UK's compliance with EU obligations, and those that apply equally across the EU. The Ecodesign and Energy Labelling Directives are the key examples of the latter. Regarding the former, the key instruments are the subsequent supplier obligations – the successive EESoP, EEC, CERT/CESP and ECO (see Table 2 for spelled-out versions of these acronyms). As mentioned above, the use of such instruments in the UK significantly preceded EU requirements (the EED was introduced in 2012, whilst the EESoP was adopted in 1994). The final key instrument in the standards and engagement pillar of policy is the Green Deal, discussed below.

Key weaknesses
The UK's policy mix contains three key weaknesses. The first, as at the EU level, is the effectiveness of the instruments under the standards and engagement pillar. For new buildings, Pan and Garmston (2012) found that of a sample taken of residential properties built between 2006 and 2009 in the UK, only a third were in compliance with MEPS for new-build properties. Indeed, awareness of regulations for existing buildings is very low, even among the construction industry (CLG, 2006).

For supplier obligations designed to tackle energy consumption and CO_2 emissions from existing buildings, a 'delivery gap' between reported and actual impact may be identified. Under these instruments, energy suppliers were assigned energy- or CO_2-saving targets to be achieved over defined periods – the majority of which were officially achieved or even over-achieved (as illustrated in Table 2). However, doubts have been expressed as to whether the imputed energy and CO_2 savings from the installed measures have been achieved in practice. The evaluation of the performance of such instruments is usually based on engineering and modelled estimates of how the installed technology will perform, rather than actual measured savings – an approach that routinely overestimates actual outcomes (Rosenow & Galvin, 2013). Although the UK compares relatively favourably with other EU member states (such as Germany) in considering some calculated evaluations (*e.g.*, in the evaluations of Efficiency Commitment by Lees, 2006, 2008) phenomena that produce such discrepancies between calculated and measured results, such as rebound effects, 'prebound effects' (coined by Sunikka-Blank & Galvin, 2012, and referring to a situation in which energy consumption prior to the introduction of a policy instrument is overestimated), technical short-comings in installed measures, and deadweight losses (from householders that would have taken the action encouraged by the policy regardless of the instrument's

existence) (Rosenow & Galvin, 2013), the evidence suggests that some discrepancy between actual and calculated savings, as used to demonstrate compliance, remains.

The current supplier obligation incarnation, the ECO, has thus far resulted in a sharp reduction in the number of energy-efficiency measures installed compared with previous instruments (particularly CERT/CESP). This stems from a reorientation towards more expensive measures such as solid-wall and hard-to-treat cavity-wall insulation, and an increase in the paperwork required for the approval of these measures. This has made it more difficult to include the new approach in existing business models. Additionally, the financial contribution required by consumers has also increased. It is estimated that in 2014 the combined ECO/Green Deal package delivered measures that will produce 74% less energy and 86% less CO_2 lifetime savings than those installed annually on average between 2008 and 2012 under the previous CERT/CESP instrument package (Rosenow & Eyre, 2015).

Although introduced as part of a policy package with the ECO, the Green Deal was not introduced in (direct) response to EU requirements, but in addition to them. As described in Table 2, the instrument was effectively discontinued in 2015. The Green Deal contained a number of innovative elements (for the UK) to encourage householders themselves to invest in regulated energy-efficiency measures. Principal among these was the availability of a loan for preapproved regulated energy-efficiency measures attached to the property rather than the householder (and thus liable for repayment by successive occupiers), with monthly repayments made via electricity bills at a level designed to be equal to or lower than the value of energy cost savings as they accrued (the 'Golden Rule'). This was intended to overcome barriers associated with capital costs, the landlord–tenant dilemma and uncertain returns on investment (if, for example, a householder were to leave before their return on investment had been achieved).

However, the instrument significantly underperformed against expectations, with only 626 householders taking part by the end of 2013, against an expected figure of 10 000. Rates of uptake did not increase substantially thereafter but did eventually gain some momentum, although the programme was abandoned in 2015 (Gillich, Sunikka-Blank, & Ford, 2016). Several factors likely contributed to this, including initial problems with the information technology (IT) administration system and the late availability of the loan facility (introduced five months after the launch of the instrument itself), but particularly the interest rate of the loan, which survey evidence suggested that at around 7.5% was considered too high by householders (Rosenow & Eyre, 2015). Other issues included uncertainty surrounding the long-term financial benefits of undertaking regulated energy-efficiency measures (Petitfor, Wilson, & Chryssochoidis, 2015), a lack of sufficient outreach to influence stakeholder participation (Gillich et al., 2016), and a possible perception that with the loan attached to the property, with subsequent occupiers liable for repayment, the sale value of the property would be reduced (Ashworth & Perera, 2015). When in June 2014 capital grants (the Green Deal Home Improvement Fund) were introduced with the intention to stimulate uptake of Green Deal finance, they were very quickly oversubscribed, with the second tranche of funding allocated within a single day, but it seems that many householders simply accepted the grant with little or no intention of supplementing this with a Green Deal finance package, as per the intended function (Rosenow & Eyre, 2015).

The second key weakness of the UK's policy mix, again aligned to the EU-level mix, is the paucity in coverage of markets and pricing instruments. The CCL and CFP/CSP (see Table 2 for spelled-out acronyms) are the UK's main energy and carbon-pricing instruments (alongside the EU ETS, applied at the EU level). Whilst the former only applies directly to the energy consumption of (all except the smallest) firms, the associated costs feed through to goods purchased by the residential sector. The latter, as described in Table 2, attempts to provide a UK-specific carbon floor price, on top of the EU ETS, which feeds through into the price of electricity. However, CO_2 emissions from remaining sources of energy in the residential sector (75% of energy consumption in 2014) remain unpriced. The CFP rate was initially intended to increase to £30/tCO_2 by 2020, and to £70/tCO_2 by 2030, but due to the low EU ETS permit price and concerns about the CSP's impacts on the competitiveness of UK industry, the CSP is currently frozen at £18/tCO_2 until 2020.

In addition, as permitted by the EU's VAT and energy directives, the UK levies a reduced-rate (5%) VAT on residential energy consumption (the standard VAT rate is 20%). The implicit annual subsidy to households that this represents is approximately £5 billion (Advani et al., 2013), although less than 30% of the implicit subsidy is taken by households in the bottom three income deciles (Preston et al., 2013). Residential energy prices in the UK currently exhibit the lowest tax and levies component in the EU (EUROSTAT, 2016). However, this is offset to some extent by on-bill levies from other climate and energy policy mechanisms (e.g. FiTs and RO/CfDs).

The third key weakness, common to all three pillars of policy, is a lack of stability and long-term credibility in the policy instrument landscape in the UK. Since its election in May 2015, the new UK government has removed or significantly amended several low-carbon energy instruments with little or no notice or indication of replacements where required to meet the UK's statutory carbon and energy targets. Regarding the promotion of low-carbon electricity, this includes removal of support for onshore wind and termination of its eligibility for the RO one year early, and an abrupt 65% reduction in the FiT (generation tariff) for small-scale (household) solar PV (< 4 kW) applicable from 1 January 2016. The Green Deal and ZCH were also both discontinued at short notice,[15] with no replacement instruments yet proposed, reducing focus on regulated energy efficiency in both new and existing residential properties. A number of other changes that fall outside the scope of this paper were also instituted.

The key reason behind such changes was the perceived high cost of these instruments.[16] Regardless of the evidence for and against the assertion of high costs, the institution of such abrupt interventions, and a lack of a forthcoming strategy with which to replace the functions of such instruments where required, has produced numerous accusations of substantially undermining investor confidence in the low-carbon sector in the future (e.g. Cuff, 2015; Gross & Watson, 2015; Grubb et al., 2015).

Suggested priority reforms

As indicated above, many of the key weaknesses in the UK's policy landscape mirror those at the EU level. As such, the suggestions for priority reform that may be derived to increase the effectiveness of these instruments are of a similar nature. Six such reforms are here proposed:

- Improve *enforcement of MEPS for new buildings*. Low enforcement (and awareness) of these requirements is a key issue in the standards and engagement pillar of policy. Such action would further lay the foundation for increased ambition, as discussed below.

- Introduce an instrument to *replace the function of the Green Deal*. The original instrument was subject to various design flaws and other barriers to uptake, as described above. However, lessons learned may be applied in order to introduce an improved instrument to fulfil the function as originally intended (e.g. Gillich et al., 2016; Rosenow & Eyre, 2015). Such a reform may also improve the functioning of the ECO (e.g. if householders were to receive a loan to finance their contribution to a measure particularly funded by ECO – currently a key barrier, as described above). This, along with the reform above, may substantially strengthen the standards and engagement pillar of policy at the UK level.

- As at the EU level, the markets and pricing pillar of policy should be extended and strengthened. This may be achieved by three key actions. The first is to extended *carbon pricing to heating fuels*, to broaden carbon pricing from 30% of residential energy consumption (electricity) in the UK to 80% of energy consumption (if a carbon price were applied just to natural gas for heating purposes).

- The second action to strengthen the markets and pricing pillar of policy is to gradually *increase the reduced rate of VAT on domestic energy consumption* (*e.g.* by 1% per year) to reduce and eventually remove the implicit energy consumption subsidy this produces. Although undoubtedly politically challenging, governments, including that of the UK, repeatedly affirm their determination to remove environmentally perverse subsidies,[17] and this is the largest such subsidy in the UK. Analysis suggests that the reform could be implemented with progressive (rather than regressive) consequences, through, for example, appropriate recycling of increased VAT revenue (Preston et al., 2013).

- The third action to strengthen the markets and pricing pillar of policy action is to restore the original trajectory of the CFP (to reach £30/tCO$_2$ in 2020 and £70/tCO$_2$ in 2030). This would strengthen the carbon price on the electricity sector, but this action seems unlikely in the absence of any strengthening of the EU ETS permit price component of the CFP.

- Introduce an instrument to *replace the function of the ZCH instrument* to restore coverage of energy efficiency in (and direct CO$_2$ emissions from) new residential properties under the strategic investment pillar of policy. This would also seek to satisfy NZEB requirements.

As many of these suggested reforms align with those at the EU level, it may be noted that reform of the policy mix at the EU level may feed through to the UK level (*e.g.* removal of the ability to levy reduced-rate VAT on residential energy consumption, or EU-level expansion of carbon pricing). However, there currently seems to be little political appetite for such EU-wide reforms (as was seen, for example, with the rejection of a revised Energy Taxation Directive proposed in 2011). Other actions may only be instituted at the national level, such as proper enforcement of building regulations (although such action may be encouraged at the EU level, through infringement proceedings), and the avoidance of abrupt interventions into the policy landscape, with unclear intentions into the medium- and long-term.

Conclusions

CO$_2$ emissions from the residential sector make a significant contribution to overall GHG emissions in the EU and constituent member states, and must be substantially reduced if the projected decarbonization of the EU is to be achieved. This paper has explored the policies that are currently in place to achieve this reduction, using the policy framework suggested by Grubb et al. (2014), which groups instruments into three 'pillars of policy': standards and engagement; markets and pricing; and strategic investment.

The key strengths and weaknesses of the policy mix at both the EU and UK levels are heavily aligned, as may be broadly expected, given the UK obligations under the EU landscape. Both policy landscapes exhibit a rich policy mix under the standards and engagement pillar of policy, with various instruments designed to tackle (both regulated and non-regulated) energy consumption and CO$_2$ emissions from both new and existing buildings (with some introduced directly at the EU level without member state-specific implementation required, such as the ecodesign and energy labelling directives). This is the first key strength of each policy mix (and the only key strength identified at the EU level). Until recently the UK went beyond EU requirements with the 'Green Deal' seeking to encourage private investment for improving regulated energy efficiency in existing buildings in addition to mandated supplier obligations. However, this instrument was effectively discontinued in 2015.

The second key strength of the UK's policy mix was (until recently) a relatively comprehensive strategic investment instrument mix, with effective renewable electricity and renewable heating support mechanisms, and until mid-2015, an instrument to promote net-zero carbon new residential buildings from 2016 (Zero Carbon Homes). By contrast, the strategic investment pillar is a key weakness at the EU level, with 'nearly-zero' energy requirements for new buildings (with low ambition in implementation by most member states), no requirement for the promotion of renewable heating and no instrument to tackle energy consumption in existing buildings from this perspective (an issue shared with the UK).

The two remaining key weaknesses across both policy landscapes are low effectiveness of the instruments under the standards and engagement pillar (despite broad coverage), and low coverage and effectiveness of instruments under the markets and pricing pillar. The first stems from low quality or under-implementation of instruments at member state level, poor enforcement and ineffective instrument design. The second stems primarily from a lack of carbon pricing on non-electricity energy products for residential consumption, and the ability to effectively subsidize residential energy consumption through reduced-rate VAT (as implemented in the UK). In addition, the UK has recently experienced significant abrupt alterations to the policy landscape, with no clear long-term strategy to replace the intended functions of instruments that have been removed, with possible impacts on investor confidence.

A range of priority reforms may be instituted to address the weaknesses identified at both the EU and UK levels. Such reforms common to both levels of governance include improved enforcement of regulatory requirements (particularly MEPS for new buildings), the expansion of carbon pricing to heating fuels coupled with the removal of reduced-rate VAT for residential energy consumption (and the implicit subsidy it entails), and reform of the strategic investment pillar to require net-zero energy new buildings and a post-2020 commitment for supplier obligations to tackle energy consumption in existing buildings. Such reforms may be introduced at the EU level for subsequent application to all member states (including the UK), or may be introduced at the UK level only. Additional reforms, such as the redesign of key 'engagement' instruments, along with the requirement (or direct encouragement) for member states to introduce renewable heat support mechanisms in the EU policy mix, and the introduction of a mechanism to replace the Green Deal at the UK level, may also be identified as priority actions.

It is clear that all these reforms would require considerable political will that is currently lacking, and is unlikely to be forthcoming in the short-term, especially as both the EU and UK seem to be on track to meet or exceed their economy-wide emission reduction targets for 2020 (EEA, 2015b). However, the EEA (2015b) project that economy-wide EU GHG emissions are likely to reach just 27% below 1990 levels by 2030 under existing measures (and 30% under 'planned measures'), falling far short of the 40% target. This places the 2050 target of 80% in significant doubt. As such, it is imperative that new policies for reducing residential energy use and associated CO_2 emissions are introduced, or the existing policy mix implemented more effectively, or both, in order to achieve the emissions trajectory for the sector illustrated in Figure 1, and in turn contribute to long-term economy-wide objectives that are at present on track to be missed. By identifying the scale of the challenge, and examining the strengths and weaknesses of the present policy landscapes at both the EU level and in a key member state, this paper hopes to make some contribution to that outcome.

References

Advani, A., Bassi, S., Bowen, A., Fankhauser, S., Johnson, P., Leicester, A., & Stoye, G. (2013). *Energy use policies and carbon pricing in the UK*. London: The Institute for Fiscal Studies.

Allocott, H., & Greenstone, M. (2012). Is there an energy efficiency gap? (Working Paper 17766). NBER Working Paper Series. Cambridge: Massachusetts, National Bureau of Economic Research.

Ashworth, A., & Perera, S. (2015). *Cost studies of buildings* (6th ed.). Abingdon: Routledge.

BPIE. (2015). *Nearly-zero energy buildings: Definitions across Europe – Factsheet* [Online]. Retrieved August 3, 2015, from http://bpie.eu/uploads/lib/document/attachment/132/BPIE_factsheet_nZEB_definitions_across_Europe.pdf

CLG. (2006). *Review of the implementation of Part L 2006*. London: Author.

Cohen, R., & Bordass, B. (2015). Mandating transparency about building energy performance in use. *Building Research & Information*, 43(4), 534–552.

Cuff, M. (2015, September 11). UK investors urge government to back renewable energy investment. *Business Green*. Retrieved January 6, 2016, from http://www.businessgreen.com/bg/news/2425520/uk-investors-urge-government-to-back-renewable-energy-investment

DECC. (2010). *The Green Deal: A summary of the government's proposals* [Online]. Retrieved December 29, 2015, from https://www.gov.uk/government/uploads/system/uploads/attachment_data/file/47978/1010-green-deal-summary-proposals.pdf

DECC. (2015a). *Digest of UK energy statistics 2015: Chapter 6* [Online]. Retrieved December 29, 2015, from https://www.gov.uk/government/uploads/system/uploads/attachment_data/file/450298/DUKES_2015_Chapter_6.pdf

DECC. (2015b). *Levy control framework cost controls* [Online]. Retrieved January 7, 2016, from https://www.gov.uk/government/speeches/levy-control-framework-cost-controls

DG Energy. (2012). *European Commission Directorate General for Energy consultation paper: Financial support for energy efficiency in buildings*. European Commission, Directorate-General for Energy, Brussels.

EC. (2011). *Energy roadmap 2050 – Impact assessment*, SEC (2011) 1565.

EC. (2014). *Energy prices and costs report*, SWD(2014) 20 final/2.

EC. (2015). *Proposal for a regulation of the European parliament and of the council setting a framework for energy-efficiency labelling and repealing directive 2010/30/EU*, COM(2015) 341 final.

Ecofys. (2012). *Towards nearly zero-energy buildings: Definition of common principles under the EPBD: Final report*. Germany: Ecofys.

Ecofys (2014a). *Overview of member states information on NZEBs: Working version of the progress report – Final report*. Germany: Author.

Ecofys. (2014b). *First findings and recommendations: Evaluation of the Energy Labelling Directive and specific aspects of the Ecodesign Directive*. Utrecht: Author.

Economidou, M., Atanasiu, B., Despret, C., Maio, J., Nolte, I., & Rapf, O. (2011). *Europe's buildings under the microscope – A country by country review of the energy performance of buildings* [Online]. Retrieved July 28, 2015, from http://www.bpie.eu/uploads/lib/document/attachment/20/HR_EU_B_under_microscope_study.pdf

EEA. (2015a). *EEA greenhouse gas – Data viewer* [Online]. Retrieved July 26, 2015, from http://www.eea.europa.eu/data-and-maps/data/data-viewers/greenhouse-gases-viewer

EEA. (2015b). *Trends and projections in Europe 2015 – Tracking progress towards Europe's energy and climate targets* (EEA Report No. 4/2015). Copenhagen: European Environment Agency.

EUROSTAT. (2016). *Energy price statistics* [Online]. Retrieved January 6, 2016, from http://ec.europa.eu/eurostat/statistics-explained/index.php/Energy_price_statistics

Gillich, A., Sunikka-Blank, M., & Ford, A. (2016). Lessons for the UK Green deal from the US BBNP. *Building Research & Information*. doi:10.1080/09613218.2016.1159500

Gillingham, K., & Palmer, K. (2013). *Bridging the energy efficiency gap: Insights for policy from economic theory and empirical analysis* (Resources for the Future Discussion Paper 13–02).

Gross, R., & Watson, J. (2015). *UKERC blog: Restoring investor confidence for low carbon power* [Online]. Retrieved January 6, 2016, from http://www.ukerc.ac.uk/news/restoring-investor-confidence-for-low-carbon-power-a-blog-by-ukerc-co-director-rob-gross.html

Grubb, M., Hamilton, I., Mallaburn, P., Ekins, P., McDowall, W., & Smith, A. Z. P. (2015). *UK energy policy: Politicisation or rationalisation?* London: UCL Institute for Sustainable Resources Research Report, UCL.

Grubb, M., Hourcade, J.-C., & Neuhoff, K. (2014). *Planetary economics: The three domains of sustainable development.* Abingdon: Routledge.

Gynther, L., Lapillonne, B., & Pollier, K. (2015). *Energy efficiency trends and policies in the household and tertiary sectors: An analysis based on the ODYSSEE and MURE databases* [Online]. Retrieved August 3, 2015, from http://www.odyssee-mure.eu/publications/br/energy-efficiency-trends-policies-buildings.pdf

Heinzle, S. L., & Wüstenhagen, R. (2012). Dynamic adjustment of Eco-labeling schemes and consumer choice – the revision of the EU energy label as a missed opportunity? *Business Strategy and the Environment, 21*(1), 60–70.

Hirst, E., & Brown, M. (1990). Closing the efficiency gap: barriers to the efficient use of energy. *Resources, Conservation and Recycling, 3,* 267–281.

HM Treasury. (2015). *Fixing the foundations: Creating a more prosperous nation.* London: HM Treasury.

HM Treasury, & HM Revenue and Customs (2010) *Carbon price floor: Support and certainty for low-carbon investment* [Online]. Retrieved December 29, 2015, from https://www.gov.uk/government/uploads/system/uploads/attachment_data/file/81273/consult_carbon_price_support_condoc.pdf

IEA. (2012). *Energy technology perspectives 2012: Pathways to a clean energy system.* Paris: IEA/OECD.

Ipsos Mori, CAG Consultants, UCL, & Energy Saving Trust. (2014). *Evaluation of the carbon emissions reduction target and community energy saving programme: Report for DECC* [Online]. Retrieved December 29, 2015, from https://www.gov.uk/government/uploads/system/uploads/attachment_data/file/350722/CERT_CESP_Evaluation_FINAL_Report.pdf

Jaffe, A. B., & Stavins, R. N. (1994). The energy-efficiency gap what does it mean? *Energy Policy, 22*(10), 804–810.

Janda, K.B., Bright, B., Patrick, J., Wilkinson, S., & Dixon, T.J. (2016). The evolution of green leases: Towards inter-organizational environmental governance. *Building Research & Information.* doi:10.1080/09613218.2016.1142811

Jordan, A., Wurzel, R., & Zito, A. (Eds.). (2003). *'New' instruments of environmental governance?: National experiences and prospects.* London: Frank Cass.

Kahneman, D., Slovic, P., & Tversky, A. (1982). *Judgement under uncertainty: Heuristics and biases.* Cambridge: Cambridge University Press.

Laibson, D. (1997). Golden eggs and hyperbolic discounting. *The Quarterly Journal of Economics, 112*(2), 443–478.

Lees, E. (2006). *Evaluation of the energy efficiency commitment 2002–2005.* Wantage: Eoin Lees Energy.

Lees, E. (2008). *Evaluation of the energy efficiency commitment 2005–2008.* Wantage: Eoin Lees Energy.

Lipsey, R. G., & Lancaster, K. (1956). The general theory of second best. *The Review of Economic Studies, 24*(1), 11–32.

London Economics, & Ipsos. (2014). *Study on the impact of the energy label – And potential changes to it – On consumer understanding and on purchase decisions* (No. ENER/C3/2013–428 FINAL REPORT). London.

Martin, R., de Preux, l., & Wagner, U. (2011). *The impacts of the climate change levy on manufacturing: Evidence from microdata* (Working Paper 17446). Cambridge: National Bureau of Economic Research.

Morewedge, C. K., Holtzman, L., & Ep_ley, N. (2007). Unfixed resources: Perceived costs, consumption and the accessible account effect. *Journal of Consumer Research, 34*(4), 459–467.

NAO. (2007) *The climate change levy and climate change agreements* [Online]. Retrieved December 29, 2015, from http://www.nao.org.uk/wp-content/uploads/2012/11/climate_change_review.pdf.

Norgaard, R. (2010). A coevolutionary interpretation of ecological civilization. Retrieved from http://neweconomy.net/webfm_send/23

OECD. (2007). *Instrument mixes for environmental policy.* Paris: OECD.

OECD. (2008). *An OECD framework for effective and efficient environmental policies.* Meeting of the Environment Policy Committee (EPOC) at Ministerial Level, 28–29 April 2008, OECD, Paris.

Ofgem. (2012). *Energy companies obligation (ECO) 2012 – 2015: Guidance for suppliers* [Online]. Retrieved December 29, 2015, from https://www.ofgem.gov.uk/ofgem-publications/75771/eco-guidance-consultation-23-november-2012-pdf

Ofgem. (2014a). *Non-domestic renewable heat incentive (RHI)* [Online]. Retrieved December 29, from https://www.ofgem.gov.uk/sites/default/files/docs/2014/10/drhi_1st_10000.pdf

Ofgem. (2014b). *Non-domestic renewable heat incentive (RHI)* [Online]. Retrieved December 29, 2015, from https://www.ofgem.gov.uk/sites/default/files/docs/2014/12/rhi_1gw_milestone_0.pdf

Ofgem. (2015a). *ECO compliance update to April 2015* [Online]. Retrieved December 29, 2015, from https://www.ofgem.gov.uk/sites/default/files/docs/2015/04/eco_compliance_update_april_2015__0.pdf

Ofgem. (2015b). *Tariff tables* [Online]. Retrieved December 29, 2015, from https://www.ofgem.gov.uk/environmental-programmes/feed-tariff-fit-scheme/tariff-tables

Ofgem, & Energy Saving Trust. (2003). *A review of the energy efficiency standards of performance 1994–2002* [Online]. December 29, 2015, from https://www.ofgem.gov.uk/sites/default/files/docs/2003/07/4211-eesop_report_july03.pdf

Pan, W., & Garmston, H. (2012). Compliance with building energy regulations for New-build dwellings. *Energy, 48*(1), 11–22.

Petitfor, H., Wilson, C., & Chryssochoidis, G. (2015). The appeal of the Green Deal: Empirical evidence for the influence of energy efficiency policy on renovating homeowners. *Energy Policy, 79,* 161–176.

Preston, I., White, V., Browne, J., Dresner, S., Ekins, P., & Hamilton, I. (2013). *Designing carbon taxation to protect Low-income households.* York: Joseph Rowntree Foundation.

Rennings, K. (2000). Redefining innovation — Eco-innovation research and the contribution from ecological economics. *Ecological Economics, 32*(2), 319–332.

Rosenow, J. (2012). Energy savings obligations in the UK—A history of change. *Energy Policy, 49,* 373–382.

Rosenow, J., & Eyre, N. (2015). *Re-energising the UK's approach to domestic energy efficiency.* ECEEE Summer Study Proceedings (1–6 June 2015), Club Belambra Les Criques, Presqu'île de Giens Toulon/Hyères, France.

Rosenow, J., Fawcett, T., Eyre, N., & Oikonomou, V. (2016). Energy efficiency and the policy mix, *Building Research & Information,* doi:10.1080/09613218.2016.1138803

Rosenow, J., Forster, D., Kampman, B., Leguijt, C., Pato, Z., Kaar, A.-L., & Eyre, N. (2015) *Study evaluating The National policy measures and methodologies to implement article 7 of the energy efficiency directive.* Didcot: Ricardo-AEA.

Rosenow, J., & Galvin, R. (2013). Evaluating the evaluations: Evidence from energy efficiency programmes in Germany and the UK. *Energy and Buildings, 62,* 450–458.

Rutherford, M. (1983). J. R. Commons's institutional economics. *Journal of Economic Issues, 17*(3), 721–744.

Simon, H. A. (1956). Rational choice and the structure of the environment. *Psychological Review, 63*(2), 129–138.

Stern, N. (2006). *Stern Review on the economics of climate change.* London: HM Treasury.

Sunikka-Blank, M., & Galvin, R. (2012). Introducing the pre-bound effect: The gap between performance and actual energy consumption. *Building Research & Information, 40*(3), 260–273.

Thaler, R. H., & Sunstein, C. R. (2008). *Nudge: Improving decisions about health, wealth and happiness.* Newhaven, CT: Yale University Press.

Tovey, K. (n.d). University of East Anglia, Carbon Reduction (CRed) project presentation. Retrieved from www.uea.ac.uk/~e680/energy/Old_modules/env2e02/powerpoint/env-2e02_section5.ppt

Waide, P., & Watson, R. (2013). *Energy labelling: The new European energy label: Assessing consumer comprehension and effectiveness as a market transformation tool* [Online]. Retrieved from http://www.clasponline.org/Resources/Resources/StandardsLabelingResourceLibrary/2013/~/media/Files/SLDocuments/2013/2013_05_EU-Energy-Lab-elling-Comprehension-Study.pdf

Wesselink, B., Harmsen, R., & Eichhammer, W. (2010). *Energy savings 2020: How to triple the impact of energy saving policies in Europe.* The Hague: European Climate Foundation.

Wurzel, R., Zito, A. R., & Jordan, A. J. (2013). *Environmental governance in Europe: A comparative analysis of the use of New environmental policy instruments.* Cheltenham: Edward Elgar.

Endnotes

[1]Much of the research reported here was carried out in the context of the EU FP7 project CECILIA2050 (see http://cecilia2050.eu/) over 2013–14. However, for this paper the research has been updated.

[2]Data for 2011 and 2012 illustrate annual CO_2 reductions of 16% and −4% (an increase) respectively on the previous year (an average annual reduction of 6%) (EEA, 2015a). Data for 2013 onwards are not yet available, and thus the presence of a long-term trend cannot be discerned (particularly due to the substantial variability seen between these two years).

[3]Although this target is economy-wide, EU ETS and transport sectors may be excluded. Other instruments such as CO_2 taxation, training and information campaigns or the creation of an 'Energy Efficiency National Fund' may also be used to secure compliance if equivalent energy savings are produced. The remaining 11 member states intend to use such mechanisms alone to secure compliance.

[4]A 'minor' credibility issue is defined as that in which 'confidence that the policy package as notified by the member state will realise 90% or more of the required target', whilst a 'major' credibility issue is defined as that in which the 'risk that the policy package as notified by the member state will realise less than 90% of the required target either due to insufficient policy savings and/or significant methodological issues' (Rosenow et al., 2015). The final member state, Portugal, has an existing EEOS instrument in place but has not notified the European Commission of its intention to use this for Article 7 compliance, and thus instrument credibility was not assessed by this study.

[5]Member states are not required to set minimum standards that are not cost-effective over the economic lifecycle of the building elements concerned, as determined using the comparative methodology framework described in Article 5 and Annex III of the directive.

[6]Data for 2013 and the EU-28 were obtained from EUROSTAT.

[7]Data source: http://www.res-legal.eu/compare-support-schemes/.

[8]DG Energy (2012) estimates that 75% of the EU's building stock standing in 2005 will remain present in 2050.

[9]This date was chosen because it was the year of the signature of the Kyoto Protocol, which set statutory GHG emission-reduction targets on industrial countries for the first time, and it was also the year of the election of a Labour government, which introduced many of the instruments to be discussed.

[10]Data source: Energy Consumption in the UK (ECUK) Statistics, table 3.35.

[11]Data for 2013, from DUKES, Chapter 3 (domestic data tables).

[12]DEFRA GHG conversion factors.

[13]Direct CO_2 emissions from the UK residential sector were 74 $mtCO_2$ in 2012, and 94 $mtCO_2$ in Germany. In proportional terms, these emissions account for 15.4% of the UK's total CO_2 emissions in that year, exceeded only by Belgium (16.4%), Ireland and Hungary (both 15.9%) (EEA, 2015a).

[14]The UK must achieve 15% of its gross final energy consumption from renewables by 2020 (from electricity, heating, cooling and transport), Under the 2001 Renewable Energy Directive, the UK was required to produce 10% of gross electricity generation from renewable sources.

[15]The ZCH instrument had been in place since 2006, for compliance to be achieved by 2016. The instrument was cancelled in mid-2015.

[16]The cost of renewable support mechanisms is projected by government to rise to £9.1 billion per year by 2020 – £1.5 billion above the limit set by the Levy Control Framework (DECC, 2015b), whilst the ZCH was removed to 'reduce net regulation on housebuilders' (HM Treasury, 2015), in order to encourage increased building rates.

[17]See, for example, the Leaders' Declaration from the 2015 G7 Summit at Schloss Elmau, which stated explicitly: 'We remain committed to the elimination of inefficient fossil fuel subsidies [. . .]' (https://www.bundesregierung.de/Content/EN/Artikel/2015/06_en/g7-gipfel-dokumente_en.html).

RESEARCH PAPER

Comparative review of building commissioning regulation: a quality perspective

Sue-Fay Lord, Sarah Noye, Jim Ure, Mike G. Tennant and David J. Fisk

Building regulations are an important policy instrument available to governments wishing to improve building energy efficiency, which should be a priority to policy-makers wishing to target cost-effective avenues in support of carbon-abatement targets. Meanwhile, building system commissioning has been recognized as a cost-effective measure to cut energy consumption, but in practice commissioning quality can deliver less-than-satisfactory outcomes. Regulation needs to better support commissioning outcomes. A five-grade commissioning scale is developed to assess the quality of commissioning and propose a common language to assist with regulation setting. Using this scale, building regulation and polices related to new and refurbished building commissioning were analysed in comparative case studies between jurisdictions England and California. This study finds that Californian regulations mandate a higher quality of commissioning and regulations that are more enforceable. The crucial elements to support better-commissioned buildings were identified as: outputs-focused regulation (not input based); regulation and process clarity; commissioning agents and building official training; as well as acknowledging the financial burden of upholding more complex building regulations. For the full benefit of commissioning to be realized, policy and regulations for existing buildings will be required.

Introduction

With climate change featuring on many political agendas, governments are seeking to reduce anthropogenic greenhouse gas (GHG) emissions in a cost-effective manner. Energy efficiency represents a unique opportunity to reduce simultaneously GHG emissions and cost. The International Energy Agency (IEA) has identified residential and commercial sectors as the largest final energy consumers with buildings being a major contributing factor (IEA, 2008, p. 7). In 2010, buildings were responsible for 32% of global final energy use (IPCC, 2014, p. 678). Clearly buildings have a substantial role to play in the efficiency agenda.

The UK has committed to reducing emissions by 80% by 2050 against a 1990 baseline (Climate Change Act 2008, Clause 1.1). Additionally as a European Union (EU) member state, the UK is required to meet 20-20-20 targets, which will see a decrease in GHG emissions by 20% and increases in energy efficiency and renewable energy each by 20% (European Commission, 2015) as well as ensuring that all new buildings will be nearly zero carbon (NZC) by 2020 (Energy Performance of Buildings Directive, 2010, Article 9, 1(a)). Similarly, California has commitment to cut GHGs and has legislated 2020 targets (California Global Warming Solutions Act 2006; California Environmental Protection Agency, 2006, clause 3855). A 2030 goal has been set for all new commercial buildings to be zero net energy (ZNE) (California Energy Commission, 2007, p. 5). Only the EU commitments are externally enforced and so UK commitments are considered more ambitious than California's.

A common method deployed to reduce energy consumption is through tightening building codes (IEA, 2008, p. 15). This approach is consistent with Belzar, McDonald, and Halverson's (2010, p. 37) US policy analysis that establishes that emissions reductions have been achieved through strengthening building codes. If governments seek to enforce provisions, thereby reducing GHG emissions through building regulations, then regulations must be strong and enforceable and coupled with inspections at appropriate intervals by qualified persons (Burby, May, & Paterson, 1998, p. 332).

Building commissioning (Cx) has been identified as a cost-effective method to avoid GHG emissions in meta-analysis by Mills (2011, pp. 154–159). This is the largest study to date of new and existing building Cx, analysing 643 commercial buildings. The study estimated the cost of emission avoidance to be –US\$110/tonne of CO_2 in existing buildings and –US\$25/tonne of CO_2 in new construction. Further, payback periods were found to be 1.1 years in existing buildings and 4.2 years in new buildings, which should be attractive even with tight budgets. This should be of interest to both governments that need to meet GHG emission reduction commitments and for clients paying for Cx services. Since Cx offers such a high rate of return, it should be a defensible option for regulation enforceable against spendthrift clients. Development of new technologies is leading to new perspectives to tackle GHG emissions. However, commissioning will remain relevant, as it is conditional to the operational success of such energy-efficiency technologies and strategies.

In the building context, commissioning now refers to a number of activities that occur at different points in a buildings life cycle (*i.e.* from construction to operation). It can be a defined stage of work within a building contract, but here it is applied more widely. There are differences in definition of commissioning for new and refurbished buildings according to location:

• The UK-based Chartered Institution of Building Services Engineers (CIBSE) provides the definition of Cx as:

The advancement of an installation from the state of static completion to full working order to the specified requirements. It includes the setting to work of an installation, the regulation of the system and the fine tuning of the system.
(CIBSE Commissioning Code M, 2003, p. 2)

• The US-based American Society of Heating, Refrigerating and Air Conditioning Engineers (ASHRAE) provide the definition of Cx as:

A quality-focused process for enhancing the delivery of a project. The process focuses upon verifying and documenting that the facility and all of its systems and assemblies are planned, designed, installed, tested, operated, and maintained to meet Owner's Requirements.
(Knebel and McBride, 2005)

Even if a building has been commissioned during construction, drifting of calibrations, ageing equipment and changes to building function can mean a building requires recommissioning (Castro & Yoshida, 2008, pp. 8–15; Lewis, Riley, & Elmualim, 2010, p. 6). In this paper, the different commissioning processes applicable to existing building will be referred to as R-Cx. R-Cx is the correct objective for long-term energy efficiency and will be discussed when applicable. However, current regulations mostly address commissioning for new system and assembly installations and will be the main focus of this paper. Unless otherwise mentioned, Cx will then refer to new and refurbished buildings.

Cx has been a feature of the commercial building industry since the introduction of building services at the beginning of the 20th century (Xiao & Wang, 2009, p. 1145). Commissioning was previously focused on setting to work of heating, ventilation and air-conditioning (HVAC), before becoming a quality process addressing all systems and assemblies (Sterling & Collett, 1994). More recently, Cx has been applied to resolve energy concerns and introduced to building regulations. Two of the first locations to introduce Cx to regulations were the UK and California. In 2002, reference to Cx was first introduced by the UK to building regulations Part L Conservation of Fuel and Power (Conservation of Fuel and Power L2, 2002, pp. 26–27). California introduced voluntary provisions in 2008 and mandatory provisions in 2010 (California Building Standards Commission, n.d., p. 1). California provides an excellent comparative case study as according to the IEA it most likely has the most comprehensive building energy efficiency standards in the world (IEA, 2008, p. 47). This study compares country and state-based cases, which is appropriate because each has the jurisdiction to establish building regulations and have made commitments to abate carbon emissions.

Despite the relevance of Cx regulations to the climate debacle, peer-reviewed research on building efficiency and regulation is scarce. Little literature is found that examines Cx in the context of regulation design, and Pitt & Sherry Consulting (2014) found that Cx is not well covered in codes and regulations. Considering the importance of the climate change agenda and the role that Cx may play, this research aims at assisting policy-makers to understand how policy can encourage better Cx, better efficiency and therefore reduced emissions.

The paper is structured as follows. The next section describes the methods used including key propositions and scope limitations. The third section proposes a model of Cx to evaluate the outcome of commissioning regulations. The fourth section inventories the regulations and policies relevant to Cx in England and California. The fifth section discusses the results of the comparison between both regulations. The sixth section gives the recommendations and conclusions of this work.

Methods

A comparative study approach was undertaken to understand how and why each regulatory authority context differs in its approach to Cx regulation (Yin, 2003). A representative selection of literature, based on keywords, source location and date range, was critically analysed to define research questions and the model of Cx established. The model of Cx is based on the different qualities of Cx, which refers to the overall benefits, such as improved energy/water efficiency or improved occupant comfort, derived from Cx. Governance frameworks for each location were constructed from literature and regulations published on online sources (Table 1).

Approved guidance and appropriate subsidiary codes were considered in light of the research questions. The intent of relevant regulations was analysed, verified and discussed through an iterative review process between the authors and key informants. Since the academic literature relating to building regulation is limited and introduction of Cx to regulations relatively recent, it was necessary to conduct a limited number of interviews to identify the most important current issues and shed light on issues not discussed in the literature or explored in regulation or guidelines. Interviews were conducted with British and Californian Cx agents (CxAs) and senior representatives from the National Environmental Balancing Bureau (NEBB), arguably the most established international certification association for training building systems experts, and CIBSE, an internationally recognized body for setting best practice and training building services engineers.

Table 1 Documents analysed

Case	Law, policy, guidance or government reports
California	California Energy Code California Code of Regulations 2013.
California	Guide to Title 24 California Building Standards Code 2010.
California	Non-residential Compliance Manual 2013.
California	California Green Building Standards Code California Code of Regulations 2013.
California	California Global Warming Solutions Act 2006.
Europe	Council Directive 2010/31/EU on The Energy Performance of Buildings (recast) [2010]. OJ L 153.
Europe	Council Directive 2012/27/EU on Energy Efficiency [2012]. OJ L 315.
Europe	Progress by Member States Towards Nearly Zero-Energy Buildings 2013.
United Kingdom	Climate Change Act 2008.
England and Wales	The Building Regulations 2010.
England and Wales	The Building (Amendment) Regulations 2012.
England and Wales	The Energy Performance of Buildings Regulations 2012.
England and Wales	Improving the Energy Efficiency of Our Buildings: A Guide to Air Conditioning Inspections for Buildings 2012.
England and Wales	F1 Means of Ventilation 2010.
England and Wales	The Energy Performance of Buildings 2012.
England and Wales	CIBSE Commissioning Code M 2003.
England	Conservation of Fuel and Power L2A 2013.
England	Conservation of Fuel and Power L2B 2013.

Finally, English and Californian contexts were compared by contrasting methods of enforcement and gap analysis between content. Findings were synthesized and policy recommendations generated.

Some limitations are acknowledged. Firstly, it was necessary to distinguish between qualities of commissioning. This research adopts Mills' (2011) findings that a higher quality/comprehensiveness of Cx will result in greater energy and water savings and that a Technical Cx approach is superior to a Process Cx approach. The distinction between Process and Technical Cx is made by McFarlane (2013) as being that Technical CxAs will personally test every building system and service when Process CxAs would samples the results of the test made by contractors. By inference this means that a Technical Cx should have a deeper understand of how other building systems may impact on the expected readings as well as be better able to trouble shoot issues. This makes Technical CxAs personally responsible for accurately documenting any inaccuracies, allowing one specialist systems expert to see a building as a system rather than the sum of the individual parts (McFarlane, 2013, pp. 33–34).

Secondly, this study focuses on the minimum required by building regulations and so therefore there are some items that are out of scope even though these items may also play a role in the final outcome. These include display certificate regulation (energy performance certificates (EPCs) and Energy Star); differences in normal contractual practices (*e.g.* which party hires the person responsible for delivering successful Cx); publically funded incentives; and certification schemes (*e.g.* Leadership in Energy and Environmental Design (LEED) and Building Research Establishment Environmental Assessment Methodology (BREEAM)). Finally, the inherent nature of case study analysis means for results to be extrapolated to other locations context specific differences will need to be considered.

Model of commissioning

The IEA (2010, p. 1) reports a lack of consistent definition between Cx processes internationally, and so to interpret the outcome of regulatory requirements it was necessary to establish a model that distinguished between qualities of Cx.

Mills (2011, p. 152) reports that Cx varies substantially in scope and ambition, which may be reliant on client budget and project size. However, a lack of consistent terminology applied in the literature may also contribute. The same study also finds a positive association between the comprehensiveness of Cx and impact or outcomes. It is acknowledged that Mills' (2011) research focused on energy consumption, and for a holistic perspective, consideration of all resources including water is necessary (*e.g.* water systems and sprinklers) (US Department of Veteran Affairs, 2013), increased lifespan and cost avoidance from equipment failure, reduced operational and maintenance costs (IEA, 2010, p. 158) and finally occupants' comfort (CIBSE, 2008), which is linked to employee productivity (IEA, 2010, p. 159).

Figure 1 is proposed as a reference point and has been synthesized from aggregating existing models (ABS Consulting, 1998, p. 17; Noye, Fisk, & North, 2013, p. 3). Additionally, findings from other sources have been integrated. This includes Claridge et al. (2004, p. 19), Liu (1999, pp. 46–56), and Xiao and Wang (2009, p. 1145), who find the highest quality of Cx occurs regularly over the lifetime of a building, and NEBB's definition of the highest quality of Cx as whole building technical commissioning (WCx):

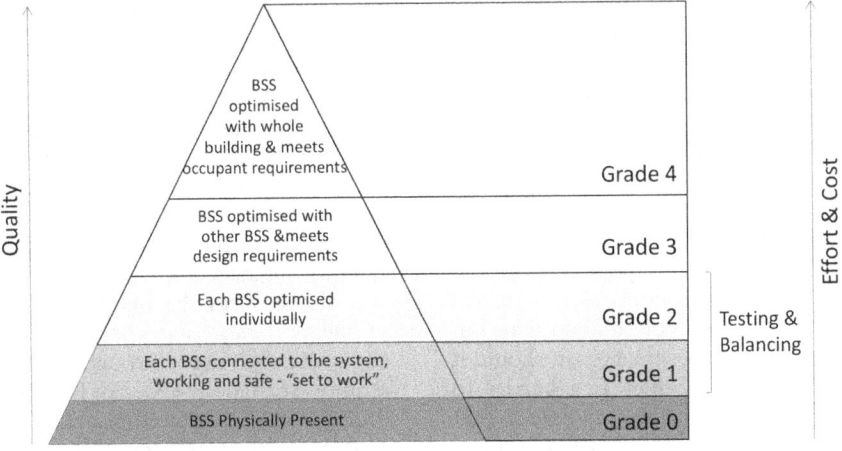

Figure 1 The quality model of commissioning

Cx of all building [systems and] services including building envelope, HVAC, electrical, special electrical (fire alarm, security & communications), plumbing and fire protection.

(NEBB, 2014, p 50)

Figure 1 explains the quality commissioning model. Each grade builds on the former in increasing quality (efficiency balanced with comfort), effort and cost. It is the role of regulation to set the minimum grade of Cx to ensure new buildings reach acceptable performance. However, it is the client's responsibility to establish which grade above the minimum they believe provides the best trade-off between cost and quality. Cx quality can also indirectly be affected by decisions made during the different stages of the construction process.

Grade 0 is the lowest level and only requires building systems and services be physically present irrespective of any other requirements. This level will be achieved for any building and therefore it is important to draft it well. For example, in the case of renewable energy, Grade 0 would make sure that the specified solar panels are attached to the roof, but would not certify that it is wired to the system.

Grade 1 ensures each building system and service is installed; its components work in isolation, is connected to the system, and is safe. It is equivalent to the British interpretation of 'set to work' and will ensure that the system turns on, but makes no claim to how well it works. For example, solar panels may be installed under shade, thus reducing energy production. If regulation were directed at producing a specified amount of renewable energy, Grades 0 and 1 would be avoided.

Grade 2 ensures that each building system and service has been optimized (usually for efficiency) in isolation of other building systems and services. Grades 1 and 2 are the equivalent of testing, adjusting and balancing (TAB) in the US. As outlined by Nakahara (2003, p. 2), the main focus of TAB is on HVAC, although testing and adjusting can be applied to any building system or service, whereas Cx focuses on the entire building and takes into consideration the interrelated nature and impact of building systems and services on each other. As such, TAB results in an operable building, whereas Cx ensures the system (building envelope, electrical, plumbing and fire systems) functions optimally. All grades to this point require the input of installers. According to interviews, installers rarely receive manufacturer training and so may not have sufficient expertise to adjust the system should it not function to specifications, which is a barrier to Grade 2 Cx. The exception being any issues that could bring unsafe plant on stream where trained engineers are required.

At Grade 2 an installer requires skills to competently test and adjust outputs (e.g. airflow, temperature and humidity) to match design specifications. Peter Kinsella, CIBSE President says:

As installers tend to focus on their own building systems and services they are unlikely to initiate adjustments required by other contractors to ensure that the building works as an interrelated system rather than the sum of the individual parts.

(Kinsella, personal communication, 2014)

This issue is extended by McFarlane (2013, p. 29) who provides:

Specialists [installers] do not understand the 'big picture' of how all of a building's various systems interact, nor do they speak the same language as other trades.

To achieve Grade 3, an integrated or systematic approach is required. For this grade, Cx also extends to other elements such as the need to provide client specification, integrated design of building systems and services, a Cx plan, operator training and documentation as well as reporting to ensure benefits are realized. As such this requires: contributions from multiple installers and a coordinator; or installers to have systems expertise; or extremely careful design, necessary because system settings are interdependent. For example, low-power LED lighting emits less heat and so HVAC may need adjustment. This grade also requires activities be scheduled across design, construction and up to handover and will likely require expertise beyond any individual. Therefore, to achieve this level a coordinator/CxA is required. In this context, a CxA is not required to have any particular skills or experience. A CxA completing Process Cx generally reviews contractor reports and witnesses samples from TAB results to verify completion.

Grade 4 is the highest level of Cx and is consistent with the definition of Technical WCx, a concept that appears to be absent from existing European literature. Interviews substantiate that this concept may be absent from the language of English practitioners. As with Grade 3, the Cx timeline extends beyond functional performance testing. However, this grade better considers the entire building as a system, requires the coordinator to be technically competent and certified (e.g. systems engineer with US-based NEBB certification), and also ensures the building suits the requirements of the occupier as it changes with climatic seasons and potential uses throughout the year. This will require regular R-Cx. A Technical rather than a Process CxA is required to achieve this level. Grade 4 recognizes the importance of ongoing adjustments after handover to ensure a building meets occupant needs, which can vary throughout the lifetime of a building and even between design and occupation.

Evidence from interviews (although no reference could be found in literature) point to Quality Cx not being performed regularly, more so in England than California.

The Quality Cx Model proposes four grades to assess the outcomes of building regulation on Cx. As the grade of Cx increases, as does the skill level of CxAs (and building inspectors) and therefore training is considered to be a substantial barrier to broad application of Cx. Similarly, the time required to deliver and therefore costs are expected to increase in accordance with quality. These attributes of the model correlate with the likelihood that Quality Cx is being performed on a broader scale, This model forms the basis of understanding how regulation in England and California differs.

English and Californian commissioning regulations

The following section analyses Cx and relevant regulations in both England and California. It identifies gaps in each context and reveals areas that have developed differently.

England
Regulatory framework
Section 1, Clause 1 of the Building Act (1984) allows for building regulations in England to be established by the UK government. As such the regulations relevant to this section are: the Building Act (1984) (UK); The Building Regulations (DCLG, 2010b) and associated amendments (England and Wales); Conservation of Fuel and Power L2A (HM Government, 2013b) and L2B (HM Government, 2013c) and associated amendments (HM Government, 2013a), also referred to as the Approved Documents; and CIBSE Cx Code M (CIBSE, 2003). Additionally, the Energy Performance of Buildings (England and Wales) Regulations (DCLG, 2012a) and associated guidance Improving the Energy Efficiency of Our Buildings (DCLG, 2012b) were analysed.

The Building Regulations (DCLG, 2010b) provides minimum mandatory standards. The approved documents are an option to people who do not wish to use their own methods to meet the regulations. However, since the regulation is silent on what occurs when a person chooses not to follow this method, it is reasonable to analyse the method outlined in the approved documents. Additionally, as the method approved by the secretary of state is that outlined by CIBSE Commissioning Code M (CIBSE,

2003), even though it was never designed as a vehicle for regulatory action, analysis of this document was also necessary.

Depth and breadth of Cx regulation
English Cx regulations apply to all newly constructed buildings. They are also triggered for existing buildings being extended or altered when changes result in new or replacement controlled services or fitting and other limited circumstances. As such, regulations apply to new construction and the opportunity to accrue benefits in existing building stock limited (Figure 2).

A lack of continuity between the Building Regulations (DCLGb, 2010), approved documents and CIBSE Cx Code M (CIBSE, 2003) were identified (Lau, 2014, pp. 13–17). There are inconsistencies between definitions of Cx and discrepancies between requirements. It also lacked clarity between distinguishing mandatory or voluntary provisions. The brevity of Cx regulation, including a failure to define Cx in mandatory regulation, affords greater latitude when interpreting the law. If Cx is not defined in enforceable regulation then there is potential to argue a lower quality of Cx meets the requirements.

Listed below are the attributes that lead the authors to believe that minimum Cx required by English regulation is equivalent to Grade 2:

- Schedule 1 Part L of the building regulations refers to providing fixed building services which – (iii) are commissioned by testing and adjusting as necessary to ensure that they use no more power than is reasonable in the circumstance (DCLGb, 2010, p. 39). Although not specific, this implies the intention of Cx is to improve efficiency ($Cx > 1$).

- The definition of Cx encompasses a wide range of activities including set to work, balancing and adjusting to ensure that building systems and services run efficiently. Failure to distinguish between processes could lead to the incorrect understanding that Cx is equivalent to setting to work ($Cx = 1$ to 2).

- Cx is required only for controlled services, fixed lighting, hot water and HVAC systems installed in new or existing buildings, however standards apply individually to each component rather than employing an integrated or systematic approach ($Cx < 3$).

- Cx does not extend to the building envelope, however thermal values and pressure testing are covered under separate regulation ($Cx < 3$).

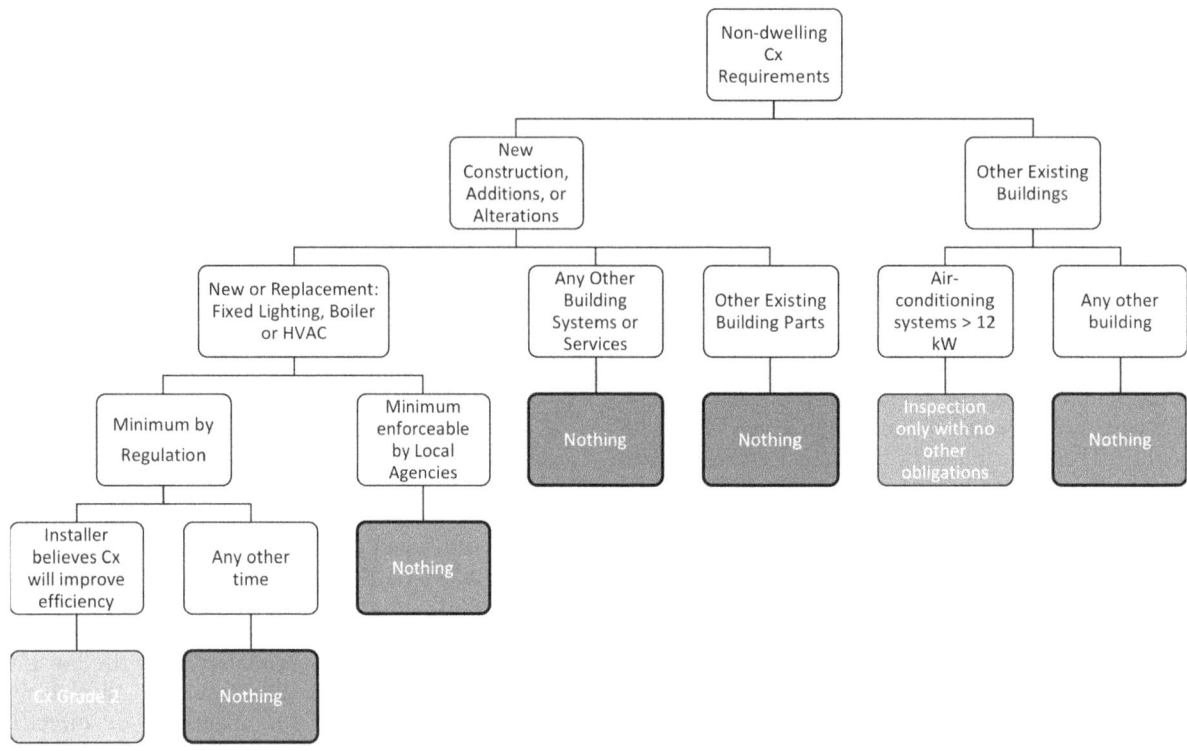

Figure 2 Limitations to the scope of building commission regulation in England and Wales

- Although some qualifications or experience are suggested no requirement for persons responsible for Cx exists (Cx < 3).

- There is no requirement for a CxA (Cx < 3).

In addition, Part 4 of the Energy Performance of Buildings requires any air-conditioning systems of 12 kW or more be inspected by an approved and qualified energy inspector every five years (DCLG, 2010a), which is a step towards regulating R-Cx. Inspections are mandatory and provide recommendations on improving efficiency; however, no legal requirement exists for recommendations to be acted upon (DCLG, 2012b, p. 8). As such, no enforceable requirement to improve the efficiency of existing buildings exists; however, reports may alert owners to air-conditioner issues and encourage action.

Cx regulations are applicable to newly installed controlled building systems and services in new and refurbished buildings and the minimum grade of Cx equivalent to Grade 2. The regulation governs inputs rather than outputs. It could be argued that a higher grade is achieved through R-Cx since air-conditioners inspected over the lifetime; however, as the regulation is aimed at reporting rather than consumption levels, this provision offers limited offset against unambitious

Cx regulation. In addition, this includes only one of the systems that should undergo R-Cx. The next step is to consider the ability of local authorities to enforce these measures.

Enforcement

Enforcement should be considered in light of: who is responsible; motivations; enforcement method; and the power of regulators to apply penalties.

In England and Wales building inspectors can outsource work to persons deemed to be 'approved inspectors' (DCLG, 2010a, p. 4). This would be normal in cases involving difficult structural engineering. The regulation requires inspectors to observe a document confirming that Cx is complete. Unfortunately no public data are available that stipulate the level of Cx training local authorities or inspectors receive. Furthermore, the regulations are silent on actions required when Cx is completed by untrained persons. However, given the level of training required to commission a building well and the level of coherence of regulations, it is unlikely local authorities or inspectors are trained to a professional level on auditing Cx. This is consistent with anecdotal reports from industry practitioners.

Another aspect is the quality of information received by local authorities. The primary method of enforcement is for each installer to submit a notice to the local authority confirming the controlled service or fitting has been commissioned in accordance with CIBSE Cx Code M (CIBSE, 2003), only necessary when the installer decides a building system and service would be more energy efficient if commissioned (DCLG, 2010b, Clause 44(2)). L2A suggests a Model Cx Plan (BSRIA BG 8/2009) be completed and submitted for all new non-dwelling construction (HM Government, 2013b, Clause 3.16), which requires details of all building systems and services irrespective of whether Cx is required by law. If required in all circumstances, local authorities would have sufficient detail to conduct a desk audit. However, in absence of a plan, local authorities have no knowledge of installed building systems and services without a thorough site inspection, which may require removing walls or fittings to gain access, particularly if Cx was not considered during design.

It is also necessary to understand the likelihood of provisions being enforced. Since enforcement is costly, unless safety is a concern, no short-term financial incentive exists for local authorities to enforce beyond the minimum requirements necessary. Clause 11 of the Building Regulations (DCLGb, 2010, p. 10) have provisions which allow local authorities the *power to dispense or relax requirements*. This clause applies to almost all provisions in the building regulations, except for those necessary to satisfy EU directives, thus local authorities have the option of not enforcing Cx provisions at all.

The final aspect for consideration is level of deterrence. Should local authorities decide owners are in breach of regulations, they may withhold approval required prior to building occupation, until satisfied (DCLG, 2010b). Failing this, local authorities may sue through the magistrates court imposing penalties of £50/day and up to £5000 in fines (Building Act 1984). As such, the power to enforce provisions is considered relatively substantial.

In theory, the Building Regulations give local authorities both the power and deterrence mechanisms to enforce Cx regulations (DCLG, 2010b), but issues no requirement for local authorities to enforce provisions. Since building regulations do not allow local authorities to insist on a model Cx plan, local authorities have no ability to assess whether a building is compliant without onerous and possibly invasive site inspections. Since there is a potential lack of training, it is unlikely that local authorities are tooled to assess compliance even if a site inspection were completed. These findings are consistent with interviews with English industry representatives reporting that Cx provisions lack enforcement.

California

After analysing English regulations relevant to Cx, this section will perform the same investigation for California as a point of comparison.

Regulatory framework

The 1787 Constitution of the United States of America establishes the rights of congress, presidents and federal courts over certain national and cross-jurisdictional matters, including federal buildings. California retains the right to create and enforce building regulations through the California Code of Regulations. Part 6 of the California Energy Code (CEC) provides regulations for minimum energy-efficiency standards and techniques for most non-residential buildings except hospitals, nursing homes, daycare centres or prisons (California Energy Commission, 2012). The CEC also addresses hospitals, nursing homes daycare centres and prisons in other parts; however, these have not been considered as part of this study. Part 11, California Green Building Standards Code (CALGreen) provides additional green building standards and Cx requirements for all new construction (California Energy Commission, 2013). A 2016 Code exists; however, this was not considered at the time the study was conducted. The guide to building energy efficiency standards Non-Residential Compliance Manual (California Energy Commission, 2015) explains how to interpret the code and, unlike Approved Documents L2A and L2B, guides are enforceable. California's building codes provide an option to meet requirements via either performance or prescriptive means, however the option for performance is not extended to Cx provisions.

Depth and breadth of Cx regulation

According to Clause 2.1 Non-residential Compliance Manual Californian Cx regulations apply to new construction and altered parts of existing buildings (California Energy Commission, 2015). As such, the opportunity to accrue benefits in existing buildings is limited. Figure 3 explains how this plays out in new and existing buildings in California.

Model energy codes are published by ASHRAE and the International Codes Council; however, California opted out (VanGeem, 2014), instead extending ASHARE 90.1-23010 to provide more stringent requirements (Ware & Bozorgchami, 2013, p. 16).

On an industry level, the US defines Cx separately from TAB (ASHRAE, 1988). This understanding is mirrored in CALGreen, which stipulates testing and adjusting in circumstances where Cx is not required (California Energy Commission, 2013, Section 5.410.4). Balancing is also required for HVAC. Since TAB in this context ensures that building systems and services have been adjusted in accordance with manufacturer

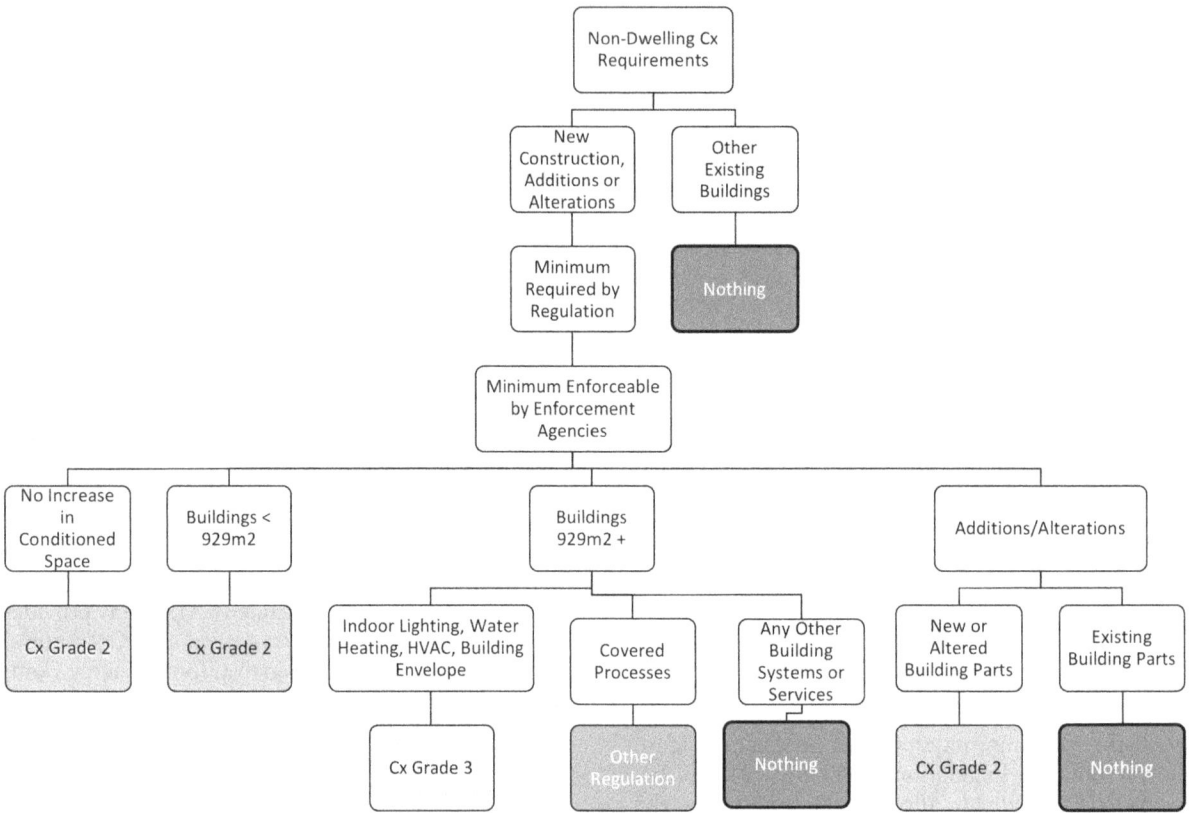

Figure 3 Limitations to the scope of building commission regulation in California

specifications TAB provisions are equivalent to Cx Grade 2.

Since TAB is equivalent to Cx Grade 2, Cx in new buildings with conditioned space larger than $929\,m^2$ is likely higher than Grade 2. Listed below are the other attributes, which lead authors to believe the minimum required by regulation is Grade 3:

- Even though a more comprehensive list of building systems and services is included (indoor lighting, water heating, HVAC and building envelope), the list is not exhaustive, notably excluding water systems (Cx < 4).

- The regulations require an individual be responsible for coordination (California Energy Commission, 2015, Clause 12.1) (Cx = 3–4).

- Regulations require Cx be considered during design and throughout construction. The regulation requires schedules for preventative maintenance and plans for future seasonal testing (Section 120.8 California Energy Code, 2013) (Cx = 3–4).

- Clause 12.1.1 of the Non-Residential Compliance Manual sets out some elements that a client may

use to assess the background or appropriateness of an assessor, however no minimum levels of experience or training are specified (Cx < 4 (California Energy Commission, 2015)).

As such Cx is equivalent to Grade 3. There are some concerns for lesser quality outcomes from a Process Cx approach, which is encouraged through the method outlined in guidance. Results from interviews highlight that Cx responsibilities may be delegated to the 'cheapest available person', raising concerns for quality due to a lack of experience or training.

Similar to England, regulations also focus on the process of Cx rather than the outcome of a well-commissioned building. Regulating outputs would demonstrate that Cx has been completed to a specified quality, but would require a much greater technical understanding.

It has been established that Cx Grade 2 applies to most new construction and Cx Grade 3 to a small subset of larger buildings with conditioned space. Despite this, concerns are raised about quality when the role of CxA is assigned to inexperienced personnel. Issues could be addressed if outcomes rather than process were regulated.

Enforcement

Enforcement has been analysed in terms of training building inspectors, the motivations of enforcement agencies, methods of enforcement and access to sufficient penalties to deter non-compliance.

Interviews point to the level of training building inspectors receive being a concern. In particular, as building regulations have become increasingly stringent, fees have not been increased accordingly. Enforcement agencies are not being compensated to provide time and training for building inspectors required to enable them to assess compliance.

That said, the enforcement framework should provide sufficient information if issues with training are addressed. The regulations provide an eight-step process enforceable at multiple points. It requires an individual be nominated as CxA and responsible for submitting documentation and personal affirmations that each step has been completed appropriately and in accordance with the regulations. CxAs are required to certify enforcement forms under *penalty of perjury* and if found falsifying documents may be fined up to US$25 000 and/or gaoled for one year (CA Penal Code § 118–131, 2005). As such the level of deterrence is considered to be high.

A well-laid-out process with sufficient checkpoints allows enforcement agencies to check whether buildings meet regulations. This is necessary as enforcement agencies may be held accountable for not enforcing provisions. Furthermore, enforcement agencies are allowed to issue more restrictive requirements as necessary due to local conditions (Section 18941.5 California Health and Safety Code, n.d.).

Californian enforcement provisions have been deemed relatively fit for purpose; however, issues exist with training building inspectors to understand how to enforce provisions and allocating additional funding for the increased costs of enforcement. The next section discusses context specific findings.

Comparison and analysis of English and Californian regulations

This investigation has revealed key differences between contexts, in particular the depth and breadth of Cx regulation and enforcement frameworks.

Existing legislation and current government mandates will see building codes largely focus on Initial Cx, which occurs during/directly after construction rather than throughout the lifetime of the building. Even though under the current framework, enforcement would be difficult, it would follow that separate regulation would address existing buildings.

When modern codes apply, the depth and breadth of California's Cx requirements are comparatively more stringent and guidance documentation far more advanced than England. This study focused on regulation rather than practice and the authors assume that commitment is similar to enforcement.

The minimum Cx required in California is Grade 2 in most circumstances where modern codes apply, except larger buildings with conditioned space that require Grade 3 Cx. Comparing with England, regulations require Grade 2 Cx, for a subset of building systems and services and only when the installer believes commissioning is beneficial. This finding highlights the importance of language and definition when specifying qualities of Cx.

The specificity of mandatory regulation is substantially different between contexts. Less specificity in English regulation allows for broader interpretation. On the one hand, the regulation could be exploited, as it is possible to meet the regulation without quality Cx, or possibly none at all. On the other hand, optionality to meet the regulations through alternative methods allows scope for innovation and so as the industry evolves (*e.g.* introduction of auto-commissioning) fewer updates will be required. The reverse can be said for California where the certainty of outcomes is greater, but policy-makers carry a greater responsibility to ensure that regulations allow for new methods.

Another outcome of new methods could be the further development of simulations to test outcomes that could put a new pressure on Cx to deliver beyond 'set to work'. There may be future prospect for systems to be designed to auto-commission from black-start. This is likely to induce a shift from building regulation to equipment standards as bearing responsibility for Cx quality. Although the technology is unproven, these issues make for potentially interesting future research questions.

Even though the level of deterrence has been determined adequate in both contexts, the final outcome will be dependent on the level of enforcement. Training of persons responsible for either declaring that Cx is complete or verifying compliance has been highlighted an issue. This is confounded by the need for greater expertise as the quality of Cx increases, as predicted by the quality model of Cx, as well as with increased project size. If either context desires a greater level of certainty that buildings comply with Cx regulations, then clear training of building inspectors and certification requirements of CxAs must be addressed. Where this results in greater burden to responsible persons, then funding and/or incentives need to be increased accordingly.

It is anticipated that the likelihood of provisions being enforced is greater in California than England. California provides far clearer regulation and guidance as well as a greater number of consistent enforcement checkpoints, whereas English regulation cannot be practically enforced without multiple physical site inspections. Also, unlike England, California requires building regulations be enforced with repercussions should enforcement agencies not comply and so the incentive to enforce provisions is greater in California than England.

Finally, Californian regulations assigns Cx responsibilities to an individual, which not only alleviates communication issues between parties and diffusion/dilution of responsibilities, but also gives enforcement agencies a central point of communication and accountability.

In summary, the comparison between English and Californian policies finds that Cx regulations and policies substantively apply to new construction, largely omitting existing structures. In cases where modern codes apply, a lack of definition could mislead the quality of Cx required. Californian regulation is far more specific and detailed than English regulation, however this comes at an increased responsibility for regulators to keep regulation up to date. Enforcement is an issue in both contexts, but for different reasons and is of greater concern in England than California.

Conclusions

In an effort to reduce anthropogenic-fuelled climate change, California and England have implemented mandatory Cx requirements in building regulation with different approaches and enthusiasm.

In this perspective, the quality model of Cx, which provides the basis to define differing qualities of Cx, has been established and utilized to inform case studies. The proposed model offers an appropriate method of distinguishing between differences in terminology used in each context but is also relevant to the international setting. Adoption will enable policy-makers and enforcement bodies to have informed conversations about Cx, thus reducing the need for specialist training. A common language will reduce complexity and improve consistency, therefore reducing barriers to entry for other countries considering methods to reduce GHG emissions.

For countries considering reducing GHG emissions through updating building regulations, the authors recommend that Grade 3 Cx in larger commercial buildings and Grade 2 Cx in all other buildings is an achievable minimum standard for any country with established building codes. In order for Grade 4 to be achieved, governments would need to be certain that a sufficient number of trained CxAs who can deliver quality Cx are available. Although it is important to acknowledge the existence of Grade 4 Cx as clients should be offered the opportunity to access best practice Cx so they can make an informed trade off between cost and quality.

Cx regulation in both contexts focus on inputs rather than outputs. If the final goal is to mitigate climate change by reducing GHG emissions then equally regulation should focus on outputs rather than inputs. To do this it would be necessary to regulate to at least Grade 3 Cx as it provides for reporting of building performance during occupation. Among others, energy and water consumption should be key indicators for outputs based regulation. As such, in an ideal world, building regulation would be targeted at reducing these throughout a buildings lifetime. Rather, there would appear to be an overreliance on equipment installed as opposed to operation with little acknowledgement of the embedded environmental cost of manufacture. The authors acknowledge that it is important to install efficient equipment. However, in many circumstances it is the way the equipment is operated which is the key to driving true gains in efficiency.

In addition, the mechanism of enforcement being scheduled for building occupation has focused legislative requirements on Cx. If outputs were regulated for existing buildings, there would be a greater incentive for building owners to invest in Ongoing-Cx or Grade 4 Cx. Of note, this point is relevant to much broader regulation than Cx alone. California has recognized this in performance-based codes, but the concept is yet to be extended to Cx provisions. Presumably this is due the complicated nature of setting points suitable for every building and type of habitation. Alternatively, if promises for auto-commissioning are realized, the focus could be shifted to specify a quality of equipment. Nevertheless, future research to identify a performance-based Cx code would make an invaluable contribution.

In the meantime, it is important to acknowledge the ultimate objective and therefore consider building regulation in light of its ability to deliver reduced consumption. There is little value in creating regulation which applies to very few buildings and so the breadth of applicability of Cx regulation to the number of new buildings is relevant.

Californian regulation is more likely to provide more consistent and higher minimum quality Cx at the expense of increased burden on policy-makers to ensure regulation remains current. The key will be in striking a balance between cost and quality/quantity that extends beyond any individual policy portfolio. Even though English policy-makers may initially save costs by writing non-specific policy, if regulation is ineffective or even prevents enforcement, emissions reductions are likely to be compromised. If this results in EU fines for failing to meet 20-20-20 targets or changes to climatic patterns, the final cost needs to be weighed against any savings. The irony in this being that even though California has invested more in building regulation, this state (or even the US as a whole) is less likely to be held to account for a failure to meet carbon-abatement commitments due to a lack of legislated external penalties. Further investigation to quantify all costs and benefits including on flow affects would assist governments to prioritize resource allocation better.

Enforcement has also been revealed a key issue in both contexts, with England suffering from multiple confounding issues. Substantial gains could be made from adopting a framework of enforcement that enables Cx provisions to be verified with minimal site inspections. As an easy first step such a framework would require building regulations be updated to allow local authorities to enforce submission of Cx plans. Another important, but arguably more contentious, step would be to mandate local authorities to uphold and enforce the provisions of the building regulations. However, it is also necessary to acknowledge the time impost on assessors, for training and assessment.

For any country wishing to implement Cx regulation, it is critical to ensure the steps of enforcement are thought through in context with regulation, policy and guidance. There is little point in regulating a minimum standard of Cx without a system that enables and supports enforcement. Only when this is in place is it reasonable to expect enforcement agencies to enforce provisions.

For existing buildings, the situation is different. Since the final opportunity for enforcement is when the certificate of classification is issued, additional energy efficiency policy will be required to encourage R-Cx through regulations. This will pose specific complications with enforcement as there can be no threat of withholding occupancy. Furthermore, the highest quality of Cx occurs periodically throughout the lifecycle of a building. It would be cheaper and arguably more effective for governments to focus on regulation which influenced operational costs rather than regular reporting/auditing. As such the authors recommend a building tax based on gas, electricity and water consumption would offer a suitable mechanism for encouraging more efficient buildings and by inference quality Cx. This would provide an offset for the cost of implementing better policy, training and enforcement.

In conclusion, if policy-makers wish to make a genuine impact on GHG emissions through building regulation, a broader perspective must be adopted. Siloed portfolios and an inability to acknowledge the flow through affects of decision making (or even a failure to act) can prevent deployment of effective building Cx regulation. A common language and a focus on outputs sets the scene for better quality commissioning linked to better emissions reductions. Broader and more specific regulation will mean better outcomes. Any regulation is underpinned by appropriate enforcement provisions and a budget to match effort. These findings should be of interest to any country seeking to improve regulatory outcomes.

Acknowledgments

The authors would like to acknowledge the helpful comments from the reviewers.

References

ABS Consulting. (1998). *Commissioning Buildings - An Opportunity to Reduce Energy Costs*. London.

Belzar, D., McDonald, S., & Halverson, H. (2010). A Retrospective Analysis of Commercial Building Energy Codes: 1990-2008. US Department of Energy. Report number: PNNL-19887. Retrieved May 2, 2016, from: http://www.pnnl.gov/main/publications/external/technical_reports/PNNL-19887.pdf

Burby, R. J., May, P. J., & Paterson, R. C. (1998). Improving compliance with regulations: Choices and outcomes for local government. *Journal of the American Planning Association*, 64(3), 324–334. doi:10.1080/01944369808975989

California Building Standards Commission. (N.D.) *The CalGreen Story*. Retrieved July 2, 2015, from: http://www.documents.dgs.ca.gov/bsc/CALGreen/The-CALGreen-Story.pdf

California Energy Commission. (2007). Integrated Energy Policy Report. Report number: CEC-100-2007-008-CMF. Retrieved May 2, 2016, from: http://www.energy.ca.gov/2007publications/CEC-100-2007-008/CEC-100-2007-008-CMF.PDF

California Energy Commission. (2012). *California Energy Code California Code of Regulations 2013: Building Energy Efficiency Standards for Residential and Non Residential Buildings*. Title 24, Part 11. California. Retrieved May 2, 2016, from: http://www.energy.ca.gov/2012publications/CEC-400-2012-004/CEC-400-2012-004-CMF-REV2.pdf

California Energy Commission. (2013). *California Green Building Standards Code California Code of Regulations 2013*. Title 24, Part 11. California. Retrieved May 2, 2016, from: https://law.resource.org/pub/us/code/bsc.ca.gov/gov.ca.bsc.2013.11.pdf

California Energy Commission. (2015). *The Building Energy Efficiency Program Title 24, Part 6 and associated administrative regulations of Part 1: The Non-Residential Compliance Manual for the 2013 Building Energy Efficiency Standards*. Sacramento. Retrieved May 2, 2016, from: http://www.energy.ca.gov/2013publications/CEC-400-2013-002/CEC-400-2013-002-CMF-REV2.pdf

California Environmental Protection Agency. (2006). *Division 25.5 of the Health and Safety Code, California Global Warming Solutions Act 2006, Assembly Bill No. 32, Chapter 488*. California. Approved by the Governor on 27 September 2006. Retrieved May 2, 2016, from: http://www.leginfo.ca.gov/pub/05-06/bill/asm/ab_0001-0050/ab_32_bill_20060927_chaptered.pdf

Castro, N., & Yoshida, H. (2008). *The International State of Building System Commissioning*. The ACEEE Summer Study on Energy Efficiency in Buildings. Retrieved May 2, 2016, from: http://aceee.org/files/proceedings/2008/data/papers/8_45.pdf

CA Penal Code § 118-131. (2005). Leg Sess. Retrieved May 2, 2016, from: http://law.justia.com/codes/california/2005/pen/118-131.html

Chartered Institute of Building Engineers. (2003). *Commissioning Code M 2003*. ISBN 1903287 33 2. Norwich. Retrieved May 2, 2016, from: http://www.cibse.org/knowledge/cibse-cc/commissioning-code-m-commissioning-management

Chartered Institute of Building Engineers. (2008). *Commissioning Guide M*. ISBN 978-1-903287-93-4. Norwich. Retrieved May 2, 2016, from: http://www.cibse.org/knowledge/cibse-guide/cibse-guide-m-maintenance-engineering-management

Claridge, D. E., Turner, W., Liu, M., Deng, S., Wei, G., Culp, C., Chen, H., & Cho, S. (2004). Is Commissioning Once Enough? *Energy Engineering, 101*(4), 7–19. doi: 10.1080/01998590409509270

Climate Change Act, Chapter 27. (2008). Enacted by Her Majesty the Queen on 26 November 2008. Retrieved May 2, 2016, from: http://www.legislation.gov.uk/ukpga/2008/27/pdfs/ukpga_20080027_en.pdf

Department for Communities and Local Government. (2010a). *Building and Buildings, England and Wales, The Building (Approved Inspectors etc.) Regulations 2010*. No 2215 of the Secretary of State made on 6 September 2010. London. Retrieved May 2, 2016, from: http://www.legislation.gov.uk/uksi/2010/2215/introduction/made

Department for Communities and Local Government. (2010b). *Building and Buildings, England and Wales, The Building Regulations 2010*. No 2214 of the Secretary of State made on 6 September 2010. London. Retrieved May 2, 2016, from: http://www.legislation.gov.uk/uksi/2010/2214/pdfs/uksi_20102214_en.pdf

Department for Communities and Local Government. (2012a). *Building and Buildings, England and Wales, The Energy Performance of Buildings Regulations*. No. 3118 of the Secretary of State made on 17 December 2012. London. Retrieved May 2, 2016, from: http://www.legislation.gov.uk/uksi/2012/3118/pdfs/uksi_20123118_en.pdf

Department of Communities and Local Government. (2012b). *Improving the Energy Efficiency of Our Buildings: A Guide to Air Conditioning Inspections for Buildings*. London. Retrieved May 2, 2016, from: https://www.gov.uk/government/uploads/system/uploads/attachment_data/file/51121/A_guide_to_air_conditioning_inspections_for_buildings.pdf

Energy Performance of Buildings Directive. (2010). *Directive 2010/31/EU of the European Parliament and of the Council of 19 May 2010 on the energy performance of buildings*. Brussels. Retrieved May 2, 2016, from: http://eur-lex.europa.eu/LexUriServ/LexUriServ.do?uri=OJ:L:2010:153:0013:0035:EN:PDF

European Commission. (2015). *The 2020 Climate and Energy Package*. Retrieved July 14, 2015, from: http://ec.europa.eu/clima/policies/package/index_en.htm

HM Government. (2013a). *The Building Regulations 2010: Amendments to the Approved Documents*. England. Retrieved May 2, 2016, from: https://www.gov.uk/government/uploads/system/uploads/attachment_data/file/430891/approved-documents-amends-list_2013.pdf

HM Government. (2013b). *Conservation of Fuel and Power L2A 2013*. England. Retrieved May 2, 2016, from: http://webarchive.nationalarchives.gov.uk/20160324025209/https://www.gov.uk/government/uploads/system/uploads/attachment_data/file/441420/BR_PDF_AD_L2A_2013.pdf

HM Government. (2013c). *Conservation of Fuel and Power L2B 2013*. England. Retrieved May 2, 2016, from: http://webarchive.nationalarchives.gov.uk/20160324025209/https://www.gov.uk/government/uploads/system/upload/attachment_data/file/441422/BR_PDF_AD_L2B_2015.pdf

International Energy Agency. (2008). *Energy efficiency requirements in building codes, Energy Efficiency Policies for New Buildings*. Retrieved May 2, 2016, from: https://www.iea.org/publications/freepublications/publication/energy-efficiency-requirements-in-building-codes—policies-for-new-buildings.html

International Energy Agency. (2010). *Energy conservation in buildings and community systems – Commissioning Overview*. 47(1). Retrieved May 2, 2016, from: http://www.nrcan.gc.ca/sites/www.nrcan.gc.ca/files/canmetenergy/pdf/Annex47-report1-Final.pdf

Intergovernmental Panel on Climate Change. (2014). *Climate Change 2014, mitigation of climate change, Working Group III Contribution to the Fifth Assessment Report of the Intergovernmental Panel on Climate Change*. Retrieved May 2, 2016, from: http://www.ipcc.ch/report/ar5/wg3/

Knebel, D. E., & McBride, M. F. (2005). *ASHRAE guideline: The commissioning process, 0-2005, American Society of Heating, Refrigerating and Air-Conditioning Engineers, Inc*. American Society of Heating, Refrigerating and Air-Conditioning Engineers. Atlanta.

Lau, S. (2014). Mind the gap – a comparative case study analysis of commissioning of commercial buildings in England and California. MSC. Imperial College.

Lewis, A., Riley, D., & Elmualim, A. (2010). Defining high performance buildings for operations and maintenance. *International Journal of Facility Management, 1*(2), 6.

Liu, M. (1999). Improving building energy system performance by continuous commissioning. *Energy Engineering, 96*(5), 46–57. doi:10.1080/01998595.1999.10530472

McFarlane, D. (2013, June). Technical vs. process commissioning. *ASHRAE Journal*.

Mills, E. (2011). Building commissioning: A golden opportunity for reducing energy costs and greenhouse gas emissions in the United States. *Energy Efficiency, 4*(2), 145–173. doi:10.1007/ s12053-011-9116-8

Nakahara, N. (2003). Study and practice on HVAC system commissioning. Proceedings of the 2003-4th International Symposium on Heating, Ventilating and Air Conditioning. October 9–11, 2003, Beijing, China. 44–61.

National Environmental Balancing Bureau. (2014). Procedural standards for whole building systems technical commissioning for new construction. NEBB. Maryland. 4.

Noye, S., Fisk, D., & North, R. (2013). Smart systems commissioning for energy efficient buildings. CIBSE Technical Symposium. April 11–12, 2013, Liverpool John Moores University.

Pitt and Sherry Consulting. (2014). *National energy efficient building project: Final Report, November 2014*. Canberra. Retrieved May 2, 2016, from: http://www.pittsh.com.au/assets/files/Projects/NEEBP-final-report-November-2014.pdf

Sterling, E. M., & Collett, C. W. (1994). The building commissioning – quality assurance process in North America. *ASHRAE journal, 36*(10), 32–39.

The Building Act, Chapter 55. (1984). Enacted by Her Majesty the Queen on 31 October 1984. Retrieved May 2, 2016, from: http://www.legislation.gov.uk/ukpga/1984/55/pdfs/ukpga_19840055_en.pdf

US Department of Veteran Affairs. (2013). *Whole Building Commissioning Process Manual*. Arlington. Retrieved May 2, 2016, from: https://www.wbdg.org/ccb/VA/VACOMM/va_cxmanual.pdf

VanGeem, M. (2014). Energy codes and standards. [Online] Retrieved July 30, 2014, from http://www.wbdg.org/resources/energycodes.php

Ware, D.W., & Bozorgchami, P. (2013). *Energy Efficiency Comparison: California's Building Energy Efficiency Standards and the International Energy Conservation Code.* *Sacramento: California Energy Commission, High Performance Buildings and Standards Development Office.* California Energy Commission report number CEC-400-2013-009.

Xiao, F., & Wang, S. (2009). Progress and methodologies of lifecycle commissioning of HVAC systems to enhance building sustainability. *Renewable and Sustainable Energy Reviews, 13*(5), 1144–1149. doi:10.1016/j.rser.2008.03.006

Yin, R. (2003). *Case Study research design and methods.* (3rd ed.). USA: Sage Publication.

Index